MW01277835

THOMAS' STOWAGE

The Properties and Stowage of Cargoes

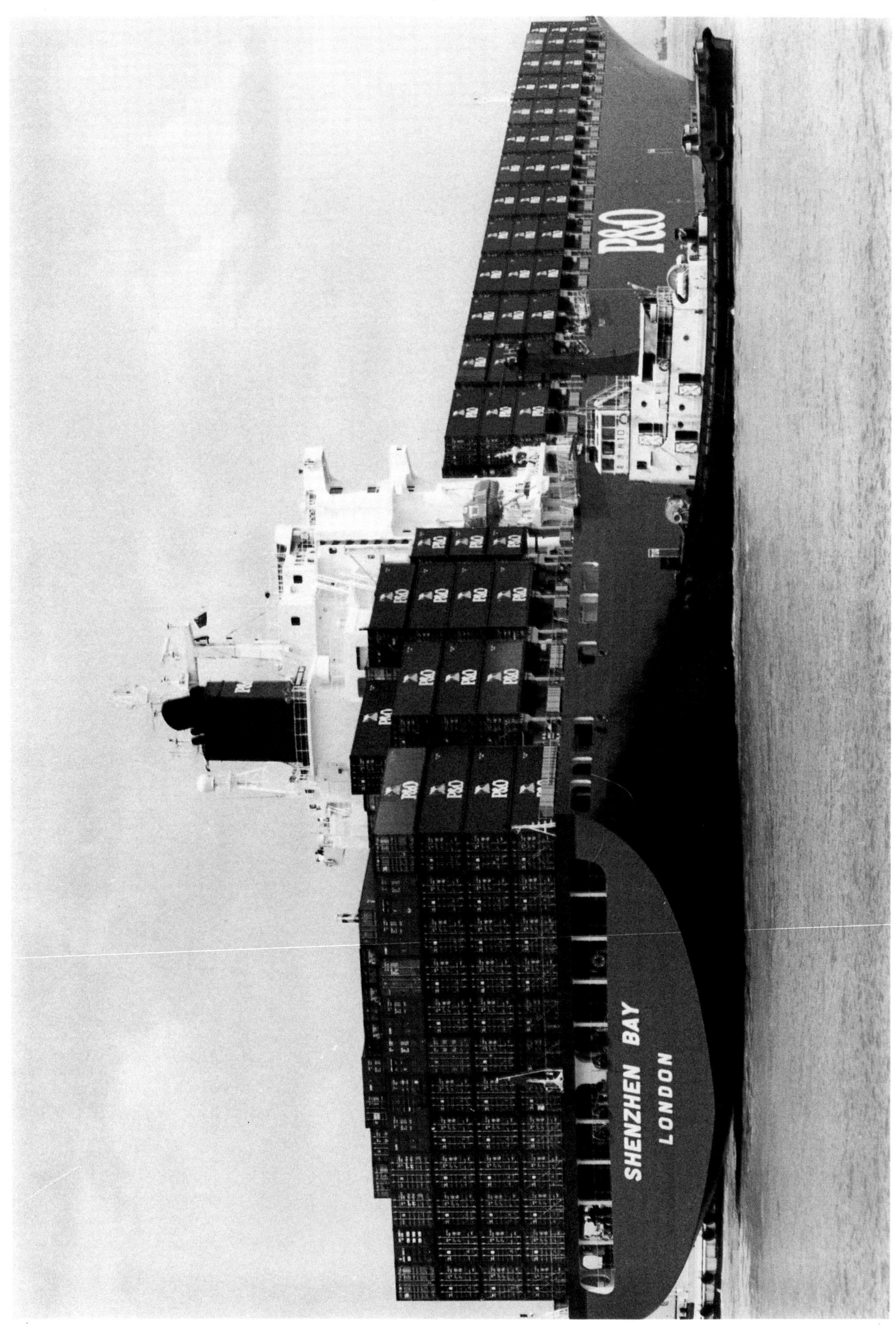

(Courtesy of P&O Containers)

FRONTISPIECE:— Container ship. 1994 built, 4236 teu capacity.

THOMAS' STOWAGE

The Properties and Stowage of Cargoes

ORIGINALLY BY
CAPTAIN R. E. THOMAS, Extra Master

RE-WRITTEN AND COMPLETELY REVISED BY
CAPTAIN O. O. THOMAS, F.C.I.T., Master Mariner

JOHN AGNEW, Master Mariner

K. L. COLE

FURTHER REVISION BY
CAPTAIN KEN RANKIN
1995

GLASGOW

BROWN, SON & FERGUSON, LTD., NAUTICAL PUBLISHERS

4-10 DARNLEY STREET

First Edition	-	*1983*
Second Edition	-	*1985*
Reprinted	-	*1990*
Reprinted	-	*1993*
Third Edition	-	*1996*

This Edition is Dedicated to the Memory of
Captain O. O. THOMAS

ISBN 0 85174 625 X
ISBN 0 85174 503 2 (Second Edition)

Printed and Made in Great Britain

PROLOGUE

That the art of stevedoring or stowing goods in a ship was well understood and practised by the ancient Phoenician seamen is testified to by Xenophon, born 43 B.C.; died at Athens 355 B.C..

Xenophon's description of his visit to and inspection of a "great Phoenician sailing vessel" brings vividly to the mind of the seamen of today how well those ancient mariners had mastered the golden rule of "a place for everything and everything in its place".

He wrote as follows:

"I think that the best and most perfect arrangement of things which I ever saw was when I went to look at the great Phoenician sailing vessel, for I saw the largest amount of naval tackling separately disposed in the smallest stowage possible.

"For a ship, as you will know, is brought to anchor, and again got under way, by a vast number of wooden implement and of ropes, and sails the sea by means of a quantity of rigging, and is armed with a number of contrivances against hostile vessels, and carries about with it a large supply of weapons for the crew, and, besides, has all the utensils that a man keeps in his dwelling house, for each of the messes. In addition, it is loaded with a quantity of merchandise, which the owner carries with him for his own profit.

"Now, all the things I have mentioned lay in a space not much bigger than a room that would conveniently hold ten beds; and I remarked that they severally lay in such a way that they did not obstruct one another, and did not require anyone to look for them, and yet they were neither placed at random, nor entangled with another, so as to consume time when they were suddenly wanted for use.

"Also, I found the captain's assistant, who is called "the lookout man", so well acquainted with the position of all the articles, and with the number of them, that even when at a distance he would tell where everything lay, and how many there were of each sort.

"Moreover, I saw this man, in his leisure moments, examining and testing everything that a vessel needs at sea, as I was surprised, I asked him what he was about, whereupon he replied: 'Stranger, I am looking to see, in case anything should happen, how everything is arranged in the ship, and whether anything is wanted, or to put to rights what is arranged awkwardly'."

PREFACE

The shipping industry continues to adapt in order to keep pace with the ever-changing face of technology and commerce. Super carriers cross the oceans with the aid of state-of-the-art computers monitoring bridge, engine-room and cargo control systems. At the same time there may also be a listing freighter sheltering in the lee of a headland with a full cargo of timber or long steel; or a master who has just been advised that he is to carry horses on deck!

The aim of the book is to give good first hand advice on all varieties of cargoes, their properties and stowage. It can be used as a reference by those not actively involved in the carriage of these cargoes and, to this end, there are new entries on the specialised cargoes of LPG, LNG and chemicals. The Commodities section has also been updated where appropriate and the entries on containers, petroleum, steels and oils and fats have been largely rewritten.

Finally, *Thomas' Stowage* would not be complete without my thanks to the experts who assisted in its revision.

KSR

ACKNOWLEDGEMENTS

The writer and publishers gratefully acknowledge the guidance and contributions given by:

CWA Consultants — Chemical Cargoes
Oils and Fats (In Association with Wolf Hamm)
H. Burgoyne & Partners — Coal
Jarrett, Kirman & Willems — Commodities
M. H. Maunder & Co. — Petroleum
R. I. Wallace & Co. — LPG & LNG
Sparks & Co., Antwerp — Steel and Iron

Holman Fenwick & Willan, Solicitors
The International Maritime Organisation
(Text reprinted by kind permission of IMO)

London Offshore Consultants
Marine Management Ltd., Malta
Ocean Fleets
P&O OCL
TT Club
Vine Gordon & Co.
Vivienne Allen
Roland Hart

CONTENTS LIST

LIST OF ILLUSTRATIONS

PART 1

SAFETY

INTRODUCTION

The majority of the world's merchant ships exist to carry cargoes on a commercial basis from one place to another. The types of ship undertaking these voyages vary enormously, and they may range in size from the smallest coaster to vast oil tankers with cargo capacity approaching half a million tonnes. In every case, the loading, distribution, stowage, security and monitoring of the cargo is of prime importance to the safety of the ship, her personnel and equipment as well as her ability to earn a profit for her Owners. In addition the cargo may represent a potential source of danger to other parcels of cargo on board, and some or all the cargo may pose a significant hazard to the environment should some disaster overtake the ship. Today, an increasing amount of the workload in planning and stowing the cargo is done ashore, often using computers, but the Master must always be aware that the responsibility for the safety of the ship remains with him. He must satisfy himself that at all times the ship is being maintained in a safe condition, and will be able to undertake a proposed voyage with no danger to her structure or her cargo.

SEAWORTHINESS

The legal concept of seaworthiness is not always the same as a layman's understanding of the word. A legally seaworthy ship may, in fact, have serious defects, but they would have to be defects that her crew could not reasonably have discovered. Legal seaworthiness is not generally an absolute standard but instead constitutes an adequate defence to liability for the Owners which will result if, amongst other things the Master and crew have exercised 'due diligence'. What comprises 'due diligence' will depend on the particular ship, cargo and voyage contemplated. However, in general terms it means the exercise of reasonable care.

Improper stowage of cargo may cause a ship to be legally unseaworthy in two ways:

(i) it may make her unfit for the voyage contemplated. For example, if cargo is inadequately secured and liable to break loose rendering the ship unstable.

(ii) it may make her unfit to receive further cargo. For example, if cargo on board is inadequately secured and liable to break loose damaging subsequently loaded cargo.

In considering how to avoid a finding of unseaworthiness resulting from improper stowage, reasonable care or due diligence must be employed, particularly in relation to:

(a) Load Distribution
(b) Cargo Securing
(c) Effect of cargo
(d) Machinery and Equipment
(e) Good Seamanship

LOAD DISTRIBUTION

A vessel's ability to return to an upright position when heeled by some external force, such as by action of waves, is a measure of her stability. The force of gravity acting downwards and the buoyancy force acting in opposition cause a righting lever which acts to return the ship when heeled.

The magnitude of this lever is determined by the position of the Centre of Gravity within the ship which is itself affected by the disposition of cargo, fuel, ballast and fresh water etc. Broadly speaking, the lower the weights in the ship the lower the Centre of Gravity; the lower the Centre of Gravity, the larger will be the righting lever at successive angles of heel (that is, the greater the ship's ability to return to the upright).

To an extent this suggests that heavier goods be stowed in the lower part of the ship with lighter goods on top. However, it should be borne in mind that very large righting levers (an excess of stability by virtue of a low Centre of Gravity) may give rise to excessive, violent rolling and consequently potential damage to both ship and cargo. Some Classification Societies restrict the maximum permissible GM. This may be known as "super stability" and restrict the deadweight intake of closeweight cargoes. Careful loading will ensure adequate but not excessive stability — said to be neither "tender" nor "stiff".

When performing stability calculations the centres of gravity of various parcels of cargo can often only practically be arrived at by approximation. It is better to err on the side of safety, assuming the centre of gravity to be *higher* than it probably is. For instance, where ISO containers are concerned, the vertical centre of gravity almost always lies *below* the mid height point of the container; if the mid point itself is used, a safety factor for the overall stow will automatically be incorporated because the actual Centre of Gravity of the ship will lie somewhere below that calculated.

Apart from stability considerations, distribution of cargo (and to a lesser extent bunkers, ballast and fresh water) can induce unacceptable bending moments, sheer forces and torque — particularly in larger vessels. Care should be exercised to ensure that any limits established by her designers are not exceeded. Instances are on record of vessels having broken in two during cargo operations; continually subjecting larger vessels to excessive loads throughout the ship's life can give rise to structural failure in a seaway. Loading the vessel with excessive weights at each end also tends to make that vessel sluggish in rising to a head sea, and liable to undue strain in heavy weather.

CARGO SECURING

Roll on-Roll off cargoes present a number of problems, such as the unsuitable state or absence of lashing points on some road vehicles, high centres of gravity on certain loads, inadequate lashing equipment etc.

Inadequate lashing of cargo on Ro-Ro's has frequently been the cause of cargo shifting and the vessel taking on a list. The dangers are exacerbated by water entering onto the loading deck as a large "free surface" of liquid diminishes the vessel's stability, potentially to a point where a vessel heeled may have insufficient righting lever to return to an upright position.

Containerised Cargo: It is often not possible for ships' staff to examine or monitor the securing of cargo within a container, although the Master has the right to open a container for inspection should he suspect that all may not be well within. Cargo which is visible — such as that on flatracks — can be examined at the ship's rail and any lashing arrangements which are suspect may either be adjusted by ship or shore staff or alternatively the container should be landed ashore.

Cargo stowed on deck requires particular care in stowage and securing, whilst at the same time affording adequate access to sounding pipes, fire hydrants etc. and to the ship's side should the need to jettison arise. Whilst at sea deck cargo should be inspected daily and lashings tightened when required.

Carriage of bulk cargo as noted above may require the use of considerable measures to ensure that any shift of the cargo caused by motion of the ship will not endanger the safety of the vessel. In the case of grain cargoes, the stowage of the grain, and the calculation of its likely effect in the event of a shift must be in compliance with the IMO grain rules which will be implemented by the national legislation of the flag state of the ship. However, other cargoes carried in bulk may produce similar results in the event of a shift, and the need to restrain cargo, particularly athwartships, must always be considered. These aspects are dealt with further in Part 2 of this book — Bulk Cargoes.

EFFECT OF CARGO

Under this heading the Master must consider the effect of any given parcel of cargo upon other cargo carried on board, as well as possible effects of the cargo on the structure of his vessel. Clear guidelines apply to the stowage and segregation of dangerous cargo, and in some cases may require particular commodities to be carried in completely separate holds. Clearly, the interaction of two cargoes will not occur if the packaging of that cargo remains intact. However, the Master must always consider the possible effect should the cargo escape for any reason and should not restrict his consideration to those cargoes which are listed in the IMDG code. An example might be fruit juice concentrates which although having no risk to the ship or personnel, can write off an entire cargo susceptible to taint even if the escape is relatively small. For this reason, when loading the cargo particular attention should be paid to damaged containers and any which are not satisfactory should be rejected.

The Master should also bear in mind the effect of the cargo on his ship's structure. An example of this is the carriage of high sulphur content coal, which under certain circumstances can lead to the formation of sulphuric acid and very greatly accelerated wastage rates in the ship's steel work. He should also be mindful of the effect of the cargo on subsequent cargoes planned to be carried in the same space, particularly where foodstuffs are involved, which may lead to claims in the future based upon ship being in an unfit condition to load.

The planned rotation or loading sequence of heavy bulk cargoes should be strictly adhered to. The master must resist commercial or operational pressure to alter this sequence.

MACHINERY AND EQUIPMENT

To ensure against the entry of sea water, rain or spray, all weather deck and hull openings (hatches, doors and ramps etc.) should be tightly secured and always maintained in good order. Manual hatch covers, if not interchangeable, must be clearly marked to show where they belong. Beams left in should be pinned.

Battening down with manual cleats can be a long job in larger ships with small crews. It must often be done at night, in inclement weather, and in similar circumstances in which the crew may not be as attentive to their duties as they should be; cleats left undone, eccentric wheels not turned up, multiple panels incorrectly aligned and cross joint wedges not hammered up are amongst the most common sources of water ingress. Permanent local damage to gaskets with consequent leakage may also occur during battening down as a result of obstructions such as lashing wires or cargo residues left on hatch coamings or between panels.

When water penetrates a hatch seal, it is usually collected in a drainage channel and discharged clear of the hatchway. Water may spill over onto the cargo in the hold below if these channels are allowed to become blocked or restricted.

Any fork lift truck or other vehicle used on board must operate in an area free of obstructions. Because of the danger from fumes, units powered by diesel and petrol engines are not usually suitable for use within the confines of a ship's hold or 'tween deck unless adequate ventilation is available. (N.B. Diesel and petrol fumes can also taint some sensitive cargoes; see part 3 — Commodities). Any bridging used to improve working areas (e.g. flared holds in bow and stern) must be of adequate strength to accept the wheel loading of the equipment and its cargo. The same strength requirement of course applies to permanent equipment such as limber boards; see also Part 4 — Damage.

SEAMANSHIP

While a vessel may be in good condition on leaving port, this condition must be maintained at sea by prudent handling and good seamanship. What constitutes "good seamanship" in this context is beyond the scope of this book, but examples include the alteration of course and/or speed to minimise damage to deck cargo and fittings in bad weather; effective load distribution and ballasting to improve stability and rolling characteristics; checking the condition of all cargo and associated lashings throughout the voyage where practicable.

FIRE

The risk of fire breaking out amongst cargo in a ship's hold, the resultant damage to ship and cargo and the risk to personnel are obviously matters which must be borne in mind when cargo is being stowed.

Smoking, lighting matches, the use of electric cables with frayed insulation etc. must never be permitted in an area in which cargo is being handled — whether that be the ship's hold, the vicinity of open hatchways, the container or the container stuffing area.

Where Dangerous Goods are concerned, fire may be the principal hazard, though not the only one. In this respect Dangerous Goods are not only a potential source of ignition but can also greatly intensify or complicate a fire and make for difficulties in fire fighting.

It is particularly important to ensure that all normal precautions against fire are carried out. These should include:

1. The provision of all fire-fighting equipment (as required by local or national regulations) in properly maintained and fully operational condition.
2. The provision of proper instructions for emergency routines and the regular practise of any such routines. Fire fighting apparatus layout and cargo plans should be readily available to both ship and shore personnel.
3. Proper care and correct operation of cargo handling gear and other machinery.

Explosions can occur in a number of substances (not only those classed as explosives) with varying degrees of violence. Where this characteristic exists, care should be taken to ensure that any recommendations with regard to stowage are fully complied with. An explosion would almost certainly be followed by a fire.

Where IMO class 1 cargoes or class 3 (low flash point) cargoes are being handled by mechanical equipment, such equipment must be made safe to operate in the vicinity of such cargo, for example electric motors must be spark-proof.

Fires and explosions often occur whenever a concentration of gases or vapours are present. It is possible for pockets of gas to form even although it is thought that proper ventilation has taken place throughout the stow. Vapours or gases from substances with a wide explosive range (e.g. acetylene) and particularly those of a density equivalent to air are most dangerous.

Certain goods, if loaded when wet, may be liable to spontaneously combust. To load such substances during or after rain is to increase the risk of fire during the voyage. Extreme insect infestation in certain bulk cargoes can form "hot spots" which may become the source of fire. A fire which has started whilst stuffing or shortly after closing, an ISO container may not become apparent until that container is loaded on to the ship and is on the high seas.

It is particularly important, therefore that all possible precautions are taken to prevent such an occurrence. Containers with combustible cargoes should be stuffed in a controlled environment and their contents properly declared. Cargoes liable to spontaneously combust "in stow" should not be unnecessarily ventilated and should be checked regularly for rise in temperature.

The securing of transport, containers and other cargo on the vehicle decks of Ro-Ro vessels must be arranged in such a way that fire-fighting equipment remains unobstructed and fully accessible during loading discharging and on passage. This includes valves, emergency pumps, etc.

ACCIDENTS

Accidents are often caused as a result of improper maintenance of cargo handling equipment. Regular inspection of blocks, sheaves, bushes, wire ropes etc. should be part of any planned maintenance programme. Any associated paperwork — such as test certificates — should be kept up to date and be available for inspection when required. Measures should be taken to avoid obstructing limit switches, e.g. when painting a crane or derrick. A careless application of paint may stop up vitally needed grease nipples.

The risk of accidents to personnel and cargo can be reduced by good "housekeeping" and proper procedures. These include:—

- Clear, unobstructed access to all cargo handling and storage areas.
- Proper lighting in the holds and other working areas.

— The removal of nails from old dunnage, or the removal of the dunnage itself to a safe place.
— Proper maintenance of equipment and compliance with Flag State and Class certificates etc.
— The provision of suitable clothing and equipment; e.g. special protective clothing should be available when certain dangerous goods are carried (see IMDG code).
— Walking boards to provide access for labour over delicate cargo, e.g. chilled foodstuffs, light carton goods etc.
— Regular inspection of cargo (where accessible) during the sea passage.
— Guard rails erected, where necessary, to protect personnel.
— Adequate barriers and controls to restrict the approach of unauthorised personnel where containers are being moved, stowed, stuffed, inspected or fumigated.
— Attention to gas-freeing and other safety precautions when entering spaces that might have a high gas content or insufficient oxygen — e.g. containers that have been carried under fumigation; deep tanks after discharge; containers that have been registered by Liquid Nitrogen, Carbon Dioxide or other refrigerant; spaces that have recently held certain Dangerous Goods etc.
— Safety helmets to be worn when cargoes are being worked on deck or down below.
— Hold ladders should be checked. Those that are damaged and dangerous (and which have not yet been repaired) must be roped off to prevent access.

PART 2

TECHNIQUES AND SYSTEMS

GENERAL INTRODUCTION

The principles of stowing the basic types of cargo, although treated as separate sub-sections, do in very many instances coincide. The techniques in handling and stowing bagged cargo, bales, cases and cartons, must be similar in most handling modes. The principles of taint, sweat, separation and ventilation remain unchanged, with slight variation of application for the different handling modes. Wherever possible cross referencing has been used to avoid repetitive text.

The different types of cargo referred to in this section are:

Break Bulk Cargo — including general information on the handling and stowage of different cargo types such as bags, bales, cases, drums, etc. Also this sub-section holds the most information on ventilation, taint, dunnage, etc.

Dangerous and obnoxious cargoes.

Livestock.

Unitised cargoes.

Containerised cargoes including ro/ro system.

Iron and steel products. Timber.

Refrigerated cargoes.

Bulk cargoes.

Petroleum, LPG, LNG, and liquid chemical cargoes.

Oils and fats.

Although passing reference may be made it is not within the scope of this book to go into detailed stowage requirements of barge carrying vessels (except in so far as the stowage of the barges can be considered general or bulk cargo stowage.)

The proper and adequate stowage of cargo, whether onboard ship, inside a container or into barges, is the result of good pre-planning and careful attention to the requirements of the trade and mode of carriage employed. Properly carried out it should ensure the following:

1. The preservation of crew and ship from danger or injury arising from the manner in which the cargo is stowed.
2. Protection of the cargo from damage, or deterioration and thus ensure sound delivery of same.
3. An economy of cargo space by which depends the earning capacity of the vessel.
4. Speed of operation at time of loading and discharge, This in turn affects the turn round time of the vessel and thus the earning capacity of the vessel during the voyage. (N.B. In the case of Ro-Ro and Container vessels, this requirement is usually of greater importance than the economy of cargo space. Indeed, efficient use of space may be sacrificed to achieve greater despatch).
5. Accessibility of cargo for each port without disturbing cargo intended for a subsequent port, i,e, the minimum number of overstows.

BREAK BULK CARGO

Introduction

Break Bulk Cargo operations includes the handling, carriage, stowage and storage of cargo in individual items, i.e. crates, cases, cartons, bags, bales, bundles, drums, barrels etc., listed in a number of Bills of Lading, each consisting of a different commodity.

The characteristics of the different cargo types (crates, bags etc.) are described under the general heading of Break Bulk Cargo, but many of the handling, stowage and storage requirements of these items are equally applicable when other modes of carriage are employed e.g. containers, Ro-Ro vessels etc.

General Stowage

Charter Parties normally call for the cargo to be loaded under the supervision and or responsibility of the Master. The Master must at all times ensure that the stowage is safe and does not endanger his ship. The charterer may have time foremost in his mind. Should the Master be disatisfied with the stow for any reason the loading should be stopped and the matter brought to the attention of the superintending stevedore and the charterers representative. Tallies and mates receipts must accurately record the quantity and condition of the cargo. To issue a Bill of Lading that does not accurately reflect the condition of a cargo is fraud. (see also Part 5 — Tallies and Bills of Lading).

The loading stevedore may have despatch at his end much more in mind than speed of operation at the other end; so that, if left without firm guidance and control by those interested in despatch at the ports beyond, he may find it to his advantage to confine to one or two holds cargo which would discharge twice as fast if stowed in three or four, especially if he is working on a per ton (or all inclusive rate) basis — the covering and uncovering of extra hatches, the employment of foremen (and sometimes gangs) for less than a full day's work etc., eating into his profits, as they must, naturally tell.

In the case of multiple port loading, the loading is done by as many stevedores, and unless the stowage is under the control of the ships' officers the tendency which exists for the stevedores in the earlier ports to pick the easy spots for their cargo may very seriously complicate the stowage at the subsequent ports, and react in a serious manner at the discharging ports as well.

Stowage should aim at distributing the cargo for any particular port equally or as nearly so as circumstances permit, in every hold, thus provision is made that all the cargo handling equipment is employed to its full capacity throughout the time the vessel is loading and discharging. It is best to ensure that the "heavy" hatch does not have to remain working for extensive periods after all other hatches are finished and have been battened down, with the added restriction, probably, of only one gang and one hook being able to get access to the space being worked.

Where foodstuffs and fine goods are stowed (that is, goods such as carpets, clothes, etc, that may be easily damaged or take on taint) only clean nail-free dunnage should be used and stowage should be found in separate compartments away from such obnoxious commodities as creosote, aniline, essential oils, petroleum, copra, hides, manures, cassia, certain chemicals, turpentine, newly sawn or most kinds of timber, green fruit, onions, etc.

Weighty packages such as cases of machinery, railway bar or plate iron, blocks of stones, ore billets, ingots or pigs of metal, etc., should always be stowed on the floor and the lighter cargo on top.

As a general rule fragile and light packages should be stowed in 'tween deck spaces — the ground floor of such being, if necessary or advisable, covered with weighty goods — where they will not be subjected to excessive top weight.

The nature of the packages sometimes calls for them to be kept in a certain position, i.e. coils and rings on the flat, etc. Avoid stowing bale and light goods on top of cargo which has life and spring, or against bulk head stiffeners, deck beams, brackets, frames, stanchions or other projections, using plenty of dunnage to protect them from contacting such projections and rough surfaces.

Each tier should be kept as level as possible (with packages of uniform size it should be perfectly level). Packages should not be stowed in such a manner or position that they tilt either way, as will occur at the turn of the bilge, or with the rise in floor in the fore part of the forward hold, etc., unless properly dunnaged or bridged.

Broken Stowage

Any break in stowage — or broken stowage — caused by the presence of pillars, stanchions, brackets, web frames, etc., for the filling of which certain packages are not available, or space which is unsuitable to receive a package of cargo, should be packed firmly with suitable dunnage, in order to prevent any movement of cargo in a seaway and to afford a stable and level platform for the next tier.

The loss of valuable cargo space, where the nature of the cargo justifies economy, is best avoided by:

1. Compactness of stowage.
2. Selecting packages which, by the nature and value of their contents and their construction, are suitable for filling broken stowage. Reels of barbed wire, bales of binder twine, coils of small wire, for example, are very useful for this purpose.
3. Always keeping a supply of such packages, or of low freighted goods, ready at hand in the holds, for use when wanted.
4. Stowing casks and drums upright rather than on their sides.
5. Nesting and/or stowing pipes "bell and cantline" (See Steel and Iron Cargoes, Tins Section). Blocking in spaces left between large cases with smaller packages. Care should be taken that these packages cannot become crushed.
6. Special selection of cargo suitable for filling beam spaces, i.e. cargo which is not liable to chafe (bale goods being very unsuitable) or liable to damage by sweat, if moist or heated cargo is carried in the same compartment, or refrigerated cargo in the compartment above. It should be borne in mind that up to 6/8 per cent of the hold capacity in 'tween decks may be contained between the deck beams.

General Cargo in Refrigerated Chambers

An important consideration when stowing general in refrigerated chambers is that of avoiding damage to brine pipes, insulation, and airtrunks. Where practical soft packages and cases should be stowed in the hatch wings. Bulky or unprotected heavy cargo should be stowed in the hatch square. These should be clean lift as dragging out from the wings and ends can cause damage to insulation.

Advantage follows if the floor or ceiling in squares of hatchways are protected by means of stout boards or sheathing. Sizeable packages, having flat surfaces (cases in preference to bales), should be selected for stowing against the pipe battens at sides and ends, so as to minimise the risk of cargo protruding into the grids and, by exerting pressure on the piping, cause the same to leak at the joints.

The permanent battens should be in good order or, in the absence of such, battens should be provided for the protection of the piping and cargo, such to be placed vertically, close spaced and "stopped" in position.

When bale, bag or other goods liable to damage from sweat or rust stains are carried, the piping as well as other ironwork should be covered with kraft paper etc.

Goods or liquids such as creosoted materials, aniline oils, essential oils, cargoes with a phenolic base, chemicals, etc., which leave behind strong odours, which safely cannot be dissipated by ventilation, should not, on any account, be stowed in refrigerated chambers. Some of the most serious claims for damage and consequent costs for getting chambers again fit for refrigerated cargoes have been through phenol taint. All dusty cargoes liable to choke air passages should be avoided.

Heavy packages should be stowed on the ceiling with dunnage or bearers and so avoid damage to ceiling and insulation by distributing the weight over the maximum surface.

Bagged Cargo

Materials

The materials from which bags are made will depend on a number of factors, i.e. the commodity to be shipped, its physical composition and its properties (e.g. moisture content, sensitivity to contamination, etc.). The principal materials are manufactured in different weights and strengths, and may be used singly or in multiples (i.e. several layers), may be sewn, glued or welded, etc. The principal materials include:

Paper (single or multi-ply) — which may be sewn or glued.

Plastic (which may be air tight).

Woven polypropylene. May also have an inner sealed bag made of polythene.

Jute, hessian, etc. (traditional materials and the type most likely to have repeated use, e.g. as second-hand bags, which having carried previous commodities may hold residual taint, moisture, or be stained).

Open mesh sacks (e.g. nets, open mesh plastic fibres, etc., for maximum ventilation).

N.B. Some bags, particularly jute and hessian, may have a re-sale value.

Commodities

The commodity itself has to be robust to withstand outside pressure and compression, for the bags will only hold the contents in one place and will not provide protection against external damage. Very often these are the sort of commodities that might be carried in bulk or mini bulk operations (see under those headings). Such commodities might typically be fertilisers, grain (rice, maize, wheat, etc.). seeds, dried fruit, sugar, cement, coffee, flour, copra, small items such as shells, resins, etc., mail, salt, mineral sands and ores, meal (fish, seed, copra, etc.), dried blood, dried milk (casein, etc.).

Characteristics

The sizes of the bags vary although it is standard practice to have such a size which, when filled with the particular product, can be handled and carried continuously by labour working throughout a shift. The most common filled weight of bags today is 50 kilograms. This weight of bags allows rapid and easy calculation of weights taken on board, into containers or to make up sling loads, etc. Care must be taken when loading large quantities of bagged cargo to allow in calculations for gross, nett, and tare (bag) weights.

The shape of the bag, and therefore the ease or otherwise with which it can be stowed, made up into slings, pallet loads, etc., varies with the type of commodity and the way in which the bag has been constructed. The bag may be "shaped", usually these are of paper or plastic construction, and form a near rectangle which is stable and has flat surfaces. Bags made out of materials such as jute are usually sewn flat at each end and, therefore, tend to have no flat surfaces, although these may be induced by the pressure during the storage or stowage cycle. If the contents are powdery (e.g. cement) the package is more likely to conform to outside pressures around it, such as weight on top, pressure from the sides, etc. If on the other hand the contents are hard and bulky (e.g. frozen offal then the bags will be of irregular shape and will probably not respond to the influence around it.

Commodities in hessian and open mesh bags may be subject to contamination from powder or small granular cargoes stowed beside or above them (see "Damage"). Cargo in polypropylene or hessian bags, if of a powder or granular nature, may itself sift though the bags with resultant risk of loss and contamination. Bags which have recently been filled tend to settle and spread outwards and this can affect stowage arrangements, e.g. containers (see "Unitised Cargoes"). Woven polypropylene bags, because of their non-absorbant characteristics, may be more suitable for moisture inherent/ sensitive cargoes (e.g. sago, coffee, etc.) than traditional hessian. These bags may have an inner layer of polythene.

Increasing use is being made of jumbo bags. These vary in size and normally consist of woven polypropylene with handling straps sewn into the bag. They may weigh several tonnes and are most suitable for hatch square stowage where they can be clean lifted upon discharge. Wing and end stowage is not recommended unless special lifting equipment is supplied. (see I.B.C.).

Handling

Different bagging materials lend themselves to different handling methods. For instance a jute bag with good "ears" on it may be more easily lifted by the human hand than, say a heavy fully filled plastic sack with nothing to grip. There is a great temptation therefore for labour to use "cargo hooks" to manhandle these into and out of their stowage positions in holds, containers, etc. Since most bagged commodities (see above) are of an easily spillable nature, and some of them are very valuable (e.g. coffee), it is important that cargo hooks should not be used for bagged cargoes except in particular circumstances — these will be indicated against the particular commodities listed in the alphabetical section.

Different bagged materials and different commodities may find themselves more suited to one handling technique than another. For instance tightly filled, bulging plastic bags may be extremely difficult to make up on to pallet loads because of the tendency to slide and the shape of the bags. Some bagged materials (multiply paper sacks) may not be safely lifted by rope slings or snotters without fear of rupturing the bags. Flat webbed slings from man-made fibres are probably best suited for slinging most bags, and the "clover leaf" sling arrangement may be used to advantage with bags that are difficult to handle, i.e. because of shape or material. These may also be used for pre-slinging requirements and unit loads (see "Unitised Cargoes"). Canvas or similar materials may be required for loading and discharging such commodities as flour, coffee, cocoa, etc., where the nature of the bagging materials and the value of the commodity makes them vulnerable to high cost damage if bags are ruptured. The care of mechanical handling equipment drivers (cranes, winches, fork lift trucks) is also required when loading or discharging this type of cargo so that damage is not incurred by swinging or rubbing against obstructions such as beams, hatch coamings, etc.

Stowage

Most bagged cargoes are liable to be damaged if stowed with moist cargo or cargo liable to sweat. They should be well protected against obstructions such as beams, brackets, stringers, etc., because as the cargo settles pressure on the unsupported or projecting part of the bag may result in tearing and spilling the contents. They will benefit by being protected by mats, paper, etc., from bare steel work and from likely sources of moisture running down bulkheads, pillars, etc., and serve to protect the bags from discolouration by rusty metal.

Where two types of bagged cargo are carried in the same space, and risk exists of one contaminating the other (e.g. bagged plastic granules over bagged rice), then proper protection should be provided in the form of plastic or similar sheeting between the different cargo types. Similar sheeting should be provided where sifting or loss of cargo might be expected — particularly in the use of valuable cargo such as coffee — so that sweepings may be collected and included in the discharge. Any such protective separation must be carefully handled at time of discharging, and any spillage collected and cleared before moving to the cargo below (see "Sweepings").

Careful tallying is essential and to facilitate this slings should always be made up of the same number of bags, as should pallet loads. Bagged cargo bills of lading should be endorsed "weight and quantity unknown" or at the very least "said to contain . . .". Slack or damage bags must be rejected as also should be damp or stained bags. The latter particularly applies to bagged sugar, cocoa, and coffee.

Some commodities are liable to rot the natural fibre bag, e.g. certain manures and chemical products. In all such cases the bills of lading should be claused so as to protect the ship from having to bear the cost of re-bagging. The ship should also be protected against any claim for loss of contents due to leakage from bags which are not sufficiently strong or not of the correct texture to prevent such loss; sometimes to prevent loss from the above causes double bags are used.

The stowage factor will vary depending on whether the bags are well filled (as for instance coffee and cocoa) or otherwise.

Bleeding of Bags

At certain ports cargo may be loaded by cutting and bleeding bags into the hold from the edge of the open hatch. This is normal practice, however, care must be taken that the bags are fully emptied of their contents before they are discarded. The bags may be bled through a wire or rope grille in order to ensure that the bags themselves do not end up in the hold. In practice this frequently occurs and it is also quite common for foreign or "field matter" to be mixed with the contents. Should the presence of bags and foreign matter be observed in the stow, Master should stop the loading and draw these deficiencies to the attention of the stevedores. He should also draw this problem to the attention of the charterers. Mates receipts should be claused accordingly.

It will be appreciated that the presence of foreign material and bags in the cargo causes difficulties at the discharge port and many claims may be raised at this time.

The ship should endeavour to check the intake by means of an accurate draft survey. Shippers will inevitably want clean bills of lading, however if there is a presence of foreign matter every endeavour should be made to have this noted in the bill of lading. Where relevant, charterers should be advised of this problem and encouraged to deal with the consequences. Shippers in certain countries may offer a letter of indemnity which is of no comfort to the ship as it is the owners who will be sued by the receivers. The likelihood of recovery from the shippers is extremely remote.

Discharging

Avoid heaving slings of bagged cargo out of the wings, or ends of holds. Such treatment will tear the bottom bags on plate butts and landings or even from splinters from tank top ceiling or other obstructions.

If the type of bag and cargo warrants it a needleman should be available for sewing up open or torn bags when loading and discharging bagged cargo.

All torn, slack or empty bags and packages should be carefully inspected while discharging is in progress and delivered (against tally) along with the cargo, otherwise a claim for short delivery is likely to follow.

Bales and Bundles

Materials

Hessian or similar strong material usually constitutes the outer cladding. Other materials include plastic; paper; woven man-made fibres; waste from the contents of the bale. For more valuable goods a secondary covering or even a tertiary one is included. Some bales, e.g. straw, hay, etc., have no covering whatsoever.

Commodities

Typical commodities that are baled will include wool, cotton, vegetable fibre (e.g. jute, kapok, hemp), paper pulp, tobacco, skins, furs, rubber, hair, cloth and other material.

Characteristics

Sizes of bales vary. Some commodities (wool and hemp) may be compressed into high density bales. It is even more important in these cases not to rupture a covering, break the binding material or in some cases, allow ingress of moisture. The resultant expansion of the commodity can make unloading almost impossible, and, in containers might distort or rupture the side walls.

Sizes and weights will, where applicable, be found under the commodity name.

Bales may vary quite considerably in shape. Some square (e.g. wool, unless double dumped when the shape becomes oblong), some cylindrical and some completely shapeless. In many cases the bales are designed as modules of the interior of an ISO container.

Hessian is a common outer covering, but in some cases (e.g. low grade skins, rags etc.) waste items of the commodity itself may be utilised. Close woven polypropylene has a non-absorbent water resistant quality which is particularly suitable for some commodities.

Handling

While cargo hooks are acceptable for raw cotton, wool, etc., they are totally unacceptable for high quality goods such as skins, furs, cotton piece goods, etc. Most baled commodities are impervious to damage from rolling or dropping from limited heights. However, it can be dangerous to drop bales of rubber due to their ability to rebound in almost any direction.

Metal or other banding is usually employed to hold the bale in compression, and should never be used as a lifting point, unless marks or labels indicate to the contrary.

Stowage

Being as a rule vulnerable to chafe, they should be well protected by matting and dunnage from sharp edges and other cargo which may cause damage. They should not be stowed between hatch beams for this reason.

Bales with torn or stained covers should be rejected, or the bill of lading claused accordingly. They should not be stowed with dusty or dirty cargo as they can mark or stain the covers.

Cases, Crates, Cartons, etc.

Materials

Cases and crates are usually made of timber, which may be plywood or thin low grade material. Heavier cases may be built up of 150 mm × 25 mm (6″× 1″) planks with strengthening pieces internally and externally. Some crates are built in a skeletal fashion to allow air to permeate through the contents, or alternatively to reduce the weight of the crate. Larger cases and crates will almost certainly have skids or bearers to allow mechanical handling equipment to gain entry, or for slings to be put in position. These skids are normally about 50 mm deep (2″). Large cases with high density goods may well be built to have certain parts specially strengthened for lifting. These will be marked on the outside and any slings, snotters, chains, etc., should be slung at those points; similarly any securing should make use of the stronger points of the case.

Cartons are usually made of single or multi wall fibre board. Very often the contents provide part of the strength and shape for the cartons, and this should be borne in mind when stowage is being carried out. Some carton material is waxed or otherwise protected against moisture absorption. Wooden crates and cases may have to be treated for Government and State quarantine immunisation.

Commodities

Cases and cartons hold a very wide range of commodities depending on the requirement of the particular cargo. Canned goods are nearly always packed in cartons. Many refrigerated goods are packed in cartons or open-sided crates. Attractive items that may be vulnerable to pilferage are often packed in strong well constructed cases.

Characteristics

Some large cases, e.g. C.K.D. (cars knocked down), may be stored outside prior to shipment. In these circumstances the crates and sometimes the contents are already wet or damp on shipment, with the possibility of introducing moisture into the hold, container, etc., to the detriment of adjacent cargo. In some instances, for example, very large cases of construction items may have to move a long way inland to remote construction sites, and so the packaging is most important for the protection of the contents as well as, sometimes, having a subsequent use as storage or even temporary accommodation. Crates or cases holding machinery may leak oil.

Drum head clamps.

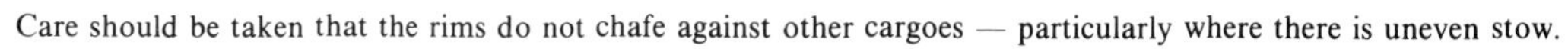

Care should be taken that the rims do not chafe against other cargoes — particularly where there is uneven stow.

Cartons will absorb moisture and gave out moisture under certain atmospheric conditions. Carton material will equilibrate with the moisture content of the air surrounding it in about 36 hours. This means that cartons of goods from a damp area may be relatively full of moisture when loaded into the hold or container. A moisture content of less than 12% in the fibre board is usually considered safe from this particular problem. There have been known instances where carton material has been actively wetted by the packers of the goods to make the material more flexible and more easy to put into shape prior to filling with canned goods. A container load or a compartment load of cartoned goods can therefore have a very high volume of moisture inherent in the cartons at time of loading, with subsequent possible problems of condensation (see "Ventilation").

Handling

Where relevant shippers/charterers instructions should be obtained as to how many high they may be stowed. It may be necessary to dunnage at intervals to avoid tier compression.

Cargo hooks should never be used on cartons. They can sometimes be used to good effect on heavy crates. Very heavy crates normally require mechanical handling equipment, e.g., pallet trucks or fork lift trucks. Cartons are best lifted on board or ashore by means of cargo trays. The slinging of heavy crates mentioned above should be undertaken with due regard to the strength points in the crates themselves. Metal strappings or banding round crates, cases and cartons should never be used as lifting points.

Stowage

Heavy cases should be reserved for bottom stowage where possible with smaller packages on top. Small, very heavy packages should not be placed on top and within the four corners of larger, lighter packages, or resultant damage may occur. Any marks or indications for stowage upright or for protected stowage should be carefully observed and adhered to. Light cases and cartons should be stowed one upon the other so that each one below bears the full weight of the one above and particularly in the case of cartons of canned goods, no overhangs should occur which might distort the cartons and rupture the contents. Special care in the case of refrigerated cargo should be taken to make sure that adequate air flow can move through and around the cases, crates or cartons as required for the particular commodity (see "Refrigeration"). Any broken or holed cartons, cases or crates should be examined at time of loading and if the contents are intact they should be mended prior to stowing. Any cases pilfered or damaged should be rejected or the bills of lading claused accordingly.

Drums, Barrels, Casks, etc.

Materials

Drums may be made out of metal, fibre board or rigid plastic. Casks are built of wood staves and bound by hoops.

Commodities

Liquids such as latex, chemicals, whisky, detergents, oils, molasses, casings, paints etc., may be carried in drums and casks. Powders, granules and other solids such as chemicals, cement, some ores, swarf, scrap metal (sometimes cast in spheres usually in second-hand drums) may also be shipped in these containers. Some of the above may be classed as Dangerous Goods, in which case the packaging requires the appropriate approvals.

Characteristics

The bilge (the part with the greatest circumference) of the barrels or casks is the part least able to support external pressure, so the weight should be taken by the quarter — which is near the ends of the barrels or casks. Metal drums may have hoops to improve their strength and to facilitate rolling and manoeuvring by hand. Light drums such as fibre board can only take top weight when in the upright position.

Handling

Normally rope slings, nets or trays for lifting on or off the ship. Metal drums may usually be gripped using drum handling attachments or mechanical equipment which grip the top rim of the drums (e.g. grabomatics). Care by the operator is often needed to avoid puncturing the drums when gripping or carrying them this way. Side clamps and barrel handlers are also standard attachments for mechanical handling equipment and are used if drums and barrels are sufficiently robust to take this type of treatment. Although often practised it is not recommended to balance drums or barrels on the tynes of a counter balanced fork lift truck for positioning and stowage. This system is prone to accidents and spillages. If drums or barrels are placed on pallets there will be a loss of space particularly if the pallet is not square and does not have plan dimensions which are multiples of the diameter of the drums. While it is not recommended, it may be necessary for drums, etc., to be dropped, e.g. while unloading a container on its trailer from ground level. In this case some form of protection should be provided — e.g. old tyres — to break the fall.

Stowage

Drums stowed on their side should not have other cargo stowed on top. Drums holding liquids should always be at the bottom of the stow with the bung or lid upmost. Barrels stowed on their sides should be "bung up and bilge free", i.e. the bilge of the barrel should not support any weight, and quoins (shaped wedges) should be used to support the weight at the quarter. As a general rule, fibre board drums should always be upright. When more than one tier of metal drums are to be stowed, particularly into the container where vibration may be experienced during inland transport, it is prudent to lay soft dunnage (wood, hardboard, chipboard) between each tier. This may not be necessary if drums are designed to nest one above the other, although in the case of containers it is usually prudent to provide intermediate dunnage. This dunnage allows the rings of the drums to bite in and grip and also protects against chafe. Since some rolling hoops on metal drums form part of the drum wall itself, these hoops may be very vulnerable to chafe, and in some instances may need protection with soft dunnage to prevent them rubbing one against the other and rupturing while in transit.

The tops of drums which have been standing out in the open may hold rain water, frost or snow trapped within the rims. This should be tipped out and if possible the damp areas allowed to dry before loading into a compartment or stuffing into a container — particularly if other cargo in that compartment can be affected by moisture.

Returned empties (that is drums, barrels, etc., which have been emptied of their cargo and are being returned to that source) should be treated as far as contamination, source of liquid, dust, etc., is concerned as though they still held the original commodities. Unless a certificate is provided to say otherwise, returned empties should always be treated as though they still hold traces of the previous contents. This means they may be a source of dust, moisture, taint, or infestation. It is most important that empties being returned after carrying dangerous goods be treated as "dangerous" unless a chemist's certificate declares that they are free of residue from the previous cargo (see "Dangerous Goods").

Discharging

Drums and casks should not be dragged out from wings or out of a tight stow using snotters or slings. Any damage sustained at this stage will be to the ship's account.

Intermediate Bulk Containers

Definition

An I.B.C. may be described as a disposable or re-usable receptacle designed for the carriage of bulk commodities in parcels of between 0.5 and 3.0 tonnes. They are interchangeable between various transport systems and the design incorporates attachments to facilitate efficient mechanical handling using commonly available equipment. Some are designed for carrying liquids. They are not suitable for pressurised commodities and emptying should be possible without the use of pressure. They differ from ISO Freight Containers in size and non-uniformity of shape. Indeed, some are designed to be modules of the internal dimensions of ISO Freight Containers (see "Containers") and are thus compatible with that system.

Materials and Construction

They may be constructed in either a flexible form, i.e. bags able to carry dry bulk or liquids, or a rigid form, e.g. of fibre board (either collapsible or non-collapsible with or without linings).

Dry bulk bags may be constructed of:

Woven polypropylene and nylon;
P.V.C. or P.U. coated polyester;
plastic and rubber.

Rigid bins may be constructed of:

Fibre board, glued or stapled;
all welded metal;
welded plastic;
a combination of the above.

Commodities

Typical commodities might include cement, china clay, sugar, plastic granules, carbon black, powdered chemicals, syrup, fruit juices, oils, detergent liquids, non-hazardous chemical liquids.

Characteristics

Construction of bags and bins are usually carried out to satisfy particular customer's requirements and the requirements of the commodities to be carried.

The size also varies, depending on the user's requirements and the commodity to be carried. Plan dimensions may be compatible with standard pallet sizes — indeed some I.B.C.s utilise pallets as part of the base. This allows stacking and handling to be complementary to the stacking and handling of other unit loads.

Bins may be square or octagonal in shape and may have lifting points on top or access for fork lift tynes at the bottom. The fibre board bin is constructed from heavy duty double walled corrugated fibre board which is sometimes glued and more usually stapled at the joints. The lining may or may not be available or required. If it is used is usually of a light plastic material of a throw away nature.

Rigid bins are often supported by a metal frame which improves the rigidity and allows tiers of bins to be stacked one on top of the other. The bags themselves are usually designed to be lifted from above — and all these systems may have facilities on the under side for gravity discharging the contents.

The lifting points which are built into the bags may consist of loops or sleeves, depending on the material and the lifting modes to be used. Alternatively a harness or clover-leaf sling may be permanently attached to the bag and provide the necessary lifting points.

I.B.C.s should not be filled with commodities which can damage the material or construction in such a way as to reduce their overall strength. The I.B.C.s should not be filled with more goods than the label or certificate indicates. Appropriate national or international regulations should be complied with, e.g. IMO dangerous goods code.

Handling

The majority of bags are designed for existing methods of cargo handling such as fork lift trucks, cranes and derricks. Single, double, or four-fold lifting points may be provided — and it is important that all those provided should be utilised whenever the loaded bag is lifted.

Bins are usually supplied with skids, fork pockets or similar, and should always be lifted and handled using those facilities correctly.

Stowage

Some I.B.C.s are constructed in such a way and of such dimensions that they are suitable for stuffing into ISO containers to obtain maximum utilisation of cube and weight. When a container is so stuffed, care must be taken to ensure all lifting points are easily accessible when stripping the container.

When stowed on general cargo ships it is advisable to keep I.B.C.s separate from other forms of break bulk cargo. They are required to be lifted and handled and positioned using mechanical handling equipment.

When I.B.C.s are damaged during handling or stowage they should immediately be given temporary repairs to preserve their integrity until the end of the voyage. I.B.C.s that are designed for one voyage only should not be re-used — particularly if damaged.

Manufacturers' recommendations for handling should be followed at all times regarding hoisting methods, stacking, securing and lifting points. They should not normally be stacked more than three or four high, often less. Stacking I.B.C.s several tiers high should only be carried out when the construction of the I.B.C. itself or the strength of the contents will permit the resulting pressure of such top weight. If stored in the open, protection against weather and sunlight should be provided.

Special lifting cradles have been designed to facilitate the multiple handling of I.B.C.s. Care should be taken to check S.W.L. of these cradles.

Deck Cargo

Introduction

A large variety of goods, because of their inherent properties — length, other proportions, weight of individual units, etc. — are carried on deck. The bills of lading must be claused accordingly, and suitable provision must be made for safe securing of these cargoes.

It may be prudent for the master to check the bill of lading to establish which Cargo Convention the goods are being carried under.

On deck only means an uncovered space, special deck houses having doors which can be continuously open (except in heavy weather) may be used. N.B. Deck houses and mast houses can be considered suitable for either "on deck" or "under deck" stowage and Dangerous Goods stowage.

Included amongst goods normally carried on deck are: (a) certain goods classified as Dangerous Goods amongst which are asbestos, compressed gases, flammable liquids, substances giving off inflammable vapours, corrosive substances, etc., for which on deck or on deck only stowage is prescribed or desirable; (b) heavy logs of timber, sleepers, props, etc., sawn timber exceeding that which can be stowed under deck as per charter-party, long structural steel and other forms of steel; (c) heavy bridge girders or sections, pipes, railway engines, boilers, pontoons, boxed machinery, small vessels, etc.

When planning a deck cargo the following points should be considered.

1. The weight involved, its location and distribution.
2. Strength of the deck and its shoring. Point Loading (max permissible weight expressed in tons or tonnes per square feet or metres) should be obtained from ships plans.
3. Stability of the carrying vessel at the time of loading and discharging the heavy lifts.

Notwithstanding that deck cargo is usually at shipper's risk, liability for loss of or damage to same may rest with the ship in certain circumstances, e.g. if due diligence and practical measures were not observed in securing and preserving the cargo; and the stowage was negligent or improper, such as by overstowing weak packages with heavy goods, etc.

Should Dangerous Goods be carried on deck, the packaging should not exceed in size or weight that which can be conveniently handled should the necessity arise to jettison that cargo in the interest of crew or ship. While it may not be possible to jettison containers laden with Dangerous Goods (although there are proprietary pieces of equipment available for doing so in special circumstances) when those commodities are carried which are so sensitive that they may require to be jettisoned, then access to those containers must be clear, doors or other fastenings clear for opening, and the containers themselves stowed at deck level so they may be easily opened and entered.

When deck cargo is carried, access to all important parts of steering gear, boats, bilge and tank sounding pipes, etc., should be preserved and where such is called for, properly made and protected gangways should be provided for the crew.

Securing

Additional ring bolts and eye plates may be necessary to facilitate the securing of deck cargo and these should be firmly rivited or welded to deck plates and beams, deck stringer plate or upper part of the sheer strake, and closely spaced.

Chocks or beds on which the packages are to rest should be positioned over the beams. When particularly heavy items are carried on deck they should if possible be placed so that the heavy part is over the bulkhead below — and the decks, where appropriate, be given additional support by shores placed under and over beams wedged up hard to fine wedges. When dunnaging is necessary under heavy cargo it should be of sufficient size and material to distribute the weight, and spread over the deck, the boards should be laid diagonally (at an angle of 45 degrees) to avoid the buckling of deck plates.

Heavy Indivisible Loads

Introduction

Heavy indivisible loads may be defined as those weights which, because of their mass and/or their shape cannot be handled by the normal gear available on board ship or on the quay alongside.

Loads of 20 and 30 tons, i.e. containers, are handled fast and continuously at specially constructed terminals. In those circumstances they can no longer be considered as heavy lifts, but in ports and on ships where specialist gear is not readily available, they may have to be handled and treated as heavy lifts.

The techniques for handling and transporting very heavy indivisible loads have changed and become more specialised with the growth of the off-shore oil industry. For this market heavy indivisible loads are commonly divided into two categories, i.e. those of above 150 tonnes and those not capable of transport on the public highway. Physical dimensions are often a more limiting factor than the weight for transportation over public highways. Most countries have very strict legislation on the width and height of loads that may be transported. The base dimensions of the load also govern the number of axles which can be placed under it and hence the axle loading transmitted on to the ground. As the result of the need to transport very heavy indivisible loads by sea, fabrication facilities have over the years been developed close to water, and in many cases shipyards have converted their facilities to build such units.

A new generation of vessels have been developed to carry these heavy indivisible loads and range from vessels with specially designed lifting equipment (derricks, cranes and shear-legs) to self-propelled flat topped platforms which are semi-submersible. The semi-submersible vessels are not usually equipped with heavy lifting gear, but rely on being able to submerge the deck below the unit to be lifted. With the ability of these vessels to raise and lower their cargo decks, rolling devices and skidding techniques may be used for loading and discharging units on to these vessels, if floating on or off is not viable.

The techniques developed for moving these heavy indivisible units have produced two areas of skill:

(a) techniques of moving the load laterally;
(b) techniques of moving the load vertically.

The vertical movement of heavy indivisible loads has developed as a natural extension of the heavy lift operations of traditional merchant shipping. The masses that may require to be moved as one unit may reach a weight of 3,000 tonnes on a single boom. Stability requirements (which become more critical when moving these types of loads) may be improved when vessels of a semi-submersible nature are used — the semi-submersible position providing a very stiff vessel. A further requirement, particularly when moving horizontal loads direct from quay to vessel, is a high speed ballast transfer pump system to control lift and trim to a very accurate degree.

The Load

At the time of booking and pre-planning the transport of heavy indivisible loads, certain information is vital:

(a) the weight and size and construction of the loads;
(b) what support points of the loads are required — or are permissible;
(c) the maximum load of each specific support point which the design would tolerate;
(d) where may the lifting attachments be made on the load and how are they to be made;
(e) what securing points are available on the load and where are they located;
(f) is the speed of movement of the load in any way critical;
(g) what additional lifting equipment is required to handle the load, what is its weight, who provides it, does it travel with the load.

Preparation

Where traditional heavy lift derricks are used which are themselves supported by the mast or samson post, these supports, unless suitably stiffened, should be themselves supported by not less than three stays; one leading fore and aft and the other leading to each side forming an angle of about 45 degrees with the former. Stays, also shrouds, should be set up taut and stretching screws securely stopped. A slack stay will not commence to function until after the mischief has been done. Careful attention should be given to the ships plans relating to the rigging and staying of the heavy lift gear.

All leads should be set up to avoid chafing the falls. Avoid slack turns getting on to the winch barrel when winding in slack wire — the presence of such is fraught with great danger when the derrick becomes loaded. Where "steam guys" are led to a single winch, care must be taken to ensure that the weight of both remains balanced, and that one does not become slack during any stage of the operation. It is sometimes the practice to overcome this problem by securing the end of the topping lift purchase to a yoke which in turn supports the end of each steam guy. Thus although the derrick may be topped or lowered a weight is maintained on the steam guys no matter at what angle the derrick is operated. In these circumstances care must be taken to ensure that the yoke is so positioned that it will operate correctly.

It is important that the vessel is trimmed to an upright position before heavy lift operations commence. In many instances vessels are now fitted with suitable fast flooding ballast transfer tanks.

The Gear

In the U.K. a heavy derrick, crane or hoisting machine with its gear, has to be tested with a proof load which should exceed the safe working load as follows:

Safe Load	*Proof Load*
Up to 20 tons	25% in excess
20–50 tons	5 tons in excess
Over 50tons	10 tons in excess

The gear should be adequate for the load, including any extra weight, i.e. for lifting tackle, which itself may well weigh many tons.

The heavy lift derrick should not be operated in a depressed condition, e.g. near the horizontal, since if the vertical component of the thrust becomes low, there is danger of the heel springing out of the shoe.

When using container cranes to handle uncontainerised heavy indivisible loads, there may be provision for a "rams horn hook" to be installed in place of the container spreader. In some instances the spreader has special lugs to which lifting wires may be shackled, and the heavy lift supported directly by the spreader. Where shore gantry cranes are used, it must be ascertained that the port of loading and port of discharge have cranes of suitable capacity, with the necessary distance between the legs (and/or a revolving spreader) to allow the load to be handled.

Tackle

Slings should be made up of the appropriate wire or chain to provide an adequate safe working load, and of the correct length so that they do not damage the cargo and do not require shortening because of too long a drift.

The two ends of slings should be connected to the block of the derrick with the eyes on the bow of the shackle and not on the pin, especially if the drift between lift and shackle is short and the spread large — which should be avoided where possible.

Slinging should be set up in such a way that the load remains level. However there may be occasions, when, to allow a particular awkwardly shaped load to enter the hatch square, the load will have to be slung at an angle. If this is the case it must be ascertained that the load itself will not be damaged by such treatment, and the slings made up accordingly.

Where goods may be become crushed, or where the length of the load is such that the angle of the sling will create unacceptable bending moments on the load, the use of a spreader, or beam may be required. As mentioned earlier the weight of any such tackle must be included in the calculations for the lifting gear.

Full use must be made of all proper lifting points on the load, and those points which are marked as suitable for placing the sling.

Vertical and Horizontal Movement

Careful winch driving is essential to avoid the gear being put to undue strain. Too great a speed of operation increases the strain on the hauling part and may be dangerous if for any reason it should become necessary to stop suddenly.

Horizontal movement may be necessary for the full loading operation (and/or discharging operation), or for stowage in the wings of the holds. For this latter purpose heavy drag links should be fitted on bulkheads and deckheads, with suitable doublers to prevent buckling to which the hauling or bowsing tackles may be attached. The use of beam grabs or clamps, may distort and disfigure beams, frames and stiffeners, and is not a recommended practice. Consideration must be given to the coefficient of friction between the moving surfaces and the effect of gravity if the move is other than on the level (e.g. if a list develops). Consideration must be given to the break out force required to overcome inertia and start the load moving, not just the (usually) smaller force needed once it is moving. It is necessary, particularly when moving loads which are not on the level, to consider the need for restraining mechanism.

The ability to move a very heavy load horizontally, can be improved by using various methods each of which reduces the coefficient of friction, the time of break out, and for moving the load.

System	*Approx. Coefficient of Friction*		*Approx. Bearing Pressures*
	Break-out	*Running*	
Primitive Slides			
Ship launching greases	0.25	0.055	1.0
Molybdenum disulphide greases	0.06	0.04	700
Sophisticated Slides			
PTFE	0.12	0.05	14/45
PTFE woven	0.08	0.055	70
Thin Film Fluid Bearings			
Air	0.005	0.001	0.035–0.35
Water	0.002	0.001	0.07–0.4
Grease	0.005	0.0015	6.8
Oil	0.0015	0.0005	6.8
Thick Film Water Bearing	0.035/0.05	0.02/0.3	0.6/1.0
Air hover systems	0.01	0.005	0.002/0.014

Stowage Requirements

The stowage position must be selected which can best support the weight of the heavy lift. There may be also a requirement to have the heavy weight in such a position in the ship that the *g* forces generated by pitching and scending will not affect the load or its securing.

The load should be positioned in such a manner that heavy point loading is not exerted on deck or tank top plating in between frames and beams. When this is not possible, bearers should be provided to spread the load. Usually the bearers should be laid fore and aft or diagonally to achieve the maximum support from floors and beams.

Support pads provided for container stowage may, occasionally, be used directly for the support of uncontainerised heavy lifts. In certain circumstances flat racks, platforms, often specially designed for heavy lifts, may be located in the container stowage positions and utilised for supporting the heavy lift.

It must be borne in mind that when heavy lifts are carried on the tank top of a cellular vessel, there is most likely to be a loss of container stowage positions vertically above that cargo.

Restraint

As mentioned earlier the cargo may have to be restrained during the loading operations — particularly if it is in a low friction device such as air or water skids. Once stowed in position lashing and securing and tomming must be carried out to prevent the slightest movement of the load. All lashings must be set up tight, wooden tomming must be secured in such a way that it cannot be dislodged by ship vibration, working and movement. As a rule of thumb a total lashing on one side of a load should have a combined breaking strain of at least 1½ times the total weight of the load to be restrained. Tomming and chocking would be extra to this.

All lashing and tomming should utilise the appropriate and approved points on a load to give the support. Where necessary extra securing points will have to be welded or rivited to ships' frames and decks.

Loading

The ship should be upright when loading (or discharging) a heavy lift and have adequate stability. The double bottom tanks must be either pressed up or dry. If ship's gear is being used it must be remembered that the effective centre of gravity of the load moves to the derrick head as soon as the lift is floated.

Lifting operations should be interrupted to carry out checks on the slinging arrangements:

(a) as soon as the weight is taken by the slings.
(b) when the load has been lifted a short distance.

The whole system should be checked through to ensure that no undue strain is being imposed on gear, equipment, or the load itself.

Clear lines of communication should be established between all those involved in the operation, with one person only in charge. It may be necessary to discontinue other operations close by on ship or shore to allow those involved with the heavy lift to hear and understand the control signals.

Discharging

Similar precautions should be taken as described under "loading". Furthermore it may be that the vessel, being at the end of a voyage or passage, may have less bottom weight in the form of fuel oil or water. This will have to be taken into account when considering the stability requirements. It should also be borne in mind that when a heavy lift is hoisted from its place of rest the centre of gravity of the weight is transferred to the derrick head. Again it is important that one person only should be in charge of the operation, and that all activities are pre-planned with appropriate stops to check equipment and gear.

Occasionally heavy lift cargo has to be offloaded straight to the sea. This might be a requirement of semi-submersible vessels, or a requirement of the cargo itself (e.g. lighters, launches, cylindrical and rectangular tanks, etc.). In such cases, and provided that the pieces are of sufficiently strong and robust construction to withstand the stresses and falls, they are carried on deck and launched over the side at destination.

In the case of cylindrical objects, which of course must be perfectly watertight and free from small parts protruding, they should be mounted on a strong launching structure built of timber, the inboard end of which rests on top of the adjacent hatch and the other on the bulwark, being adequately supported at intermediate points. The frame of the structure should dip about 2 degrees in the direction of the ship's side.

So mounted the discharging is a simple process and is affected by lifting the vessel in the desired direction by means of water ballast or fuel oil. Tanks not in use for that purpose should either be empty or pressed up hard and it should be borne in mind that when the weight drops off the vessel it will roll sharply in the opposite direction.

For lighters, rectangular tanks, etc., a similar structure is built, extending a little beyond the bulwarks.

On top of the main structure a movable cradle like arrangement is built to the shape of the craft so as to support it at three or four different points, according to its length. This is mounted on launching or slipways formed integral with the main structure with suitable means of locking it into position. The craft is securely lashed by wire lashings to prevent movement during transit and is launched overboard sideways by giving the ship the necessary lift.

After Discharge

It sometimes becomes necessary to unship stanchions when stowing heavy lift cargo. All such, forming as they do important parts of the structural strength of the vessel, should be properly restored and secured in their original position before stowing any cargo on the decks above.

Where as sometimes happens, it is not possible or convenient to restore all stanchions into the position without seriously embarassing the stowage, heavy shores of timber, laid on stout timber fore and aft bearers, should be fitted and tightly wedged into position to carry the weight of the deck above. Neglect to do so will likely result in the decks being set down.

Special Cargo ("Specials")

This term is usually applied to attractive goods for which special stowage, supervision, and careful checking and tallying is considered desirable. This may be on account of the value: the ease with which attractive articles can be removed (e.g. small items easily pocketed) or consumed (e.g. liquor). Precious or valuable goods would be treated as bullion — with the appropriate special strong room stowage (see below).

Typical articles which might be described as "specials" include: beer in bottles or cans; bottled spirits; wearing apparel; fancy goods and jewellery; furs; laces; mail; portable electronic equipment; toiletries such as packaged perfumes, etc.

This cargo should be received and in turn kept in a lock up or special cargo locker or cage ashore. If circumstances or quantity permits, it should be carried onto and off the ship. An officer should be personally tallying or in attendance — and handling by night avoided — if possible.

Every vessel engaged in general cargo trade should be provided with a special cargo locker, of substantial construction, fitted with reliable locking arrangements and well lit. Its location should be remote from crew's quarters, and ventilators leading to the locker should be fitted with iron bars.

It is false economy to be too sparing in regard to the size of these special compartments and even if all the space provided is not required on most occasions, it should be an easy matter to fill unwanted space with suitable ordinary cargo.

Where "lockups" on the wharf are not available, safety of this class of cargo demands that it be put on board under lock and key, as soon as received. However that may be difficult, if not impossible, to arrange in the case of a vessel receiving cargo for multiple ports where only one special cargo compartment is available — as the goods destined for the last port must be stowed first and so on. This, added to the fact that all specials are not delivered to the dock until near completion of loading, necessitates this class of cargo remaining on the dock and being exposed to pilferage for a longer period than actually should be necessary.

This difficulty is easy to overcome in Break Bulk Cargo ships by designing a special cargo locker: (a) in the upper 'tween deck between hatches; (b) divided into four or more compartments formed by means of heavy steel netting suitably stiffened and framed with light angle bars, the doors to be of the same construction with suitable locking arrangements; (c) each compartment to be entirely separate from the other, having its own separate door; (d) the doors being inaccessible unless the hatch covers on that deck are in position, as opposed to the customary arrangement of doors in the wings where they may be tampered with; (e) arranged in such a way that special cargo may be received at the appropriate locker at any time and in any order; (f) when discharging, the door of one compartment only need be open thus reducing opportunities for pilfering and eliminating risk of over landing or over carrying.

Bullion and Specie

Bullion should not be received on board unless the vessel is fitted with a proper strong room or safe with suitable capacity.

Strong room should be fitted with two locks having different keys, one in the custody of the Master, the other in that of the Chief Officer or other responsible person.

Gold and silver bullion, i.e. uncoined gold and silver respectively: shipped in ingot or bar form, packed in strong well made boxes which are usually fitted with strong rope brackets for handling; very rarely, silver is shipped in large ingots unboxed.

ISO Containers, if used for the carriage of bullion and specie, should be of all steel construction. A careful inspection should be carried out before presenting the container to the shipper to ensure that the container is in a structurally sound condition, with door locking rods and cams working correctly. Special locks and seals to shipper's requirements may be affixed in addition to those customarily used for the trade. Because of its volume/weight ratio, bullion should be stowed as near as possible to the side and end walls where the best floor strength prevails. Careful chocking and bracing is essential to prevent any movement, while at the same time having regard to the requirement that individual boxes or crates may need to be opened and contents checked at any stage during transit.

Receiving

The utmost precaution should be observed at every stage of receiving, stowing and delivering this very valuable cargo; each operation should be personally supervised by the ship's officers, assisted when necessary by responsible members of the shore staff. If putting bullion on board by winch, it should be slung by means of special closed nets to which a small buoy is attached by a long line — care being taken to avoid breaking the seals. If carried by hand over open gangways, nets should be spread under the same, or the gangway closed temporarily for boarding. Use special bullion bags or strong canvas suitably roped — two men to a bag — but if such is not available and the cases have to be carried by hand the men should not be permitted to carry them on their shoulders — the only safe way is to carry them with both hands in front. An officer should tally bullion at the rail and its progress to the bullion room should be carefully watched. Tallying check again in bullion room before stowing, recording every mark and number, and also examining the seals. Mates' receipts should specify all marks and numbers and bills of lading suitably annotated.

Delivery

Delivery should never be made by Ship's Officers to anyone except on a written order from the Master, owners or agents (clearly specifying marks and numbers to be delivered) who will have satisfied themselves as to the bona fides of the consignees, etc., and only against a clean receipt clearly specifying marks and numbers, seals intact, etc. Any disputes should be settled before delivery.

Masters of vessels taking in bullion abroad should satisfy themselves that the bill of lading contains a clearly worded clause exempting the vessel and owner from all risk of loss, damage or theft, howsoever arising, whether through the fault of a servant of shipowner or not. Failing such clause in the bill of lading the bullion should not be carried unless the entire risk is covered by special insurance policy.

The keys of the bullion room should be kept in the safe in the Master's, Chief Officer's or Purser's room or equally safe place and should never be allowed to get into the hands of anyone except the senior officers.

Precious Stones, etc.

This includes pearls, gems, currency notes, bonds, travellers' cheques, postage or revenue stamps, high class jewellery, etc. All of these items, because of their size and value, are very tempting to thieves. Travellers' cheques are particularly vulnerable because of the case with which they may be exchanged for cash.

The packages should not be brought on board until shortly before the vessel sails. Delivery should be made to the Master, Chief Officer or Purser on board who should carefully examine each package to see that it is in sound condition and all seals intact (defective packages should be rejected) and note every mark, number, address, etc. Receipts and bill of lading should be clearly endorsed "nature and value of contents unknown said to be . . .".

Custody — immediately the packages pass into the custody of the Ship's Officers they should be safely locked up in the safe or specie room where they should remain until the vessel reaches its destination. The key of safe or specie room should not leave the custody of the Master or other responsible officer in charge of their custody.

Delivery — consignee should be called upon to take delivery immediately the vessel reaches her destination. This should be effected on board the vessel unless the bill of lading makes a special stipulation to the contrary, as is sometimes the case, when delivery has to be made at a bank, etc. In such a case the officer entrusted with delivering the package should be accompanied by an armed police guard and should make the journey in a vehicle.

Delivery should only be made on an order specifying numbers, marks, addresses, signed by owners, agent or Master, who have satisfied themselves of the bona fides of the consignee.

A clean receipt specifying seals intact, etc., should be demanded and the bills of lading surrendered; any dispute should be settled on the spot.

The loss of a package or part of the contents of any of these sort of commodities would be such a serious matter that no attempt should be made at carrying the same by any vessel that is not equipped with a thief proof safe suitably placed, or a specially built specie room of steel plating with thief proof locking arrangements. Neither should such be shipped without it being agreed beforehand that the bill of lading should be clearly worded so as to protect the owner and vessel from the consequences of loss, damage or theft, howsoever arising, whether through the fault of the shipowner or the fault and/or by the privity of his servants or not, unless such risks are entirely covered by a special policy of insurance held by the owners or agents, in which connection it should be noted that most protection and indemnity clubs will not cover any risks of this class unless they have approved in writing of the contract of carriage and of the spaces, apparatus and means used for the carriage of such valuables.

Ventilation

Ventilation of cargo spaces may be necessary to:

Remove heat.
Dissipate gases.
Supply dry air to help prevent condensation.
Remove taint.

Not every commodity requires or even benefits from being exposed to ventilation during transport. Cargo loaded in bulk carriers may only benefit from surface ventilation, i.e. a change of air between the underside of the deck or hatch above and surface of the commodity. It follows that if there is a problem within the stow such as pockets of moisture, ventilation will have little effect.

Heat may be generated by live fruit, wet hides, vermin, and commodities liable to spontaneous combustion.

Gases which may require dissipation include inflammable and explosive gases such as those given off by coal; vehicle exhausts on Ro-Ro ships; CO_2 and ethylene from ripening fruit and vegetables.

Condensation appears in two basic forms: ship sweat and cargo sweat. The former appears as tiny beads of moisture condensing on the ship's metalwork, and typically, might occur on the side of the hold when the sea temperature fails and reduces the adjacent metal temperature to a value below the dew point of the surrounding air. Cargoes sweat forms on the surface of the cargo when its temperature is below the dew point of the air adjacent to it.

The removal of existing or residual taint by means of ventilation may be carried out in conjunction with an ozonating unit. It may be required to reduce the probability of sensitive cargo becoming affected by the taint, or to improve working conditions for labour.

The proper ventilation of holds therefore, is indispensable to the correct carriage of some goods, and may assist in the preservation of the ship's structure itself. Sometimes with coal cargoes and those which give off inflammable and explosive gases, or are liable to spontaneous combustion, it may be absolutely necessary for the safety of crew and ship.

A well designed system of ventilation — natural or mechanical — aims at inducing a constant circulation of air throughout the holds and the mass of cargo by which heat, moisture, fumes, vapours, gases and odours given off by the cargo are discharged into the open air and replaced by air of a dew point suitable to the prevailing conditions.

Three systems of ventilation are in general use:

(a) The natural ventilation of ordinary cargo compartments, supplemented in certain cases by modified mechanised air circulating systems.
(b) The temperature control system of circulating air in insulated compartments, controlling its temperature and carbon dioxide content given off by certain commodities (see "Refrigerated Cargo").

(c) Mechanical ventilation in cargo spaces and control of humidity in the compartment. The principle of most of these types of mechanical ventilation is to reduce the dew point of air in ships' compartments by de-humidification to the point at which ships' sweat and cargo sweat cannot occur. No attempt is made to change the temperature condition in the holds, the whole object being to control the dew point of the air surrounding the cargo.

Correct ventilation can keep cargo in good condition in the great majority of circumstances if properly employed. No system of ventilation can prevent damage if the Ship's Officers do not know when and when not to use it. Also, of course, stowage of the cargo must be correctly planned with regard to the various classes of commodities of which it is composed, together with proper dunnaging and positioning to ensure that the ventilation reaches all the necessary cargo.

Open-top, Open-sided and Flatrack containers may sometimes be stowed with cargo requiring ventilation while on board, e.g. fruit, vegetables, spices, etc. If stowed on deck then it must be ensured that adequate ventilation can reach the contents while properly covered against rain and spray. Moreover the container should be stowed on deck in such a position that the tilt is protected against damage from wind and waves during the passage. Below decks the stowage position should be so situated as to take maximum advantage of any mechanical ventilation installed, and care should be exercised against cross-taint, e.g. a container of wet hides might leak strong-smelling brine.

When to Ventilate — Some Guidelines

Ventilate if the dew point of the air is lower than the dew point of the cargo space. The changing air will ventilate the cargo space and there should be little danger of condensation.

Restrict ventilation if the dew point of the outside air is higher than the temperature of the cargo. To ventilate under such conditions would cause danger of condensation on cargo as the outside air with higher dew point comes into the cargo space.

When the vessel is fitted with a mechanical ventilation system, the air should be re-circulated within the cargo space when the introduction of external air is restricted. Despite this guidance special attention must, of course, be given to conditions where cargoes are giving off fumes, etc., where according to certain circumstances ventilation may be necessary despite the possibility of condensation. The aim should always be to avoid stowage of cargo requiring different ventilation treatment within the same compartment. Similarly two compartments with different ventilation requirements should effectively be sealed off from each other.

It is often required that cargo normally of a very dry nature, e.g. milk powder, casein, etc., has to be stowed in the compartments adjacent to hard frozen chambers. Experience has shown that the most effective method of carrying cargo under these conditions is to restrict the ventilation completely during the passage. In some refrigerated vessels, decks or bulkheads are fitted with thermal injection. The principle of such being to provide a heat input equal and opposite to the heat loss from the refrigerated compartment. The fitting of this equipment is of considerable value in ensuring safe out turn of cargoes stowed under such conditions, but does not warrant departure from the basic principles of ventilation.

The importance of maintaining a dry ship and dry cargo cannot be over emphasised. The cargo should be protected from all unfavourable exposures of temperature and moisture from the point of origin to the delivery, and the shipowner for his own protection should know more thoroughly the history of the products delivered to his charge and of the condition that the products are received. Typical of this category is the carriage of perishables in open top and open side containers. It is often the practice for the tilts of these containers to be left open while stowed below decks to improve the prospect of ventilation. These covers should be replaced and secured at time of (or prior to) unloading. Similarly when loaded on board the ship's staff should check the appropriate covers are drawn back to take maximum advantage of below deck ventilation. In cases where terminal staff have been adjusting covers for the same purpose during the container dwell time at the terminal, it is important that ship's staff and terminal staff liaise to maintain the weathertight integrity of these containers while at the same time affording the contents maximum ventilation.

The responsibility of the terminal operator involves protection of the cargo impounded on the receiving pier in so far as practical conditions permit. The stowage, subject to the same limitations, should provide the most advantageous condition in the ship. Through an understanding of what takes place within a loaded ship, the officers can reduce sweating hazards and provide intelligent operation or sealing of ventilation systems.

There are two alternatives in meeting the problem of cargo sweat: the control of the dew point in all parts of the loaded cargo space. The tempering of the cargo to keep it above the dew point of the air to which it may be exposed while at sea or at warm ports of call or at its destination. As an example canned condensed milk is sometimes shipped in a warm condition out of temperate latitudes so that cargo sweat has less chance of occurring during the passage through warmer climates.

Ship sweat may be prevented by air conditioning systems efficiently designed and efficiently operated. Air conditioning plants are not the panacea for all ills and unless the cargo being served by them is delivered in temperatures above the dew point of the air at the port of delivery, there is no guarantee that condensation will not occur.

The temperature of the air both outside and in different cargo compartments should be regularly taken and recorded. Attention to the ventilation should always be entered in the Deck Log. It is of utmost importance that attention to the ventilation of cargo be clearly and regularly recorded in the Mate's Log with special mention to be made of any opening of hatches, etc. The absence of such records has on many occasions involved the owners in heavy claims for alleged neglect.

Except with cargo which gives off highly inflammable or explosive vapours or gases such as petroleum spirits, etc., a certain number of hatch covers, should, when conditions of sea and weather permit be kept open at each end of each hatch leading to holds containing goods which give off heat, moisture, strong odours, etc., such as fruit, jute, seeds, copra, sugar, hides, coal, gambier, etc.

With cargo which sweats profusely such as rice, maize, jute, etc., in order to avoid heavy condensation which might be produced by heavy downward rush of cold air, it is best to open the hatches gradually.

Dunnage

Dunnage may serve the following purposes, according to the nature of the cargo carried:

1. To protect it from contact with water from the bilges, leakage from other cargo, from the ship's side or from the double bottom tanks.
2. To protect it from moisture or sweat which condenses on the ship's sides, frames, bulkheads, etc., and runs down into the bilges.
3. To protect it from contact with condensed moisture, which is collected and retained on side stringers, bulkhead brackets, etc.
4. To provide air courses for the heated moisture laden air to travel to the sides and bulkheads along which it ascends towards the uptakes, etc.
5. To prevent chafe as well as to chock off and secure cargo by filling in broken stowage, i.e., spaces which cannot be filled with cargo.
6. To evenly spread out the compression load of deep stowages.
7. To provide working levels and protection for the cargo on which labour can operate and serve as a form of separation.
8. Provide access for cooled air round or through the cargo for temperature controlled requirements (see "Refrigeration").

Throughout Part 3, "Commodities", recommendations are given regarding the use of dunnage, matting, etc., to protect the cargo from contact with metal decks, bulkheads, beams, etc. However, modern fast vessels, with reduced time on passage and fewer stanchions and other metal obstructions, combined with high cost of labour and materials, frequently omit the use of such dunnage especially where bagged commodities are concerned. Whilst the results generally turn out to be satisfactory, it does not ensure that all reasonable precautions for the safe carriage of cargo have been taken. The connsequence therefore of any claim against cargo out-turn will perforce still be judged against whether all reasonable precautions have been taken or otherwise.

Permanent Dunnage

The floors of many ships, the double bottoms of which are intended to carry fuel oil, may be completely sheathed with 635 mm or 760 mm planking laid on battens so as to provide an air space so that any leakage from below or from cargo above and condensed moisture coursing down the bulkheads is drained into the bilges.

Ship's sides may be fitted with dunnage boards or spar ceilings of about 50 mm thickness, usually 150 mm to 180 mm in width, spaced close enough (should not be more than 230 mm) to prevent packages from protruding into the frame spaces, and to interfere with ventilation or coming into contact with frames or shell, and thus be damaged by condensed moisture coursing down the same.

For ordinary cargoes, the permanent ceiling or dunnage is sufficient, provided it is dry, clean and free from oil stains or otherwise in a condition likely to cause damage to a cargo — like bag or baled goods — in close contact with it. Where the position of permanent dunnage is otherwise, dunnage and matting are necessary. In this connection it is as well to remember that a cargo such as grain, in close contact with the ceiling the surface of which may appear to be dry, may draw moisture from the timber by capillary action and so be damaged.

In the bilges, however, and on top of cement caps, on stringers and brackets it is always necessary to lay dunnage as the condensed moisture runs down the former or is retained by the latter.

With the ever increasing cost of dunnage material and attendant labour a number of types of permanent dunnage are now in use particularly in insulated compartments (see also "Refrigeration"), viz:

(i) Permanent Collapsible Dunnage (P.C.D.). This is 50 mm × 50 mm or 75 mm × 50 mm suitably varnished or painted timber, the lengths of which can be easily collapsed together. The fitting is secured to the bottom of the inboard or outboard bulkhead of chambers with a swivel fitting, so that the dunnage when not used can be housed in a flat and vertical position. Not suitable for holds or large decks where cargo has to be dragged over it.

(ii) Aluminium Strip Dunnage. Closely corrugated aluminium made in easily portable dimensions 3 m × 500 mm. Suitably strengthened so that fork-lifts can operate over it. This dunnage can normally be left under general cargo and is easily cleaned. With this method squares of 'tween deck hatches have to be battened in the conventional manner.

(iii) Steel Grating Dunnage. Similar in principle to (ii) but more permanent in character and is being fitted over lower hold tank top ceilings. Suitable for fork-lift work.

(iv) Permanent Gratings. Frequently found in small lockers or in vessels with a limited number of refrigerated compartments. They can be cumbersome to handle and therefore difficult to clean.

(v) ISO Containers. Insulated and refrigerated Containers usually have extruded aluminium "T" section floors built in. (See also "Containers").

All the permanent types of dunnage described must prior to loading be closely inspected, damage made good and thoroughly cleaned.

Dunnage Materials

Many different sorts of wood and materials are in use for and make good dunnage, but in all cases it is necessary that it be sound and dry, clean and free from oil, grease or creosote stains, or matter likely to develop maggots, etc. Many authorities prohibit the landing of bark covered dunnage (i.e. Australasia). There have been numerous cases of vessels being delayed during discharge, when local authorities have detected insect ridden dunnage and have instructed the vessel to be fumigated, and/or the offending dunnage to be taken ashore and burned. Oily or greasy dunnage should never be used with dry goods, while nothing will excuse the use of wet dunnage.

Bamboo — Loose or in Bundles

If loose sign for quantity unknown. If in bundles they should not, unless permitted by Bill of

Lading, be cut open. Make excellent dunnage for dry goods liable to heat and throw off moisture. Stow on ends, sides and bulkheads, to facilitate heated air to rise.

Battens

Of 50 mm × 50 mm (2″ × 2″) or 76 mm × 76 mm (3″ × 3″) used largely with refrigerated cargo.

Boards

Generally of rough 150 mm × 25 mm (6″ × 1″) or 100 mm × 50 mm (4″ × 2″); largely used for 'tween deck dunnage, for laying over bulk and as a platform for the necessary tiers of bags to secure same, for making a platform over ores, wet goods such as oil barrels, jaggery, etc.; also for laying on bulkheads, spar ceilings, etc.

Coir

In dholls, sometimes used in the Indian trade. The greasy kind is very useful with drums of oil, etc., but it should not be used with dry goods.

Dunnage Bags

These may be re-usable and disposable. Both are inflated with compressed air. Different sizes and different materials allow almost any cargo to be restrained. They incorporate a valve to allow the quick release of the compressed air at port of discharge, where the bags should be carefully emptied and retained for future use. Disposable bags may be made of paper with plastic linings. Used on the same principle as the others, the means of collapsing them is to puncture the bag. The technique for securing in both cases is to work the cargo from bulkhead or other secure position on either side and the resultant gap remaining in the middle of the cargo is filled with dunnage bags which exert pressure on both sides, thereby restraining the cargo and providing a wide area of pressure. Ideally suited for restraining cargo that is fragile or difficult to restrain in any other way.

Hardboard, Chipboard, etc.

Usually in sheets 2,440 mm × 1,220 mm (8″ × 4″) or 1,830 mm × 915 mm (6″ × 3″). May be used to protect bagged and baled cargo, etc., from contact with ships' metalwork and other cargo. May also be used for separation. Chipboard may be used in lieu of plywood as walking boards.

Laths

To lay between tiers of green fruit or cartons of refrigerated cargo so as to assist in the circulation of air. May be of wood or polystyrene. Usually 9 mm (⅜″) thick. 25 mm or 50 mm (1″ to 2″) in width.

Paper

Kraft or similar paper is frequently used to good effect in protecting cargo from contact with bare metal, and in separating one cargo from another. Care must be taken in ensuring that the paper does not shift out of position after being laid, particularly when the practice of "dumping" bagged cargo into an open hold is carried out.

Rattans

In bundles. Those of superior quality are seldom, if ever, permitted to be used as dunnage. Bundles must not be broken up. Rattans absorb moisture and form air courses, and are so ideal dunnage for floors, sides and bulkheads, with cargoes which are liable to heat and sweat such as pepper, dry rubber (though not on floors), sago, tapioca, etc. Stow on ends at sides and bulkheads and for cargo specially liable to heat, such as pepper it is a sound practice to lay a tier between the bags at about mid height.

N.B. When different consignments of the same class of cargo to be used as dunnage are shipped, they should be stowed apart — in different compartments if at all possible — in order to avoid mixing and confusion on delivery. With horns, hooves, bones and similar goods, Mate's receipts and Bills of Lading should be endorsed "weight and quantity unknown, all on board to be delivered".

Sawdust

Comes in very useful to absorb the drainage from certain cargoes such as gambier, creosoted goods, etc., but the use of such with linseed or other seed oils is very dangerous, in as much as sawdust, when impregnated with vegetable oil, is very liable to spontaneous combustion.

Weight of Dunnage

As full cargoes of some commodities require a very considerable amount and weight of dunnage in loading, it is well that, when a vessel is chartered on a deadweight basis, it is borne in mind that the "weight of dunnage used and necessary for stowing cargo is to be counted part of the deadweight called for by the charter".

N.B. Many traded classes of goods are shipped at a low rate of freight on the understanding that they can be used as dunnage and/or for filling broken stowage, but with certain exceptions their use as such must be confined to clean cargoes. In all such cases the Bill of Lading should contain a clause authorising their use for such purposes the Mate's receipt to be endorsed to like effect.

Dunnage with Refrigerated Cargo

Dunnage used with refrigerated cargo serves a twofold purpose, i.e. providing channels for thorough circulation of the cooled air, and, as with other cargoes, even distribution of superimposed weight.

To ensure the former, dunnage is to be in line with the air flow; the amount used being dependent on the system of refrigeration being installed (see "Refrigeration").

ISO containers usually require no dunnage for refrigerated cargo, as the floors, walls and ceilings have their own built in battening (see "Containers").

Dunnage over Refrigerated Cargo Space

There are many instances where it is necessary to load general or chilled cargo over a hard frozen refrigerated space. Adequate dunnage should be laid in order to ensure that the cold air will not strike through to the cargo stowed in the compartment above. On unsheathed decks it is advisable to have at least 127 mm (5″) of dunnage, i.e. 50 mm × 50 mm (2″ × 2″), crossed and overlaid with 150 mm × 25 mm (6″ × 1″) boards, and as additional insulation a sprinkling of approximately 50 mm (2″) of odourless sawdust. On sheathed decks, approximately 76 mm (3″) of dunnage is sufficient, provided that the insulation beneath is known to be sound. Hatch coamings, iron ladders, etc., should be well covered. The most effective way of avoiding damage over any hatch coaming by condensation is to box the coaming with dunnage wood and fill in same with sawdust.

DANGEROUS GOODS

Introduction

Most countries have legislation for the safe carriage of Dangerous Goods. Dangerous Goods are defined as those classed in any such acts, rules or bye-laws or having any similar characteristics, properties or hazards. Legislation covers the classification, packaging, stowage (including permissible proximity and positioning) of Dangerous Goods during transport and storage.

The handling and carriage of Dangerous Goods must be carried out in full compliance with the laws of the country from which the Dangerous Goods are being shipped, the laws of the country in whose vehicle or ship it is moved, the laws of any country through which the goods will transit, and of course the laws of the country of destination.

So the classification, packaging and stowage of Dangerous Goods must be in accordance with any legislation which may be enforced in: (a) the country or origin; (b) the country of destination; (c) any country which it has entered; (d) the country under whose flag the carrying vessel operates.

IMO

The International Maritime Organization has produced the code for the carriage of Dangerous Goods in ships (the IMDG Code). This code is based on the report of the United Nations Committee of Experts on the Transport of Dangerous Goods, which also forms the basis for legislation and recommendations for transport of Dangerous Goods by other modes, e.g. ADR, RID, IATA — road, rail and air. The IMDG Code now has wide universal acceptance and has been adopted with legal standing in most maritime countries. It does form the basis for international movement of Dangerous Goods (with regard to classification, documentation and stowage), which is particularly important where shipping companies are operating within international consortia. It is important to maintain a library of the organisation's publications including IMDG Code 1994 consolidated edition et seq.

Classification

The IMDG recognises nine broad classes of Dangerous Goods. For the correct classification and labelling of Dangerous Goods the IMDG Code should be referred to. The fact that a substance may not be listed in the IMDG Code should not be taken as evidence that it is non-dangerous. It is the particular properties of each individual commodity which must be taken into account.

Class 1 — covers diverse hazards ranging from safety class ammunition to those which have a mass explosion risk. Commodities in this class, with the exception of safety class, are usually subject to extremely stringent national legislation and port bye-laws.

Class 2 — gases — compressed, liquefied or dissolved under pressure. According to their properties or physiological effects, which may vary widely, gases may be explosive, inflammable, poisonous, corrosive or oxidising substances, or may possess two or more of these properties simultaneously. Some gases are chemically and physiologically inert. Such gases, as well as other gases may be regarded as non-toxic but nevertheless be suffocating in high concentrations.

Many gases of this class have marked narcotic effects which may occur at comparatively low concentrations, or may evolve highly poisonous gases when involved in a fire.

Gas cylinders, although a particularly strong type of package, are potentially extremely dangerous if involved in a fire where pressure build-up may cause them to explode.

Some substances in this class, under transport conditions, are liable to polymerise, i.e. combine or react themselves, so as to cause dangerous liberation of heat or gas, possibly resulting in pressure on the receptacle. These substances should not be transported unless they are properly inhibited or stabilised; this is usually indicated with appropriate reference name.

Class 3 — flammable liquids. The main danger associated with the carriage of substances of this class is the escape of inflammable vapour (certain of which could be toxic), this being especially prone to substances having a low flash point which are naturally volatile. This vapour might either form a flammable mixture with air, leading to an explosion, or catch fire through becoming ignited by a spark or flame. It is, therefore, necessary to stow the substances well away from naked lights, fires and any source of heat and packagings should serve to protect the contents against external sources of ignition. It should be noted that inflammable substances may be either miscible or immiscible with water — an important point when fire fighting is necessary.

Class 4.1 — flammable solids — solids that readily ignite. Some may explode unless kept in a saturated condition with water or some other liquid, i.e. loss of liquid might make the substance become dangerous. Keep away from any source of ignition.

Class 4.2 — flammable substances either solid or liquid liable to spontaneous combustion. Such substances should be carefully watched for any unexplained rise in temperature. Those which ignite immediately in contact with air being especially dangerous. Items such as vegetable fibre should be kept free from contamination by oil or water. They should not be loaded — if so contaminated as the self heating may commence some days or weeks later.

Class 4.3 — substances either solid or liquid which in contact with water emit flammable gases. All substances in this class must be kept absolutely dry. In some cases the gases evolved may be toxic and some are liable to spontaneous ignition due to heat liberated by the reaction.

The characteristics of each substance in this class should be studied closely and no cargo likely to interact packed in the same compartment or container. Special attention should be paid to fighting fires involving substances of this class, where the use of water, steam, or water-foam extinguishers might worsen the situation.

Class 5.1 — oxidising substances. The substances in this class, while themselves not combustible possess the property of making combustible material burn more easily and of giving off oxygen when involved in a fire, thus increasing its intensity. Some mixtures of these substances with combustible material are easily ignited, sometimes even by friction or impact. Such mixtures may burn with explosive force. There will usually be violent reaction between oxidising substances and strong liquid acids, and highly toxic gases may be evolved. Toxic gases may also be evolved in certain substances which are involved in a fire. Packages from which there is leakage should be refused for shipment. Before loading substances of this class, attention should be paid to the proper cleaning of the storage space into which they will be loaded. Particular attention should be paid to the removal of all combustible material which is not necessary for the stowage of the cargo.

Class 5.2 — organic peroxides. Both oxidising agents and inflammable, will burn readily — sometimes with explosive force. All may decompose with heat and, in general, are the most unstable of substances. They must be kept away from other hazardous cargo and sources of heat and be in readily accessible positions.

Some evolve oxygen naturally and are usually packed in receptacles which are provided with means of ventilation. Where this is so the packages must be handled and stowed at all times in the proper upright position with the ventilators all free. They must never be rolled. No deviation from the receipt and despatch procedure is permissible. Many of the commodities in this class are required to be carried under temperature control conditions, in which case all shippers' instructions and legal requirements must be complied with.

Class 6 — poisonous substances. Poisonous substances must be handled with care, having regard to the toxic effect they produce. These are liable to cause death or serious injury to human health if swallowed, inhaled, or by skin contact. Nearly all toxic substances evolve toxic gases when involved in a fire.

Breathing apparatus and protective clothing should be readily available in case of damage to packages.

Class 7 — radioactive substances. The care and handling of radioactive substances varies widely. Very stringent precautions are taken to ensure the safe packing of radioactive substances and these are all within internationally agreed standards.

The initial booking stage, prior to acceptance, should include careful study of all the relevant port regulations. Proper documentation of goods in this class is of the highest importance and staff should seek guidance either through consulting the appropriate statutory regulations, or the authorities concerned whenever necessary.

Any instructions/requirements stipulated by the Lines and the authorities (including those of way ports) should be adhered to.

Class 8 — corrosives. Substances in this class are solids or liquids possessing in their original state, the common property of being able more or less severely to damage living tissue. The escape of such substances from their packaging may also cause damage to other cargo or equipment.

Many substances are sufficiently volatile to evolve vapour irritating to the nose and eyes. A few substances may produce toxic gases when decomposed by high temperatures.

In addition to a direct destructive action in contact with skin or mucous membrane, some substances in this class are toxic. Poisoning may result if they are swallowed, or if their vapour is inhaled, some of them may even penetrate the skin.

All substances in this class have a more or less destructive effect on materials such as metals and textiles.

Class 9 — miscellaneous dangerous substances. This class contains substances which, although dangerous, have not been allocated to any other class. It includes substances which cannot properly be brought under any of the more precisely defined classes because they offer a particular danger which cannot be properly covered by the regulations for the other classes, or which present a relatively low transportation hazard. It should not be automatically assumed that goods in this class are "less hazardous".

Dangerous Goods may be shipped under certain relaxations or even exemptions provided they comply with the provision of Section 18 of the general introduction to the IMDG code, and the relevant competent authorities have accepted this section. It is primarily intended to cover the carriage of very small quantities such as are found in laboratory chemicals and medical preparations. The U.K. has adopted this section.

Labelling

(a) Packages — All Dangerous Goods' packages offered for shipment must be correctly labelled (or stencilled) with the appropriate Dangerous Goods labels and show the correct technical name as used in the shipping documents (in addition it may also be necessary to fix labels as required by certain national regulations). This label should indicate the principal hazard and where one or more secondary hazards exists, additional label(s) indicating that hazard — but with the numeral on the label deleted — should also be affixed. Any discrepancy in correct labelling should be noted and rectified and the reason investigated.

(b) ISO Containers — Four Dangerous Goods' class labels should be affixed to the container, one on each side including one on the right hand door. The label on each side should be positioned so as to be clear of the container doors when opened and secured back. The label on the front end of the container should be positioned so as to be clear of the towing vehicle if possible. In addition one Dangerous Goods' label should be affixed to the door, and this should be fully completed with the correct technical name(s) of the substances in the container together with the U.N. number and number of packages — and any other information considered useful. This Dangerous Goods' label should be completed using a waterproof medium, e.g. a spirit pen.

All Dangerous Goods' labels should be removed from the container (or masked) as soon as the container can be considered non-hazardous.

Packaging

Packagings used for Dangerous Goods should be able to successfully withstand the tests as prescribed in IMO Annex 1. While the performance tests should be applied to all new types and designs of packaging, satisfactory practical use may be accepted as equivalent evidence in existing types and designs.

Packages of all types should be properly closed and lids, bungs and other closures be in place and tight. Although it is impossible to see the inner receptacle of the package, it should be ascertained that the outer one is in order and that this has not been damaged and that any of the absorbent packaging (if used) is not missing. Any strong smells indicating any possible leakage should be carefully investigated and if the cause is not readily found, further advice should be sought. Stains on a package particularly of a recent origin, should be investigated.

Untoward corrosion or pitting of steel or iron drums should be investigated if it appears active. Drums or other receptacles which are obviously second hand and any which have recently been painted or sprayed should be inspected for corrosion, especially at welded or soldered joints. Dented drums particularly if dented near the joints, rolling bands or spigots, should be carefully examined if the dent occurs while packing the container or loading aboard the ship and the drum concerned should be set aside and loaded last, again being examined immediately before being loaded.

Wooden barrels (hogsheads, casks, etc.) are particularly susceptible to damage and should be most carefully examined to see that at no stage have the heads sprung. Stowing on end, the bung should be specially well examined.

Plywood and fibre wood cartons with plastic interiors used for crystals and powders should not be loaded if torn — with the risk of leaking contents which are often of an obnoxious nature. This also applies to multiwall paper bags and sacks.

Cylinders and tanks of inflammable and poisonous gases should have the protective hoods for valves properly fitted, and any showing signs of damage should be refused for shipment.

Repairs — it is important that any leaking packages of Dangerous Goods are repaired, recoopered. etc., by the manufacturers who will be skilled in the handling and packaging of their own particular commodity.

Booking

Included in the documentation at the time of booking, there must be the number and kind of packages; the correct technical name of the product; the class to which it belongs; the flash point — if any; the U.N. number; weight and measurement of the goods. A statement that the goods are packed in the manner approved by the appropriate authority and adequate to withstand the ordinary risks of transport.

In the case of some goods special certificates are required stating that the product has been weathered or dried for a certain period, or that it contains or is immersed in another substance which reduces its hazard. Where such a certificate is required it is important to see that it has been issued.

A Dangerous Goods Container Packing certificate must be signed (where applicable) by the person responsible for packing a container on completion of stuffing.

Handling Precautions

Provided the correct procedures are followed and care has been taken in the handling of Dangerous Goods the risk, though present, is small. However, short cuts, rough handling of the goods and a careless attitude are a sure way to cause accidents.

Dangerous Goods must be correctly documented and handled from the time of booking until the time of final delivery.

All Dangerous Goods packages should be inspected for signs of damage, leakage, or any other unsatisfactory state liable to increase their hazard, prior to being stowed in the compartment or container.

Packages of Dangerous Goods should never be dropped or thrown down, and the use of hooks, bars, etc., kept to a minimum. Mechanical loading with fork lift trucks should be strictly controlled and any rough handling avoided.

It is dangerous to take short cuts with safety regulations, and all those persons handling Dangerous Goods should be fully aware of the hazards involved.

Dangerous Goods' packages which have been wetted by rain, and received in a wet condition with frost or snow adhering, should be effectively dried before loading. They should not be loaded unless it is certain that the enclosures are intact and no moisture has entered the receptacle. This is vitally important in the cases of goods in Class 4.2, 4.3, 5.1, 5.2, 6, 8 or 9.

Packages which are provided with means of ventilation (e.g. certain of Class 5) should be kept upright during handling. On no account should such commodities be rolled in the handling and stowing operations.

Fibre board kegs and plastic lined paper bags are frequently used for chemicals in powder or crystal form, often those which will taint other cargo. Some solids will liquefy or soften with the rise in temperature, increasing the risk of contamination by contact and allowing the tainting odours to be more readily released.

Particular care should be taken to see that these are not damaged during the loading. The importance of rejecting any dangerous cargo accidently damaged is stressed. If, in the course of handling or stowing a compartment or container, any sign of leakage should be noted, particular packages should be located and rejected if of a dangerous nature. Smell or fumes should be viewed with suspicion. Odourless fumes which irritate the throat or eyes may be a warning of leakage.

Stowage

Before loading any Dangerous Goods into a compartment or container, it should be ascertained that the place is suitable for that particular cargo, and that it is in a good dry condition.

All Dangerous Goods should be tightly stowed and well secured against any movement including chafe. Securing materials used should be compatible with the goods themselves.

Drums should be stowed close and compact, bungs or closures uppermost. Rolling hoops should not override. Avoid roll tiers if possible. Adequate soft dunnage (e.g. timber) should be laid between tiers — even, in some instances, though the drums are designed to "nest". With drums of sensitive cargo, e.g. chlorates, bromates, chlorites and substances in other classes such as nitrocellulose, it may be necessary to dunnage individually between each drum.

Whenever Dangerous Goods and general cargo are stowed together within the same compartment or container, the dangerous goods should be stowed (where possible) for best accessibility and to facilitate inspection — e.g. in the doorway of the container.

The presence of even one package of Dangerous Goods in a container at once renders that container hazardous and subject to Dangerous Goods' legislation.

On Deck Stowage

All goods stowed on deck must be properly secured, having regard both to the nature of the packages and the weather conditions liable to be experienced. Adequate security can be obtained by means of temporary structures made by using bulwarks, hatch coamings and bridge bulkhead, the structure being closed by means of portable angles bolted to bulwarks and hatch stiffeners. The cargo so stowed should be further secured by means of overall lashings or nets. Unless so stowed or secured by some equally satisfactory methods bulky packages should be lashed individually, preferably with wire rope lashings.

Stowage should be such as to provide safe and satisfactory access to the crew's quarters and all parts of the deck required to be used in the navigation and necessary work of the ship, and sufficient space for the crew clear of the goods concerned. Where Dangerous Goods are stowed in the wells they should not be stowed above the height of the bulwarks, and cargo which, by reason of its nature, is liable to damage tarpaulins or hatch covers should not be stowed on the hatch if such hatchways have tarpaulin covers.

Where deck cargo of an inflammable nature is carried special precautions must be taken to prevent smoking or the use of naked lights in the vicinity of the cargo, and notices should be displayed to that effect. Substances liable to give off inflammable or poisonous vapours should be stowed away from intake ventilators. Where petroleum spirit and other liquids subject to the same conditions of carriage as petroleum spirit are carried as deck cargo on one side of a ship, means should be employed to prevent any leakage crossing to the side used for crew's access.

In ships carrying passengers, Dangerous Goods may not be stowed in any part of the decks available for passengers or near passenger accommodation.

In the case of combustibles care must be taken to avoid the risk of ignition which may arise from electrical short circuits or old electric cables.

Segregation

Greatest care should be taken to ensure that incompatible substances are never stowed together in the same compartment or container. IMO classifies such substances according to the principal hazard, but not all substances of a particular class are necessarily compatible (e.g. Class 8 where a violent reaction may take place between acids and alkalis).

Careful consideration should be given to all other commodities (also their packaging) to be stowed with hazardous substances to ensure against dangerous interaction occurring. For instance non-hazardous cargo packed with straw, wood wool or other combustible materials should not be stowed with Dangerous Goods. Substances which react with water must not be stowed with items having a water base.

Foodstuffs should not, as a general rule, be packed with Dangerous Goods.

Miscellaneous Safety Precautions

It is both illegal and dangerous to take short cuts with safety regulations. Proper care in the initial handling of Dangerous Goods can prevent accidents arising at a later stage.

Naked lights and smoking should be prohibited if in or near a Dangerous Goods' handling area at all times (particularly when inflammables are being handled or stowed).

If possible, Dangerous Goods should be handled and stowed during daylight hours; if not, adequate lighting must be provided during the operation. N.B. The colours of some labels appear to change in the artificial light.

Ambient temperatures in relation to the flash point should be taken into account — particularly in hot weather/tropical climates.

If spillage occurs it should be carefully dealt with having regard to the dangerous nature of the substances, i.e. it should not be allowed to spread via footwear, wind, etc., and spillage should not be replaced in the appropriate package without the knowledge and advice of the manufacturer. Transport drivers should be advised of the dangerous nature of the goods carried and any necessary action which may need to be taken. Information such as: IMO class; flash point, if any; whether miscible with water; whether explosive; whether toxic — and under what circumstances; whether corrosive; the effect of heat; action in case of fire, spillage, etc.; should all be given to the driver.

The IMO publication — Emergency Procedures for Ships Carrying Dangerous Goods — should be made full use of, with a copy of this book being available on each vessel. It should be the aim of each ship or operator to persuade the shipper to submit the appropriate emergency schedule number with the other information at the time of making the booking. See also IMDG Code 1994 — Consolidated Data.

OBNOXIOUS CARGOES

Introduction

Sometimes it is a requirement of law, but always it is prudent, to ascertain the correct technical name of any chemical to be transported, together with its properties. Chemicals of a hazardous nature will be subject to the Dangerous Goods' legislations operating in the country where it is packed, the country where it will be unpacked, and the countries through which it must pass (see also "Dangerous Goods").

Dirty/obnoxious cargo may be described as any cargo liable to affect other cargo or equipment by its dirty or obnoxious nature, or which may cause discomfort or raise objections from personnel involved in the handling and stowage operation.

Characteristics

Heavily infested goods (e.g. bales of rags); very odorous commodities liable to cause strong residual taint (e.g. essential oils); cargo that gives off fine penetrating dust, thus damaging or affecting adjacent cargo (e.g. carbon black); chemicals that give off fumes — which may affect the eyes and nose, but which otherwise may have little odour (e.g. formaldehyde).

Handling and Stowage

Consideration must be given to the possibility of spillage, no matter how well the material is packaged. If the material is of an obnoxious nature then it must not be carried in the same compartment or container with substances which will take hurt if a spillage occurs. For instance poisons must never be packed with foodstuffs. It is good practice when stowing bagged powders, which may sift, to place plastic or similar material underneath the cargo (lining the floors and the wall of the container) to protect other cargo against the effects of any sifting or spillage. It also reduces the amount of cleaning required to be done at time of discharge.

Liquid chemicals in tanks, drums, or bottles, must occupy the lower portion of any mixed stow. Some chemicals require to travel and be stored under temperature controlled conditions — either to prevent the temperature of the substance rising too high, falling too low, or both. In these circumstances full consideration has to be given to the consequences of any failure in the refrigerating machinery if appropriate. Too great a rise in temperature may make a substance become obnoxious or even dangerously unstable.

Typical Dirty and Dusty Commodities

Many commodities may be objectionable to some degree — and the list below is not exhaustive. They may be dusty, or smelly or stain other cargo with which they come into contact. On no account should delicate commodities or foodstuffs be packed in the same compartment or container with any of these. Some will render the compartment or container unfit for normal use unless it has been specially cleaned. If essential oils are spilled within the compartment it may be impossible to remove the all-pervading smell without stripping and rebuilding.

Certain of the commodities listed may be a health hazard and some are Dangerous Goods (see also "Dangerous Goods").

Asbestos powder, Asphalt, Ball clay, Blood (dried), Bone meal, Borax, Camphor oil, Carbon black, Chlorides, Cement, Disinfectants, Dyes, Fertilisers, Fish meal, Fish oil, Formaldehyde, Graphite, Glass fibre, Glycerine, Greases (various), Hides, Lubricating oil, Lime, Molasses, Oakum, Oils (especially essential oils), Oxides, Ochre, Paints, Polishes, Skins, Slag (basic), Snowcem, Swarf, Sulphates, Tallow, Tar, Titanium white.

LIVESTOCK

Introduction

The transport of animals is subject to legislation in many countries. Where risk of disease may exist this legislation is rigorously enforced. In most cases the legislation not only covers the importation of animals, but also the transit of animals, through a port. For instance it may not be possible to carry certain livestock because of the national regulations of way ports that the ship may call at. A typical case in point here is the Australian requirements regarding the African horse fly; the regulations are such that if a vessel has passed within 50 miles of the coast of Africa, then any horses carried on that vessel for Australia might not be an acceptable import to that country.

Over and above regulatory requirements, there is the need for humane and hygienic treatment and conditions for animals transported by land and sea. Guidance for individual species' requirements may usually be obtained from local zoological societies and similar organisations. A large part of the ensuing recommendations has been derived from some such source.

Categories of animals may vary enormously from full shiploads of, say, sheep, to a single domestic pet (e.g a dog) which requires no more than the attention of an individual member of the crew during the passage.

Animal rights groups are exerting considerable pressure on shipowners to ban the transportation of live animals in ships not fitted for the purpose. They have been largely successful with ferry operators carrying animals in transporter lorries.

Regulations

The most stringent regulations apply to even-toed ungulates, e.g. pigs, cattle, etc. As mentioned above many countries have very strict requirements, most particularly in:

U.K. — Ministry of Agriculture and Fisheries.

U.S.A. — Department of Agriculture (Bureau of Animal Industry).

Republic of Argentina — Division of Animal Industry of Department of Agriculture.

Commonwealth of Australia — Department of Transport Marine Standards Division, etc.

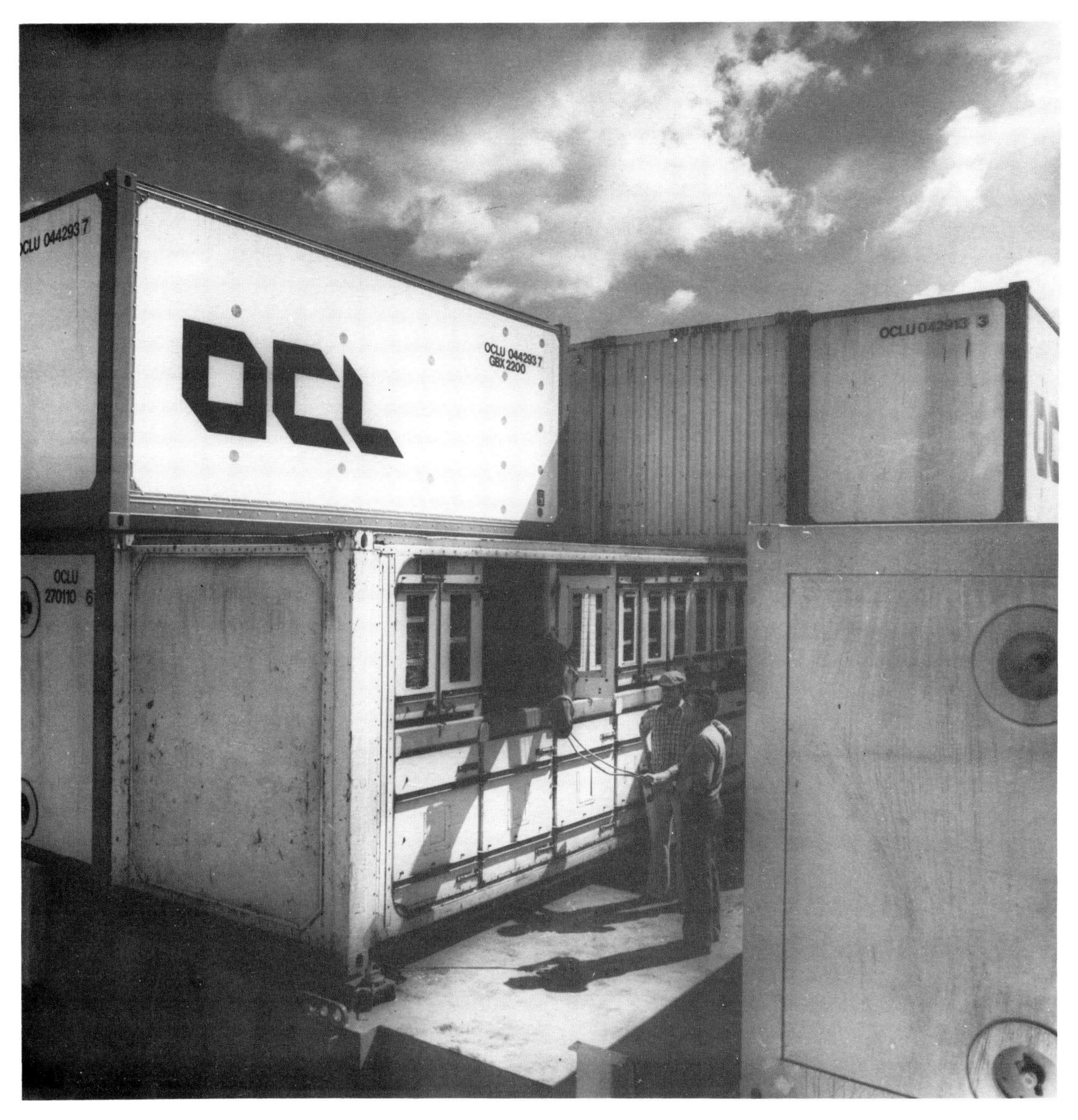

(x) The carriage of livestock shown here demonstrates container versatility. *(Photo: Overseas Containers Ltd)*

Careful study of these regulations — and any that may apply from other countries — should be carried out prior to shipment of any animals from one country to another. It is also imperative that the correct documentation is prepared, e.g. import certificates, health certificates, veterinary certificates, etc., prior to shipment.

Typical of the areas covered in the regulations over and above documentation will include: the quantity of any particular animal permitted to be carried at one time in the space available; the size of pens or cages; the strength requirements of fittings, etc.; the food requirements; fresh water services; access; ventilation; lighting; fire fighting appliances; whether animal stalls may be carried more than one tier high.

Stowage

Crates and cages should be stacked with proper protection against both heat and cold. Where possible provision should be made to set up awnings on deck to give shelter from the sun, rain and wind for those cages and animals which require it. The close covering of cages with tarpaulins in hot weather will cause undue suffering to the occupants of cages. Cages can be stacked on hatches where they are less exposed to bad weather and there is less danger of seas and spray reaching them. If, however, the ship is due to call at intermediate ports where loading or unloading will take place, enquiries must be made beforehand as to which hatches will be in use so as to avoid shifting the boxes.

Where animals are carried in freight containers, particularly on cellular vessels, care must be exercised to ensure that those particular containers are protected from extremes of heat, cold, rain and spray. Where animals have to be taken out for exercising — e.g. dogs — stowage positions should be such that this may easily be achieved. To provide adequate ventilation it is almost certain that adjacent container slots will need to be left vacant. This vacant space may under certain circumstances be utilised partially as storage area for food and equipment, and access for the handlers.

In the case of larger animals the stowage position (whether in containers, or as general cargo in crates) should be such that access can be achieved should the animal need veterinary attention during the voyage, or for removal of the carcass should the animal die. This last is most important on vessels without gear, and could mean the butchering of the animal to dispose of the carcass.

It is important to make allowances on cellular container vessels, when animals are carried in freight containers, to allow sufficient container space for the carriage of the necessary food and bedding. Such containers should have their doors readily accessible at deck level, unrestricted by lashing arrangements.

Food, Water and Bedding

An ample supply of food and bedding for the voyage must be placed on board at the port of embarkation, unless firm arrangements can be made to replenish supplies en route, and a margin should be allowed for delay. Proper arrangements should be made for stowage of food on board, where it can be reached easily during foul weather. Only sound food should be given and this, as well as bedding, must be stored in a dry place on board.

Full written instructions for feeding, watering and cleaning should be fixed to the front of each box and full power should be given to the man in charge to replenish stocks of food if necessary at ports of call.

All drinking vessels should be kept carefully scoured, and only fresh water used.

Suitable food for ruminants consists of lucerne, or clover hay, meadow hay, grain, chafe, crushed oats, maize, or other grain. Also appropriate food concentrates in the form of pellets or cake supplied under proprietary brand names.

In the case of carnivorous animals it may be possible to store the frozen meat in the ship's stores, if there is appropriate separation or barriers for hygiene requirements. Where this is not possible a refrigerated container capable of carrying at correct temperature would be ideal.

Refrigerated carriage may also be necessary for fruit, eggs, milk, etc., for the animals. Milk can most probably be carried as condensed tinned milk.

Carnivores need be fed only once a day, preferably in the late afternoon. Ruminants and all other animals should be fed twice a day — early morning and early evening.

Animals should, as a general rule, be watered two or three times a day, but more frequently in hot weather. If the drinking vessels are removed after the animals have drunk, there is no risk of the water being fouled nor can the animals knock against them and be frightened. Carnivores should be provided with shallow metal receptacles. Ruminants should be watered in the same manner as horses, small donkeys or goats, according to size.

Straw or hay makes suitable bedding for most animals, and it should be removed every morning. For kangaroos, oat-chafe is most suitable. Soft hay should be used for all animals provided with sleeping boxes. Bedding should be changed as required, and careful attention paid to the possibility of livestock eating the bedding and suffering accordingly.

In the case of ruminants the container space used for food and bedding very often exceeds that required for animals and should be planned accordingly.

Ventilation

It is most important that all animals have adequate ventilation — except in the case of some reptiles. Ventilation must be such that the animals are not exposed to strong winds and spray in the process of being ventilated! In some cases, e.g. 'tween deck stowage, and some container stowage, mechanical ventilation may have to be provided.

Lighting

Adequate artificial lighting will have to be provided particularly in 'tween decks and possibly in containers. Lighting will allow the handlers to better attend to their charges, and may also provide a soothing effect on the animals themselves. More careful attention to lighting and heating maybe required in the case of small animals, birds and reptiles.

Cleaning

The type and size of animal will dictate the frequency of cleaning of pens and cages and containers that is required. Adequate services should be provided in the vicinity of animals to allow proper cleaning to take place, e.g. hose connections, power points, etc. It is important too that adequate drainage is available and such that any cleaning water does not blow back on board in the process of draining.

All boxes and cages should normally be cleaned out at least once every day. As salt water is injurous to many animals it should be used with discretion for washing out large cages. Disinfectants if used at all, must be used very sparingly, with great care. All the smaller cats should be provided with shallow trays containing earth or sawdust, which must be changed daily.

Equipment

Appropriate medical supplies for each type of animal and length of voyage should be placed on board prior to departure from the port of loading. Appropriate humane killers — as prescribed by regulations, and the type and size of animals — should be placed on board at port of loading. Other equipment such as buckets, brooms, shovels should be supplied in adequate quantities to avoid having to use ships' supplies. In many cases ships do not carry sufficient of this type of equipment to allow their use for other than ships' work. Stowage for this equipment should be supplied and, in the case of medical supplies, humane killers, etc., should be safe from unauthorised access.

Handlers

Where ten or more head of large animals are carried, a handler (or handlers) may need to travel with the animals. This should be adequately investigated in advance, with handlers of suitable expertise and experience, and the appropriate arrangement made for transfer from the vessel at the port of final destination. It is important, too, that the necessary accommodation is available for such handlers — and of course victuals, bedding, etc. For small animals, and small quantities of animals, members of the crew might reasonably be expected to act as attendants. For hygiene purposes such animals should not be allowed into the accommodation.

MAMMALS

Small animals such as dogs, cats, monkeys, etc., may be a subject of interest to crew and passengers. Precaution should be taken to prevent any teasing taking place or ad hoc feeding with titbits, etc. Some individual animals become extremely nervous if peered at by a procession of strange human beings.

Crates, pens, stalls, etc., must allow adequate room for the animals to move, while providing appropriate support against the rolling and pitching effects of the vessel. Adequate head room, particularly with regard to horses, is most important. Typical height for that requirement would be:

Minimum Head Clearance

cattle	1,981 mm (6′ 6″)
Horses (not ponies)	2,286 mm (7′ 6″)
large dogs	1,220 mm (4′)
medium sized dogs	915 mm (3′)
small dogs and cats	610 mm (2′)

Other dimensions can also be extremely important; certain species of antelope when taking fright exhibit the reflex action of trying to bound away, and it has been known for these animals to break their neck if restricted by, for instance, the wall of a cage.

Heights, lengths and widths of stalls, crates or pens with regard to mammals are frequently governed by legislation in countries of origin or destination.

Certain animals are dangerous if approached too closely by those who are not their regular handlers. Appropriate notices should be displayed for crew, passengers, etc., where this applies.

Different types of mammals have different requirements:

1. Camels — may weigh from 810 kg (16 cwt) to 1,270 kg (25 cwt). Carried on deck at shipper's risk, clause Bill of Lading "ship not responsible for mortality". For short passages in fine weather latitudes they are just tethered to a line spread fore and aft along the deck. Some of these animals are very vicious and given to biting any stranger within reach. Wounds so inflicted are very apt to become septic, and however slight, should be given early and careful treatment. Camels should be well watered before embarkation. For other than short passages, allow ten gallons of water, three or four pounds of grain, in addition to green foodstuffs per day.
2. Cattle — the voyage itself may form part of the quarantine period for the country of destination. It is important therefore that no contamination, e.g. from personnel. etc., should take place at way ports during the voyage. Vessels are usually exempt from all claims with respect to mortality, and bills of lading should be claused accordingly.

   Attendants are usually supplied and their wages paid by the shippers. Charter party or contract should embody a provision to enable the owner to recover all expenses incurred in respect of these men — usually the victuals are provided free of charge by the ship — including expenses for repatriation which may amount to a large sum. They sign on ship's articles and are in all respects to be subject to the discipline of the ship to the Master's authority.
3. Horses and Mules — usually subject to statutory regulations of the country of origin and/or destination. The average weights of horses: heavy horses, 712 kg (14 cwt); cavalry, 560 kg (11 cwt); light horse, 406 kg (8 cwt). Animals suffering during the voyage from broken legs or other serious injury must be slaughtered by direction of the Master, hoof or other marks noted and the incident recorded in the log. Slaughter should be carried out using an approved humane killer. The vessel is usually exempt from all liability in respect of mortality or injury to animals, and the bill of lading should be claused to that effect. Wet, mouldy or loose hay should never be accepted, no matter by whom supplied, and the bales should be sufficiently well bound to keep them intact while being handled. Times of feeding and watering are usually prescribed by the shipper whose representative generally proceeds in the ship. Usually horses and mules are fed

and watered three times a day — morning, noon and evening. Owing to their heating properties it is not customary and neither is it wise, to feed oats until the voyage is well advanced and then not heavily. Dead animals should be got overboard as soon as possible, as they quickly fill with nauseous gas offensive to both man and beast, and every effort should be made to avoid entering port with carcasses of animals recently dead on board. Where a choice exists, horses should not be positioned facing over the side of the vessel.

No attempt should be made to reduce the scantlings of fittings to horse boxes, stalls, pens, etc., below those recommended or required by law. The weight of an animal such as a horse or cow in a heavy seaway can put enormous pressure or battering ram effect on all such fittings.

4. Sheep — until recent years sheep were carried in small numbers in essentially similar conditions to other agricultural stock. However, the carriage of sheep, particularly from Australasia to the Persian Gulf, has become a major trade and in the main converted vessels now carry as a specialist operation live sheep of up to and in excess of 100,000 head.

   This clearly requires different techniques in feeding. Prior to the vessel's arrival, sheep are transported close to the loading place and their feeding is converted until they are able to thrive on pellet food. Once this has been done, they are ready for loading. Mortality rates of under 1–2% have been frequently achieved during the passage.
5. Elephants — carried on deck at the shipper's risk. Bill of Lading to be claused "Ship not responsible for mortality". They may also be tethered in a suitable position on appropriate vehicle decks on Ro-Ro ships. A fully grown animal is three tons and over and varies from 2,286 mm (7′ 6″) to 2,743 mm (9′) in height. Special slings must be used for lifting these animals. Allow 115 litres (25 gallons) of water and 270 kg (600 lbs) of green foodstuff per day for each full grown animal.

UNITISED CARGOES

Definition

A grouping together of two or more items (usually of a homogeneous nature) and securing them with banding, glue, shrinkwrap, slings (e.g. clover leaf), to form a unit which, together with a base (skids, pallets, etc.) or permanent slings, allows mechanical handling equipment (e.g. tynes of a fork lift truck) to lift and transport the unit.

N.B. While ISO containers are a form of unitisation, they are dealt with separately in this chapter.

Advantages and disadvantages:

Ease of tallying.
Reduced breakages.
Reduced pilferage.
Faster speed of working between ship and shore (and on ship or shore).
May make more effective use of vertical storage space in sheds and holds by stacking units 4, 5 or 6 high (possibly with the need to incorporate shelves or racking).
Reduced labour requirements when handling between interfaces.
The disadvantages of unitisation may include:

Loss of space below decks where the shape of the vessel is not compatible with the shape of the unit.
Loss of space caused by the shape of the package being unitised, e.g. drums on pallets.
Collapsed or crushed units require labour intensive efforts to rectify, handle and store.
An element of extra cost involved in the pallet; slings; skids; shrink-wrap; strapping, etc.

Pallet Sizes

The overall height of a pallet may be from 100 mm (4″) to 150 mm (6″), depending on the construction (single-deck, double-deck, etc.) and the use to which it will be put (single trip, pallet racking, etc.).

The deck (or plan) size depends on a number of factors:

Size of cargo to be palletised.
Type of cargo to be palletised (e.g. drums, bags, etc.).
The use to which it will be put: e.g. closed circuit pool, one trip only, inland/ocean transport requirements, etc.

ISO have set some standard sizes:

1,000 mm × 800 mm (40″ × 32″)
1,200 mm × 800 mm (48″ × 32″)
1,200 mm × 1,000 mm (48″ × 40″)
1,200 mm × 1,600 mm (48″ × 64″)
1,200 mm × 1,800 mm (48″ × 72″)

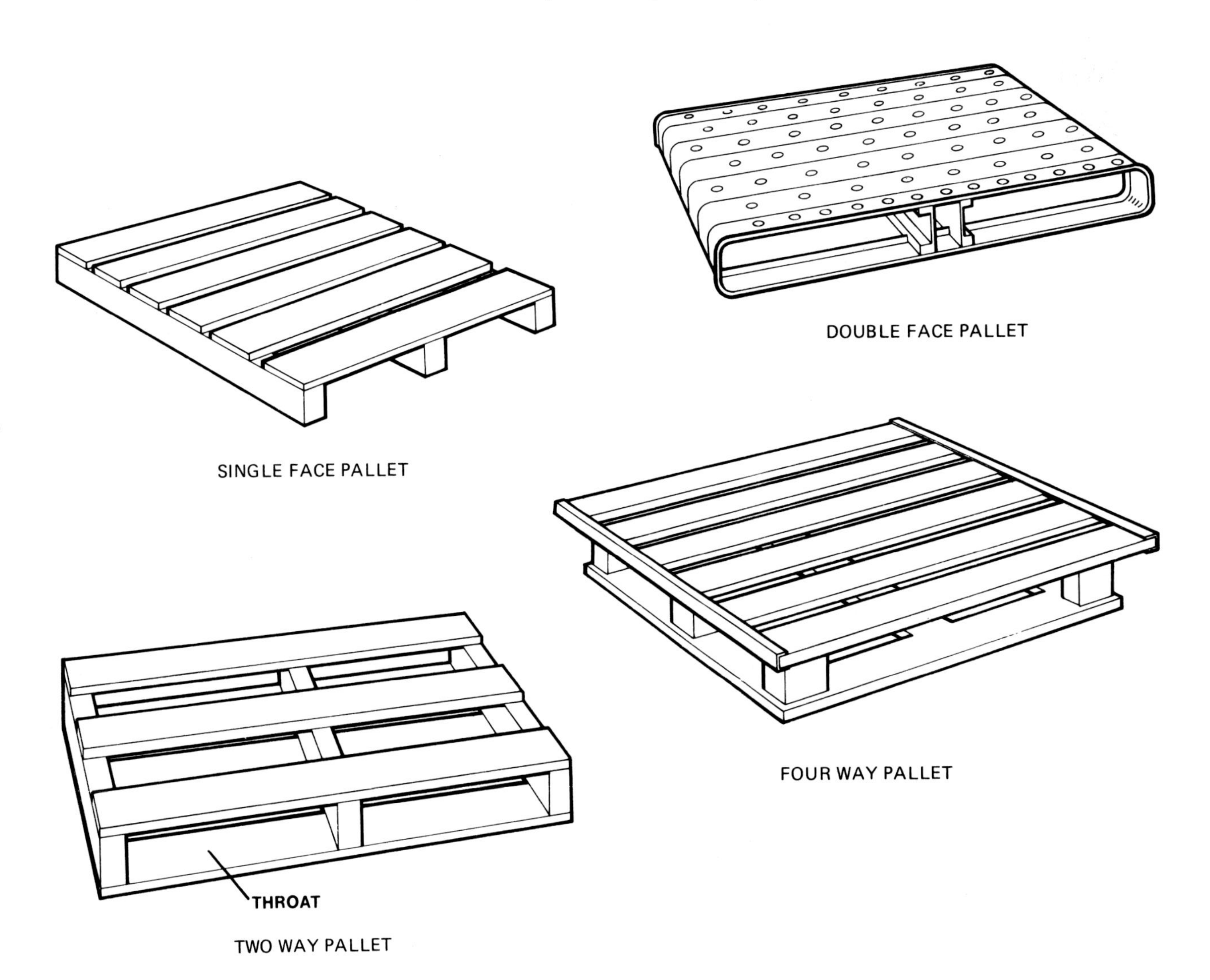

DOUBLE FACE PALLET

SINGLE FACE PALLET

FOUR WAY PALLET

TWO WAY PALLET

British Standard sizes which are recommended as being suitable for ISO containers (from BS2629, part 2) include:

1,000 mm × 900 mm (44″ × 35·5″)
1,100 mm × 1,100 mm (44″ × 44″)
1,100 mm × 1,400 mm (44″ × 55″)

Other sizes in common use, particularly in "closed circuit" systems such as stevedoring and warehouse operations:

1,500 mm × 1,200 mm (60″ × 48″)
1,170 mm × 1,170 mm (46″ × 46″)
(particularly Australasia)

Typical pallet sizes to suit sizes of bags, filled:

Bag Sizes	*Pallet Sizes*
762 mm × 380 mm (30″ × 15″)	1,140 mm × 1,140 mm (45″ × 45″)
800 mm × 400 mm (32″ × 16″)	800 mm × 1,200 mm (32″ × 48″)
609 mm × 400 mm (24″ × 16″)	1,200 mm × 1,000 mm (48″ × 40″)

Slip Pads

Sometimes known as slip sheets. These may be made up of a single sheet of fibre board or plastic, of the same plan dimensions as a pallet, but with an exposed "lip" which allows the special fitting on a fork lift truck to grip and pull the slip sheet onto a flat "spade" attachment. Thus the unit — made up on the slip sheet — is handled in all respects as a pallet, except for the special FLT attachment necessary.

The advantages of this system include:

Saved space; as against pallets or skids. Sometimes particularly important, e.g. low 'tween decks, containers, etc.
Pads may be less expensive than pallets.
Less storage area required when not being used.

The disadvantages include:

Pads may be easily damaged; if the "lip" is ripped off then the mechanical handling equipment has nothing to grip.
Requires a flat surface without obstructions on which to operate.
Requires special equipment to operate and handle.
May (usually) only be approached and handled from one side; i.e. the side on which the lip protrudes.

Making Up the Unit

The cargo should, as nearly as possible, exactly fit the pallet or slip sheet. Cargo overhanging the deck of a pallet may cause the following problems:

Put excessive pressure on the lowest tiers of cargo, particularly that which is in contact with the edge of the pallet. The *g* forces generated during transport can force the edge of the pallet deck to penetrate the cargo — particularly bags and cartons — with resultant damage to the contents.
Make the securing of the load to the pallet difficult or even impossible.
Risk the pallet load fouling adjacent cargo or units, with the load becoming wedged or jammed in the stow.

Similarly a unit module which is too small for the deck of the pallet will make the securing of the cargo to that pallet less effective, and reduce the ability of ship's bulkheads, container walls, or other cargo to support the pallet loads laterally in the stow.

The height of a pallet load should be such that the pallet and cargo combined leaves a clearance of at least:

50 mm (2″) between top of cargo and door header (or ceiling) of a container.
150 mm (6″) between the top of the uppermost pallet and the lowest deck-head obstruction in a 'tween deck or hold.

Method of Securing

The cargo should be secured onto the pallet by one or more of the following methods:

Strapping: Man-made fibre or steel. Banding should pass under the top deck of the pallet, to remain clear of mechanical handling equipment requirements. Some cargo may require protection from the point loading of the strapping, which may be achieved with scrap dunnage, waste cardboard, etc., or using proprietary equipment specially designed for the purpose.

Steel strapping may not be suitable for securing cargo that may alter shape or contract, e.g. bags, telescopic cartons etc. Man-made fibres, particularly nylon, may stretch and allow rigid items to move in some instances.

Nets: Usually of polypropylene, and used to secure awkward shaped packages onto a pallet.

Glue: For securing regular stows of cartons or bags. Should be strong in sheer strength, but with little resistance to a vertical force when dismantling the pallet load. A 5 per cent dextrine solution can be a suitable mixture for this purpose.

Shrinkwrap: A plastic (transparent) cover over the pallet load which is heat shrunk into place. Certain types of plastic wrapping may be put around the cargo without the need of heat. May be used for most goods and provide protection against dirt, pilferage, etc.

Locking Pattern: Unit loads may be built up on pallets with plan dimensions suited to the modules being handled. In this way layers may be built up each differing from its neighbours and providing restraint in a similar manner to a brick wall. Alternate layer patterns are reversed to "tie in" the layer beneath.

The Effect of the Pallet

The pallet, because it raises the cargo approximately 150 mm (6″) off the deck of the hold, has the effect of raising the centre of gravity of the cargo which may, in certain circumstances, have to be allowed for.

Pallets use more space than conventional dunnage, perhaps 10 to 14 per cent.

Some overhead space may be lost if the height of the pallet load does not form a module of the height between decks or in the holds of the ship — or the internal height of the container. If the pallets are being made up on the quay this height may be adjusted (in some instances) to suit the particular height requirements of the ship. 'Tween decks may be designed for "user friendly" palletised cargoes.

While pallet loads may be made up in such a way that the units are suitable for carriage under refrigeration, the following will need to be taken into consideration:

The air-flow through the cargo (if required) or around the cargo is not impeded by securing materials or the positions of the pallet loads.

If two-way entry pallets are being used, then the pallets must be so positioned that the correct air-flow is not blocked by the supporting members.

The deck of the pallet may have to be slatted or perforated to allow free movement of controlled temperature air. This is particularly important for cargoes (e.g. fruit) in packaging designed to allow vertical through-movement of air.

Handling and Stowage

A flat unobstructed area, e.g. deck space, is best suited for positioning and stowing pallets. They should never be dragged into nor out of the stow — such treatment will collapse the pallet (particularly one-trip pallets) and make subsequent handling and storage difficult if not impossible.

Vessels which have not been built with pallet handling in mind, e.g. without flush 'tween deck hatch coamings, etc., may need extensive preparation work carried out before loading commences, e.g.:

Temporary wooden ramps to surmount hatch coamings.

The removal of deck obstructions, e.g. ring-bolts, stanchions, etc.

The bridging and squaring off of rounded bilges, excessive shear, etc.

Temporary strengthening of hatch covers, tank tops, 'tween decks, etc., with metal plates to allow for axle loading of fork lift trucks.

Point loadings may be very high when a fork lift truck is under load, and considerations should be given to the Classification Societies' point loading figures (usually expressed in tonnes per square metre).

A table of truck weights and loadings is given below:

Type of Truck	*Total Weight of Truck and Load (tons)*											
	3	4	5	6	7	8	9	10	11	12	13	14
	Minimum Deck Thickness (irrespective of beam of longitudinal spacing) (inches)											
Two wheels at Fork end	0.30	0.35	0.40	0.44	0.47	0.50	0.53	0.55	0.59	0.61	0.63	0.65
Four wheels in pairs at Fork End	0.26	0.30	0.32	0.35	0.38	0.41	0.44	0.46	0.47	0.50	0.51	0.53

The stowage of the ship has to be planned with the strength and size of the units carefully considered. Two-way entry pallets and skidded units should be stowed so that the truck always has an entry to the cargo available. This is particularly important where the vessel is discharging at more than one port, and is often referred to as the "flowline" of the pallets. This "flowline" should be marked on the cargo plan with arrows.

If a solid stow is obtained immediately beneath the hatch square it is necessary to have key pallets clearly marked on the stowage plan to indicate which have to be lifted out first to allow access to the remainder of the stow and introduce the appropriate mechanical handling equipment. These pallets may with advantage, under most circumstances, be left with slings in place at time of loading to aid the process of discharging.

Any dunnage used with palletised (or other unit) loads should be limited, and lengths kept short to prevent obstructing or fouling mechanical equipment, cargo in the stow, and handling operations. Sometimes it is necessary to place sheets of plywood, chipboard or fibre board on edge in the stow to prevent the cargo on one pallet "locking" with the next and so obstructing the unloading operation. It is important when operating below decks with mechanical handling equipment that the decks are free of obstructions: loose dunnage, rubbish, etc.

Making up Dockside Pallets

Where pallet loads are made up on the dockside for unit loading aboard ship, care must be taken that bills of lading are not mixed on the pallet. Labelling of the pallet load should be carried out on at least two sides.

The mixed loading (or discharging) of breakbulk cargo and unitised cargo will completely negate the purpose of mechanised working, i.e. the fast handling of cargo with mechanical equipment and reduced numbers in each gang. It is also more likely to result in cargo becoming jammed and damaged.

Cargo into Units

When cargo is not suitable, because of size, shape or other reason, to be palletised, it may sometimes be made up into a unit load suitable for handling with mechanical equipment. This could be achieved in a number of ways, e.g.:

Securing skids or bearers to the bottom of a suitably sized package or case. This would require at least two such "skids" spaced to allow fork lift tynes entry. The minimum depth of the skids should be 50 mm (2″), but more suitable 65 mm (2½″).

Making up the packages that form the unit in such a way that space is left between the packages themselves for fork lift tynes to gain access. Typical commodities that are suitable for this treatment are ingots, bricks, etc.

Pre-Slinging Techniques

Pre-slinging is normally used to facilitate the speed of cargo loading and/or discharging. Also to open up the stow in way of the hatch square (see over).

Multiple sling discharge.

Pre-slung reefer cargo.

It necessitates the positioning of sling loads of cargo in the stow at time of loading, leaving each unit with the sling still encompassing the load so that the port of discharge has only to hook on and lift out.

With open hatch ships and twin hatches, the advantages to be gained by using pre-slinging techniques can be great. Traditional hatch arrangements with deep wings are not so suitable since the sling load cannot be landed directly from the hook to its stowed position.

A certain amount of space is lost (perhaps 10 or even 15 per cent), but this is usually more than made up for in the speed of cargo handling obtained. The space lost is less than that for palletised or skidded cargo, and the cost of slings is probably competitive with the cost of pallets.

Materials

Any normal sling material may be used for a pre-slinging operation:

Chain slings.
Fibre rope.
Steel wire rope.
Flat synthetic fibre.

Also, under certain circumstances, nets, baskets or trays.

It is common for the following types of commodities to have the following types of sling:

Bagged Goods

Clover leaf slings, of natural or man-made fibre ropes and strops.

Reels of Paper and Newsprint

Flat synthetic strops, or endless slings often used in pairs.

Sawn Timber

Wire or flat synthetic choker strops. N.B. Bearers and dunnage must be used between units when using flat synthetic slings to avoid severe chafing damage in bad weather.

Cartons and Cased Goods, Bales and Pulp

Flat synthetic slings — also rope or wire slings.

Steel Products

Various, depending on weight, unit shape, and number of potential lifting points. Round steel piping is often pre-slung using two flat choker strops in a bridle. Steel plate, on the other hand, will usually require wire or chain slings.

Handling

When loading it must be ensured that the sling eyes are left in an accessible position on top of the load ready for immediate hooking on at time of discharging.

When discharging care must be taken that eyes from two adjacent units are not hooked on together in the belief that they are from the same unit.

It is important that during handling operations a pre-slung load remains intact and does not become broken up due to careless handling. If pre-slung loads do break up then there is immediately a labour intensive operation required to move that cargo, and much of the advantage gained from pre-slinging is lost due to higher costs in labour and time.

It is important that a series of pre-slung units be made up of the same number of bags, cartons or whatever, so that tallying is made easier. Bills of lading should not be mixed in one pre-slung load.

The drift of the sling, when handling fragile items such as cartoned goods and small crates, etc., should be sufficient to prevent pinching and crushing of the sling loads.

Slings should be of adequate strength and colour codes checked so that safe working loads and limitations may be observed.

Small bags, e.g. of salt, may be made up into loads contained within large specially constructed bags. This pre-slinging technique is practised widely in the West African trade. It should not be confused with mini-bulk (but see also that section).

CONTAINERS

Introduction

Although there are many earlier examples of 'containers', the maritime container which is so familiar today first began entering deepsea service in the mid 1960s. The US company Sea-Land Service introduced the first genuine containerships onto the North Atlantic trade in 1966 and by the end of the decade, there were purpose-built container vessels operating on most of the world's prime trade routes. Very soon, the major trades between developed countries were fully containerised, i.e. breakbulk vessels were withdrawn almost completely and what non-container capacity remained tended to be provided aboard roll-on, roll-off/container ships.

Trades to/from developing countries began to containerise in the 1970s, firstly by carrying some containers on conventional vessels, then with the introduction of multipurpose roll-on, and roll-off/container vessels, and ultimately with cellular and non-cellular full container vessels.

Amongst the advantages cited in favour of containers are:

(a) Reduced ship time in port.
(b) Better berth utilisation.
(c) Improved trans-shipment and intermodal operations.
(d) Reduced time between producer and consumer.
(e) Less physical handling of cargo — less damage.
(f) Good security — less pilferage.
(g) Protection against weather and detrimental atmosphere.
(h) Quality control to improve shelf life of perishables.
(i) Improved safety to personnel, cargo and equipment.
(j) Reduced tallying costs.

The relative importance of each varies according to particular circumstances. For the shipowner, the increased efficiency compared with breakbulk operations soon rendered the conventional cargo vessel totally uneconomic, except in very specialist areas of employment. It was not simply a matter of higher vessel operating costs; also to be considered was the fact that higher value cargoes, paying better freight rates, gravitated towards containers, leaving the breakbulk vessels with only the low value, low rated cargoes.

For the shipper, the shipowner's customer, the advantages were even more significant. The attraction of the customer towards a system which conveyed cargo from door to door with minimal risk of damage or loss has proved to be the real driving force behind containerisation. Consequently containers today are to be found in considerable numbers all over the globe, even in countries where the landside infrastructure is clearly inadequate or where stevedoring charges are still modest.

The elimination of the breakbulk vessel from the volume trade routes explains why, today, cargoes which, on the face of it, are ill-suited to transport within closed boxes are nevertheless being containerised. There is no alternative shipping method available.

To distinguish between the loading of a container on to a ship or unloading of it from a ship, or the packing and unpacking of that container with cargo, the words stuffing and stripping are used with regard to the actual cargo work (sometimes known as vanning and de-vanning).

Cargo in a container very often has the same requirements and same characteristics as a cargo loaded into a locker, 'tween deck, or hold of a ship. For this reason many of the remarks on general stowage apply equally well to containers.

The ISO Container

The first container shipping lines took into account various factors when choosing the dimensions of their containers but the most important consideration was always the regulations governing road transport in the countries where the containers would be used. Next in importance was the nature of the cargo to be carried on particular trade routes, especially the relationship between weight and volume. So it was that the US shipping companies who pioneered containerisation developed 20 ft, 24 ft, 27 ft, 35 ft and 40 ft containers.

To meet the industry's needs for some degree of uniformity, the International Standards Organisation now recommends a series of external and internal dimensions for containers, together with gross maximum weights which the containers may achieve (complete with cargo). Under normal circumstances these sizes and weights are adhered to, but there are exceptions, the most common of which are:

Length: The US company Matson Line still operates 24 ft long containers but Sea-Land's 35 ft containers have now disappeared. Today the deepsea trades are dominated by containers either 20 ft or 40 ft in length. Recently some lines have begun the introduction of 45 ft containers but these are still restricted to the major trade routes such as those across the Pacific and North Atlantic.

In North American domestic traffic, 48 ft and even 53 ft long containers are being employed but, given current road regulations elsewhere in the world, these overlength containers are unlikely to spread into the maritime sector.

In Europe, the 30 ft container is still found in North Sea trades but the intermodal metric 'swapbodies' — 7.15 m/7.45 m/7.82 m/13.6 m in length — are not suited for carriage by container vessel.

Width: The dominance of the 8 ft wide container was unchallenged until the later half of the 1980s when the emergence of the 2.5 m (8′ 2½″) wide container began.

In Europe, the standard container had always been at a disadvantage when competing for traffic with road vehicles in that its internal width was insufficient for two standard Euro-pallets to be stowed side by side. The 2.5 m wide box addresses this problem and has proved extremely popular with European shippers.

While early 'pallet-wide' containers could not be carried in cellular vessels, most newer designs do permit loading into cellular holds and this has resulted in 2.5 m wide boxes being ordered by some deepsea carriers. As with the 45 ft container, it remains to be seen whether their use will become more widespread.

Height: 8 ft 6 in has become the standard container height and early 8 ft high containers have now almost disappeared. However, to meet the needs of certain shippers, some carriers introduced 40 ft long, 9 ft 6 in high containers and these 'hi-cubes' are now very common, especially on the East-West trades linking North America, Europe and the Far East. In many countries though, the overall height of a 9 ft 6 in container on a road vehicle or rail wagon creates difficulties and in the European shortsea trades for example, a compromise height of 9 ft is often preferred for hi-cube requirements.

Weight: ISO recommended maximum gross weights of 20,320 kg/25,400 kg/30,480 kg provide an easy to remember 20 tons/25 tons/30 tons approximation for the maximum permissible loaded weight of 20 ft/30 ft/40 ft containers. In the 1980s though, the 24-tonne 20 ft container became the norm and 20,320 kg 20 footers are fast becoming rare.

Table 1. — External dimensions, permissible tolerances and ratings for series 1 freight containers.

Freight container designation	Length, L mm	tol. mm	ft in	tol. in	Width, W mm	tol. mm	ft in	tol. in	Height, H mm	tol. mm	ft in	tol. in	Rating, R (gross mass) kg	lb
1AAA									2 1896°°)	0 -5	9 6°°)	0 -3/16		
1AA									2 591°°)	0 -5	8 6°°)	0 -3/16		
	12 192	0 -10	40	0 -38	2 438	0 -5	8	0 -3/16					30 480°°)	67 200°°)
1A									2 438	0 -5	8	0 -3/16		
1AX									<2 438		<8			
1BBB									2 896°°)	0 -5	9 6°°)	0 -3/16		
1BB									2 591°°)	0 -5	8 6°°)	0 -3/16		
	9 125	0 -10	29 11¼	0 -38	2 438	0 -5	8	0 -3/16					25 400°°)	56 000°°)
1B									2 438	0 -5	8	0 -3/16		
1BX									<2 438		<8			
1CC									2 591°°)	0 -5	8 6°°)	0 -3/16		
1C	6 058	0 -6	19 10½	0 -1/4	2 438	0 -5	8	0 -3/16	2 438	0 -5	8	0 -3/16	24 000°°)	52 900°°)
1CX									<2 438		<8			
1D									2 438	0 -5	8	0 -3/16		
	2 991	0 -5	9 9¾	0 -3/16	2 438	0 -5	8	0 -3/16					10 160	22 400
1DX									<2 438		<8			

In certain countries there are legal limitations to the overall height of vehicle and load (for example for railroad service).

Table 2 — Minimum internal dimensions and door opening dimensions for series 1 freight containers.

Dimensions in millimetres

Freight container designation	Minimum internal dimensions			Minimum door opening dimensions	
	Height	Width	Length	Height	Width
1AAA	Nominal container external height minus 241 mm	2 330	11 998	2 566	2 286
1AA			11 998	2 261	
1A			11 998	2 134	
1BBB			8 931	2 566	
1BB			8 931	2 261	
1B			8 931	2 134	
1CC			5 867	2 261	
1C			5 867	2 134	
1D			2 802	2 134	

The strength of a container relies mostly on the floor and corner posts. The walls, doors and roof do supply some strength but are very susceptible to damage from point loading — e.g. cargo stowed hard against it. The strength and weaknesses of a container must be borne in mind when stowing with other types of cargo, and particularly when securing.

Materials used may be steel, aluminium, grp/plywood, or a combination of these three. The inside floor of the container, supported by the cross bearers, is usually of timber construction. Aluminium containers usually have the walls lined partly or totally with plywood. Nearly all containers have securing points for the cargo in the floors and on the walls.

Many containers are built with fork lift pockets to ISO dimensions. Depending on the size and type of container there will be a standard distance between the pocket centres, width of pockets, clear height of opening and height of pockets above lowest point of container.

Types of Container

Of the many different types of ISO container the most common is the Dry Box (or General Purpose) container. This usually has doors at one end only, has a relatively low tare weight, a relatively low capital cost, and can carry a surprising variety of different commodities.

There are however many other types of container, generally known as specials.

In the deepsea trades, early experience showed that specials suffered from low utilisation rates and high re-positioning costs leading most carriers to restrict their operation of specials to only three categories: the open top and half-height; the platform and flatrack; and the refrigerated container or reefer. A fourth type, the tank container, is generally shipper-owned or leased; very few shipping lines now provide tanks except where there are dedicated traffic flows.

The open top container has doors at one end, portable roof supports (roof bows) and a tilt or cover for weather protection. Some open-tops have a rigid removable roof rather than a canvas tilt. This type of container is principally used for very heavy cargoes; cargoes that can only be loaded with an overhead crane; or for over height cargoes.

Although the open top can be used as an alternative to a dry box container, perhaps for positioning purposes, many shippers are reluctant to load them in case the tilt should prove not to be watertight.

The half-height container, like the open-top fitted with a tilt cover, is intended for the carriage of high density cargoes such as ingots but since such cargoes can easily be loaded into dry boxes or onto flats, many operators have ceased to offer this limited-use special.

The platform flat or the flatrack will be used for cargo which cannot be accommodated within a dry box or an open top.

The platform is basically the floor of a container with container fittings at each corner but not corner posts. 20 ft × 8 ft platforms are commonly used on ro-ro vessels as large pallets to unitise breakbulk cargo but they are rarely employed for door-to-door movements involving cellular ships since they cannot be stacked or top lifted by a conventional spreader when loaded with cargo. Container operators prefer the flatrack, a platform with either fixed or folding end walls.

The flatrack, when loaded with 'in gauge' cargo can be handled as a standard box and be incorporated within container stacks ashore and in cellular and non-cellular holds. With out-of-gauge cargo, either overwidth, overheight or both, special care needs to be taken regarding the stowage location.

Unlike a standard box container, the flatrack is designed for heavy cargoes producing concentrated floor loadings. Usually heavy-duty lashing points and slots for side stanchions are provided.

Sets of folding flats may be locked together for empty handling as a single 'container'. This provides economic advantages over fixed end flats but first cost and maintenance costs are higher.

Flatracks and platforms may also be used in cellular vessels grouped together to provide 'tweendecks for the underdeck stowage of items too large to be conveyed on a single unit.

The reefer. Refrigerated containers fall into three categories: those with an integral diesel generator to power the refrigeration unit; those reliant on an external source of electrical power; and port-hole insulated containers.

When used to carry temperature-controlled cargo, the electrically powered container is connected to the ship's own power supply and usually carried on deck. The diesel generator is also normally carried on deck or in a well ventilated area on ro-ro ships.

Port-hole containers are carried below deck connected via air ducts to the ship's central refrigeration plant. It is possible to fit 'clip-on' reefer units to port-hole containers in which case they can be treated as per the standard reefer type. This is particularly important when pre- or on-carriage can not be effected quickly.

It is important to note that reefer containers are designed to maintain temperatures and that cargo should be pre-cooled to the appropriate temperature before loading. They are not designed to 'pull down' cargo from ambient temperatures.

Tank containers are used for the carriage of a wide variety of liquids ranging from edible products — e.g. fruit juice, wines and spirits — to hazardous and non-hazardous liquids including liquefied gases. They may be heated or non-heated.

Most tanks are owned/operated by shippers and/or specialist tank operators and are carried by shipping lines as shipper-owned equipment. However some carriers do provide tanks for certain traffics, e.g. Scotch whisky or rum.

In the shortsea trades, a wider variety of equipment is offered to shippers reflecting the need to compete with road transport for specialist traffics. The following specials may however be found being operated in some deepsea trades too.

The open sided container is ideal for cargoes that need to be loaded from the side — e.g. long lengths — and for cargoes that need ventilation — e.g. fresh vegetables or livestock. Cargo side restraint may be provided by bars or a metal grill and usually a canvas tilt cover provides weather protection.

The side-door container carries both a weight and cost penalty but is favoured by shippers who require both side access and the security of a box container. Popular with rail operators since it can be loaded whilst still on a rail wagon.

The coil carrier is a derivation of the flatrack with special fittings designed for the carriage of steel coil.

Finally, there are a group of container types which are derivations of the dry box. They can be found in both deepsea and shortsea employment.

The dry bulk container resembles a dry box but will have loading hatches in the roof, doors at one end, and a discharge hatch either in the wall opposite the doors or in the doors themselves. It may be used for dry bulk cargoes and for general cargo. NB: Requires tipping trailer to operate it for discharging bulk.

The ventilated container has been developed for the carriage of hygroscopic (high moisture) commodities such as coffee or cocoa beans. The full ventilation galleries along the top and bottom side rails are designed to prevent the ingress of water and so these boxes can also be used for general cargo.

The fantainer or fan box is intended for cargoes prone to condensation problems. It may be purpose built but it is just as likely to be a converted dry box. A circular ventilation hatch, which can be sealed when not in use, is fitted with an electric extractor fan.

The garmentainer is usually a hi-cube 40 ft fitted with internal rails for carrying hanging garments. Care should be taken that the rails are not overloaded or collapse may occur.

The Stowage of Containers on Board

It is usually the practice in full container vessels for all pre-planning of the stowage positions of containers to be undertaken ashore. With the short turn round time of a ship, constantly changing information, containers arriving up to the last minute, and the requirement for an overall knowledge of future container movements (particularly empties), a centralised unit which has information available for the whole round voyage can often provide more effective planning. This stage of the planning usually gives guidelines for local planners at terminals and ports so that they may position individual containers. These guidelines would have due regard to stability, dead-weight, rotation of ports, movement of empties, forecasts of future cargo, the requirements of special cargo such as dangerous goods, over-heights (or other out-of-gauge cargo), uncontainerisable, refrigerated. As mentioned in the preamble to Part 1, such activity does not detract from the Master's responsibility for the safety of his vessel. It is important, therefore, that Ships' Officers pay particular attention to the condition of containers coming aboard, noting any damage; also the stowage positions and labelling of Dangerous Goods containers, the securing of cargo (where it can be seen, e.g. on flat racks), the declared contents of refrigerated containers so that the correct temperature setting may be checked and/or set tank containers should be scrutinised for any sign of leaks or damage to valves. Ships' Officers should liaise closely with shore planners to obtain the latest information on container stowage positions, their weights and where applicable their contents, so that stability and other calculations can be monitored and checked prior to putting to sea.

Containers of cargo that come aboard without seals (or locks) on the doors should immediately be queried with the shore staff; a fresh seal put on and a note made of the number and the circumstances.

If containers are loaded on to ships with other cargo types, such as unitised cargo or break bulk cargo, great care must be taken to prevent undue strain being put on the sides, ends or roofs of the containers. If a container has to be over-stowed with break bulk cargo (not a recommended practice) only the very lightest of cargo should be used for this purpose. The stowage of Ro-Ro ships sometimes requires that a container is blocked in and used as a natural break, or separation to other cargo. In this case it is often possible to leave the container unsecured, relying on the surrounding cargo to restrain it. Again it must be emphasised that the cargo should be of such a nature that it cannot damage the sides or ends of the container.

Containers are best stowed in the fore and aft line so that the cargo inside them gets the benefit of the restraint of the side walls. Containers stowed athwart ships will possibly result in the face of the stow collapsing and being thrown against the doors, when the ship rolls, with resulting danger or damage when the container comes to be unstuffed.

Where containers are stowed more than one tier high, the planner must endeavour to achieve homogeneous blocks of containers, e.g. avoid mixing 8 ft 6 ins and 9 ft 6 ins high containers indiscriminately in the stow. In any case the stowage arrangement must be such that adequate securing of the containers may be achieved. Where it may be possible to stow a 40 ft container on top of two 20 ft containers (if position and securing arrangements permit), however two 20 ft containers should never be stowed immediately over one 40 ft unit — unless a specially constructed frame or platform takes the weight.

The weight of a container is supported through its four lowermost corner castings. Calculations of deck loadings must take this into account. Some vessels — e.g. certain Ro-Ro vessels — have specially strengthened points on deck or tank-top to support loaded containers. Care must be taken to ensure that loaded containers are properly positioned in respect to these strengthened areas and do not exceed maximum weight limits.

Containers, e.g. those with side tilts which can be damaged by high winds or heavy spray, should not be stowed on the outboard side of the stack on deck. When animals are carried the stowage position should be such that the crew can gain easy access to them for feeding or watering or, if the worst occurs, for removing the animal carcasses at sea (see "Livestock").

Container Handling

Cranes and derricks — These require an overhead connection to the container usually supplied by a "spreader". This is necessary to supply vertical lift at each corner casting to prevent damaging the container. Twist locks enter the appropriate apertures in the corner castings and when turned 90 degrees they engage. Spreaders may be operated by the crane driver direct or they may be

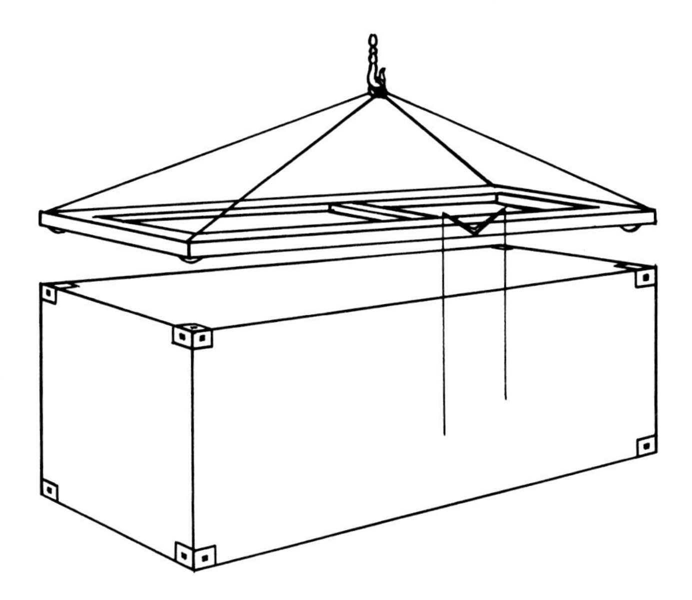

Semi-automatic, manually operated 'twist-locks'

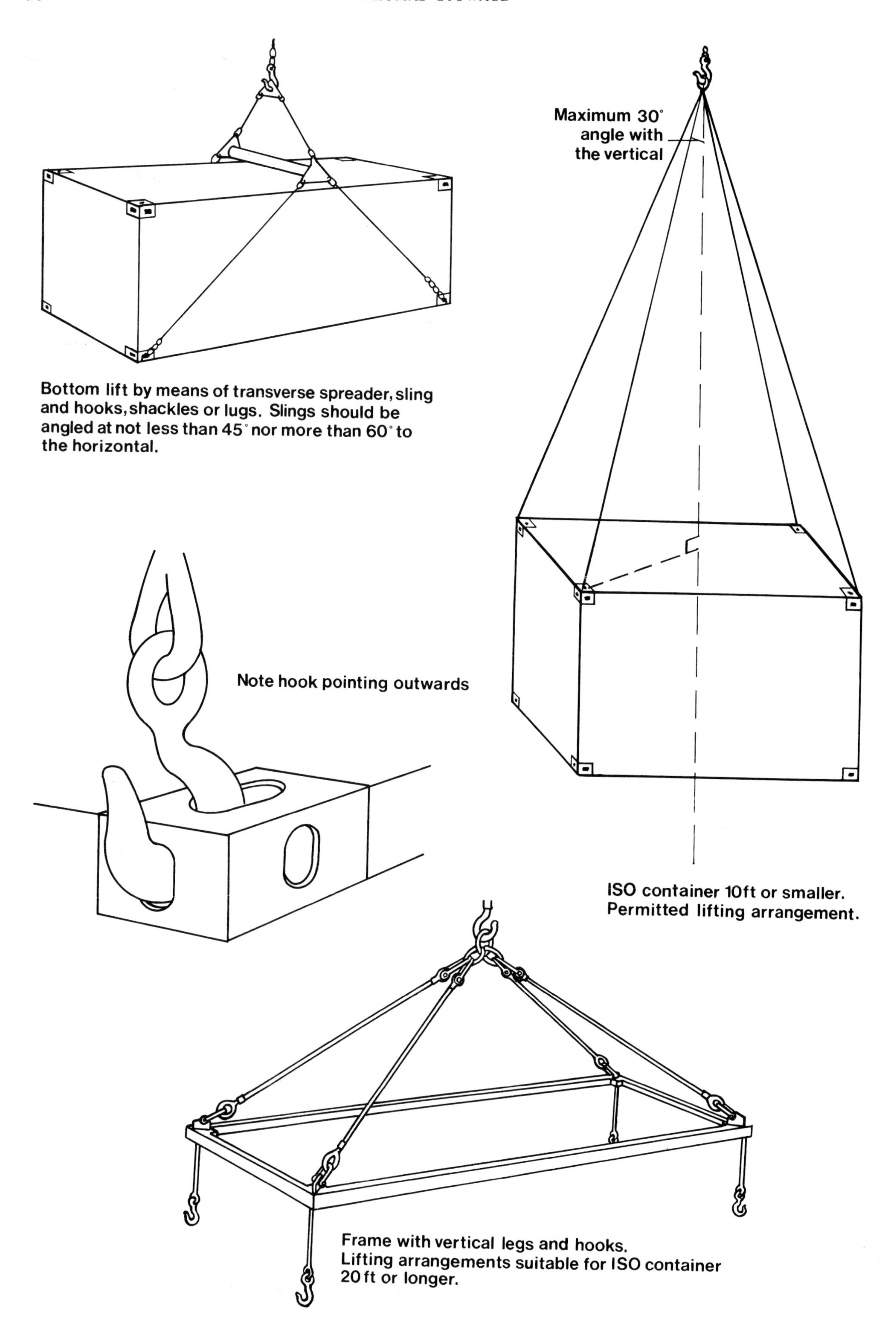
Maximum 30°
angle with
the vertical
Bottom lift by means of transverse spreader, sling and hooks, shackles or lugs. Slings should be angled at not less than 45° nor more than 60° to the horizontal.
Note hook pointing outwards
ISO container 10ft or smaller.
Permitted lifting arrangement.
Frame with vertical legs and hooks.
Lifting arrangements suitable for ISO container 20 ft or longer.

"semi-automatic" and operated by a lever on the side of the frame (often with a line hanging down for ease of access). A frame may be used with four hooks hanging vertically which are manually inserted into the correct corner castings, in which case the hooks should be pointing outward from the ends of the container to gain maximum support from the seat of the hook and make for ease of unhooking when the container is in position. 20 ft containers and above should never be lifted by direct wire slings from top corner castings without some spreader device to prevent the wires pinching and therefore damaging the container.

Fork lift trucks — These must be of sufficient capacity to handle the container if it is loaded. Mast heights must be suited to operating with overhead obstructions in the case of Ro-Ro vessels. Smaller capacity fork lift trucks may be used for empty containers. If the container has fork lift pockets then the tynes of the fork truck may be used for direct lifting. Spreaders may also be fitted for top lifting containers as also may be side frames. Side frame operation may not be suitable over uneven terrain, and care must be taken to select frames suitable for the containers to be handled, e.g. with bearing points at or near the side rails or corner castings. Empty containers may be lifted by end frames; but this is only for empty containers as it puts a great strain on the container itself.

Container terminals usually employ sophisticated equipment such as straddle carriers, gantries, etc., for movement of containers into and out of the stack. Low profile straddle carriers may sometimes be used to carry containers on and off Ro-Ro vessels.

There are certain makes of self levelling spreaders and spreaders that maintain a parallel line to the ship as a swinging crane or derrick puts the container on board. Some spreaders are able to rotate a container through 360 degrees; these being of greater weight usually detract from the maximum load that the crane is able to lift.

Certain terminals have spreaders capable of twin-lifting 20 ft containers. Such equipment can usually only operate if containers of the same height are paired. Two containers at a time, one on top of the other, may be lifted by fork lift truck (if of adequate capacity) in certain circumstances (e.g. the weather deck stowage on board Ro-Ro vessels with no overhead gear) to obtain a three-high stow. However, the lifting of two containers, coupled together by twist-locks or similar, by overhead gear is dangerous — unless the system has been properly tested and approved.

Container Securing

Care should be taken by ship's staff to ascertain the point loading on deck and heavy containers are given adjacent deck stow. There may be occasions where planners seeking to minimise moves may give heavy containers top stow. This may cause stability problems and excessive lashing strains when rolling, pitching, or working in a seaway.

The securing of containers on (or if applicable below) decks is the responsibility of the ship's staff, though not necessarily carried out by them. It is important that all rod and wire lashings are sufficiently tight but not too tight to strain fittings, containers, etc. The correct bridging pieces, twist locks, etc., should be checked in position between tiers. It should be clearly ascertained and understood which way twist lock handles are put for the locking position, i.e. the locking position should be in the same direction for all twist locks on the ship so that a quick glance is all that is needed to ensure that the locks are indeed engaged.

In cellular vessels containers are put down cell guides and landed one on top of the other. No further securing is required in these circumstances. Some ships exist with guides above deck for similar purposes. Containers stacked one above the other without the benefit of cell guides must be secured one to the other with twist locks, and/or a combination of locating cones, bridging pieces, lashing rods, wires and shores as specified by builder or supplier, to prevent any form of shifting.

The blocking in of containers in a stow with unitised or general cargo has already been mentioned. To facilitate this and to spread the load on the container sides or walls large inflatable dunnage bags might be used to advantage.

Containers that are not blocked in as above must be properly secured with wire rod or chain lashings to prevent any movement and to reduce the strain of racking on the container. When lashing containers in conventional stow or on deck, particular attention must be paid to securing the *bottom* of the container as well as the top corners. This is particularly important if two or more containers are stowed in a vertical height. It is also important that only the corner castings are used to secure the container. A wire lashing for instance passed over the top middle section of the container does not secure it adequately, and can only damage the container should it move. Ships that have been designed to carry containers either below or on deck, and do not have cell guides, will have deck fittings suitably placed for corner castings to be held. These must be used to full advantage, and any recommended lashing pattern provided by the builders or architects should be strictly adhered to particularly where containers of mixed heights are concerned. Ro-Ro ships that may carry cargo either on wheels or landed on the deck will often have securing points that can be used for either type of cargo. Proprietary products are on the market that will allow a deck fitting to be used one voyage for Ro-Ro lugs, and on the next voyage to insert container securing cones.

Container Stuffing

The stuffing of the container may take place at:

The customer's premises.
An inland terminal or depot.
A groupage agent's premises.
The sea terminal or port.
On board ship (in special circumstances).

Prior to stuffing the container, it should be checked for fitness, i.e. that it is suitable with regard to type, cleanliness and repair, for the cargo it has to carry and the voyage to be made. These checks will include:

External

No holes or tears in the walls or roof.
No broken or distorted door hinges or locks.
The roof closures of open-top containers to be sound (e.g. no tears in tilts) and well fitting.
All out-of-date labels removed.
The temperature setting (of mechanically refrigerated containers) correct for the cargo to be loaded, with recording chart in place.
Loading, discharging and relief valves on tank containers operating correctly and properly shut where necessary.

Internal

Cleanliness: there should be no remnants of the previous cargo, e.g. no dust, sweepings, grease, liquid, etc.
Dryness: the interior should be free from any sweat, frost, etc., which might adversely affect the cargo to be packed.
Infestation: the container should be free of any signs that vermin or pests are present or have been present.
Taint: particularly if delicate goods or foodstuffs are to be put in the container, there should be no residual taint, also free from odour of disinfectants or fumigants.
Watertightness: this may best be tested with the "light test", i.e. entering the container, shutting both doors, and looking for any ingress of light.

The degree of fitness that the container has to achieve is largely dictated by the type of cargo to be carried. The alphabetical list of commodities in Part 3 will indicate the requirements for each type of cargo, and where possible the most suitable types of container.

Particular care must be taken in the case of Dangerous Goods, to which the IMDG Code Section 12 refers (see also "Dangerous Goods").

It is most important that the cargo inside the container is restrained from any movement. Proper restraint will prevent:

Damage to the cargo from falling or sliding around during the container's movements on land or sea.

Damage to the cargo or packaging from chafe — usually caused by vibration on road or rail.

Damage to the container floor or walls.

Danger to personnel opening the container doors at time of unpacking or for the customs inspection.

In extreme cases very heavy cargo, which is not adequately restrained, might break through the container walls and affect adjacent cargo or even the safety of the ship itself.

Restraint can be provided by:

Using an appropriate interlocking stow of the packages.

Pre-planning the stow so that if it does not fill the container it fully covers the floor and is level in height throughout.

Where possible and appropriate leaving void spaces down the centre of the stow so that cargo may be restrained by being supported hard up against either side wall, and filling the appropriate void spaces with dunnage or waste material, e.g. spare packaging material, old tyres, inflatable dunnage bags (which must be compatible with the cargo carried, e.g. with respect to taint).

Shoring and blocking the "face" of the stow (near the doors) with timber or inflatable dunnage.

Using the floor and wall securing points to lash with rope, wire or nets. N.B. If nails are used to secure timber shoring to the container floor, the nails must not penetrate the full depth of the floor, and they should be easily removed.

The cargo may damage itself or adjacent cargo, if not properly stowed, by:

Mechanical damage, e.g. crushing, ec.

Cross-contamination, e.g. taint, spillage or leakage, migration of dust or debris, infestation, movement of moisture, i.e. condensation.

Such damage may be guarded against by:

Stowing light items (of low density) over heavy items (high density).

Keeping items such as crates separated from cargo such as paper sacks.

Providing dunnage (timber, hardboard, etc.) between tiers as required.

Stowing liquids on the floor, dry items above. N.B. Drums and barrels should be stowed with bungs or closures uppermost.

Stuffing moisture inherent items in separate containers to those requiring dry stowage.

Stuffing items with a strong odour in separate containers to those susceptible to taint damage (or which might affect the packaging). N.B. Commodities with properties that might leave long-lasting residual taint (e.g. Phenolic based substances) may have to be stuffed into special containers that are reserved for dirty cargoes (see "Obnoxious Cargoes").

Strict observance of Dangerous Goods rules and regulations (see "Dangerous Goods").

Other areas of damage that have to be guarded against at time of stuffing or during transit include:

Uneven weight distribution; the cargo weight should be so distributed that the centre of gravity remains as low as possible and near the centre of the container. No more than 60 per cent of the weight should be in one half of the container.

When very heavy high density cargo is stuffed (e.g. lead ingots) the cargo should be positioned hard up against the side walls, and the weight distributed as extensively as possible to obtain full benefit from the inherent floor strength. Some such cargoes may require dunnage or similar to support the weight and spread the load.

Cargo susceptible to damage from fluctuations in temperature may have to be carried in insulated containers under temperature control conditions, or given stowage and stacking position adequately protected from direct sunlight (see "Refrigeration").

Cargo requiring ventilation may have to be carried in containers fitted with mechanical ventilation, or in open-sided or open-top containers. The stowage pattern of such cargo should be arranged to allow the proper movement of air through the cargo to achieve the ventilation required (see "Ventilation").

Refrigerated cargo must be stuffed in such a manner that the air or gas may move around and/or through the cargo sufficiently freely to maintain the required temperatures (see "Refrigeration").

Before stripping (de-vanning, or unpacking) a container, the following should be carefully checked:

Any notices or labels (e.g. Dangerous Goods labels), or notices concerning the contents specially if bulk has been loaded.

Where there is risk of gas being present (e.g. when using liquid nitrogen as a refrigerant) then doors should be left open for some minutes to allow any gas to dissipate before personnel may enter the container.

The right hand door should be opened — cautiously — first. This is to guard against the risk of improperly secured cargo falling out and injuring personnel.

The seal should be intact and not have been tampered with. A note should be made of the number for future reference.

The external condition of the container should be apparently sound. Any damage that may have affected the contents should be noted.

When the container is empty of cargo the interior should be checked and any residue removed.

Modification of G.P. Containers

The use of "specials" adds extra cost in capital outlay; maintenance and positioning. Advantage may be gained, therefore, by the operator able to use General Purpose (Closed Box) containers for specialised cargoes.

Techniques in this direction have been developed for the carriage of dry bulk and liquid bulk using General Purpose containers.

The principle requires the installation of a temporary bulkhead in way of the doors to support the load, and, for some types of dry bulk, the lining of the container with plastic or similar. Liquids are contained in a bag made of Polyurethane, Nitrile, Butyl, Platilon, Hytrel, Hyperlastic or Nylon Film depending on the compatibility with the product to be carried. Some of the products that have been carried in bulk, using a General Purpose container include: Malt, Grain, Seed, Polythene Granules, chemically inert powders, Brake Fluid, Detergents, Fruit Juice, Wine, Oils (non-hazardous), Sodium Silicate, Fatty Acids, Maple Syrup, etc. (See also Appendix 11).

Where bulk is being loaded in a General Purpose container, it must be ascertained that the container is structurally fit for this type of treatment. Some plywood sided containers, for instance, may suffer such extreme side wall deflection that the container may foul adjacent units or even the cell guides. In severe cases the container may rupture.

When to Clean

One or more of the following circumstances may make it necessary to consider cleaning the interior of a container between cargoes:

In the event of spillage — of either liquids or solids.

To remove a source of taint — which might affect future cargo.

To remove infestation — which may not always be apparent at time of inspection.

To remove the residue of a previous cargo, e.g. bulk, for quarantine or other reasons.

As a safety or good operating practice to remove old dunnage, packing materials, residue from chafed packages, etc.

N.B. Some of the cargoes which might permanently affect the interior of a container are listed in the section on "Obnoxious Cargoes".

Different types and levels of contamination from residues or spillage require different methods of cleaning. As mentioned above, the degree to which the interior is cleaned must depend to a certain extent on the commodities likely to be carried — and even on the trade.

Removal of loose solid residue by means of a simple sweep with a broom.

Cold fresh water wash, using soap or detergent.

Hot fresh water wash, using soap or detergent.

Steam clean.

Replacement of highly contaminated parts (e.g. the wooden floor) which involves repair rather than cleaning methods.

Fumigation

Containers may be fumigated for the following reasons:

When empty of cargo — to destroy residual infestation from previous cargoes.
When stuffed — to fumigate a particular cargo, e.g. malt.
To comply with quarantine requirements, e.g. Australian Plant Quarantine Regulations.

Normally carried out by specialist Companies well experienced in fumigation requirements, containers are usually shipped after having been cleared of any residue of fumigation. Special occasions may arise, however, when containers have to be shipped while under fumigation. In such instances full agreement and understanding must be reached between ship and shore staff; correct stowage — with appropriate ventilation — provided; warning notices, labels and restriction of access to unauthorised persons must be clearly marked. Containers under Methyl Bromide or Phosphene gas fumigation are dangerous, and must be carried strictly in accordance with appropriate legislation and reference to the authorities.

Fumigation methods:

(a) Methyl Bromide may be injected into the container via special inlets on the top side rails or other suitable point. Alternatively it may be introduced via the doors. In any case it should be carried out by authorised and experienced personnel, with proper protective clothing and breathing apparatus. The containers should have warning labels clearly discernible and should be fenced off from all unauthorised approach.
(b) Phosphene may be introduced in the form of pellets, which emit gas as a result of reaction with the moisture in the air. This gas is even more dangerous than Methyl Bromide, and residue of pellets must be removed from the container before cargo handling personnel are permitted to enter.

See also "Pests and Vermin".

ROLL ON-ROLL OFF

Introduction

Roll-on Roll-off (Ro-Ro) techniques include the handling of cargo on to or off a ship primarily by horizontal (or near horizontal) movement. This means that the cargo may be handled on vehicles or trailers which remain with the cargo during the sea passage, or alternatively may be taken on board by wheeled vehicles and the cargo itself stowed as General or Unitised cargo. (Sometimes, in exceptional circumstances, e.g. the movement of very heavy loads, the use of air or water skids may be adopted). The reverse procedure is used at the port of discharge.

Ro-Ro operations may be divided into three broad groups:

1. Short-Sea — With the sea-leg of transport chain of short duration, cargo usually remains on the trailers or vehicles.
2. Medium Sea — Medium length voyages, where cargo may remain on trailers or be lifted off and stowed without its wheels for the sea leg of the voyage, or a combination of both.
3. Deep Sea — Long ocean passages, Ro-Ro cargo is usually delivered on board, removed from its wheels, and stowed in a similar manner to a general cargo vessel — space requirements taking preference over speed of operation.

Operational Efficiency

Good planning and proper supervision are of a paramount importance.

Since the time taken to manoeuvre the load into position on a Ro-Ro vessel can take up to 50 per cent of the vehicle cycle time, it is important that proper means and techniques are provided to allow vehicles or their cargo to be manoeuvred into position with the least possible delay or complication.

The loading time of a Ro-Ro vessel is very often dependent upon the time taken to manoeuvre and secure the cargoes, trailers, and vehicles.

The speed with which a vehicle can complete its cycle (and sometimes its ability to operate at all) is affected by:

(a) the gradient of each ramp;
(b) width of each ramp;
(c) bends and turns to be negotiated, and blind corners;
(d) the speed of operation of elevators or other similar handling equipment;
(e) the "vehicle envelope";
(f) the change in gradient;
(g) the organisation and traffic flow.

The angle of the ramps — particularly the ship to shore ramps which are subject to a greater flow of traffic as well as to external weather conditions — can affect the speed and sometimes even stop operations. A gradient of one in ten may reduce the cycle time of vehicles by up to 8 per cent when measured against the speed obtained with purely horizontal access. It is important therefore that where possible vessels use the appropriate trimming tanks to adjust the threshold heights of their ramp to enable the latter to have the least gradient possible. Too steep a change in slope (e.g. in excess of 7 degrees) may sometimes cause the towing unit to become uncoupled from the trailers that they are pulling.

Where the width of a ramp may seem sufficient for two lane vehicular traffic, cycle time may be affected adversely if too many vehicles have to pass each other on the ramp. Traffic flows can sometimes be improved to reduce this.

The "vehicle envelope" is the total space required by the whole vehicle and its load as it negotiates a change in gradient, without fouling any obstructions above, e.g. entrance, deckhead, trunking, etc.

A change of slope if too severe, can ground vehicles either at the centre or at their extremity. Vehicles thus grounded may suffer damage themselves and cause damage to the ramp. Typical maximum angles when setting up equipment that involves a change of slope, would be 6 to 7 degrees, and about 6 metres between each such change of slope.

The use of good experienced drivers can greatly improve the speed of cargo handling in a Ro-Ro vessel. This is particularly apparent if vehicles have to be reversed up the ramp because of insufficient manoeuvring space on the vessel, or to position the trailers so that they may be pulled out directly at the port of discharge.

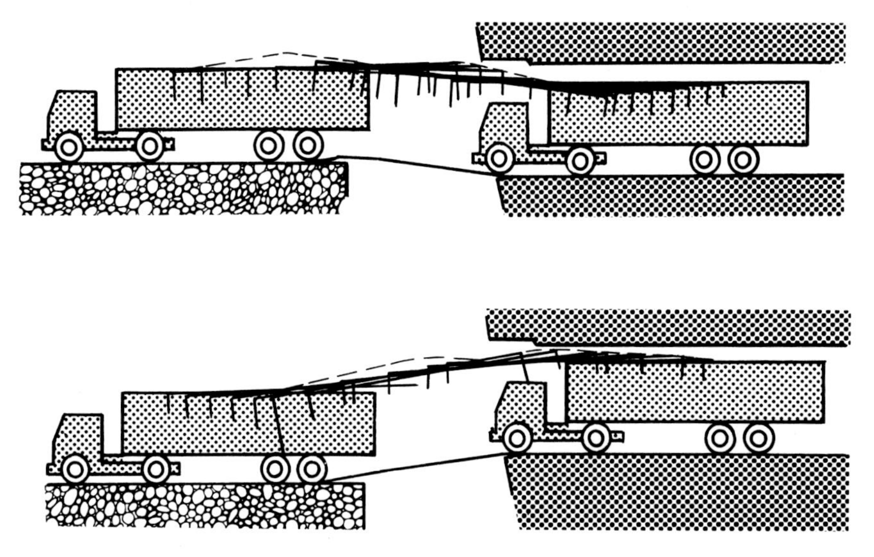

'Vehicle envelope'– the extra vertical space required by a vehicle negotiating a change of slope.

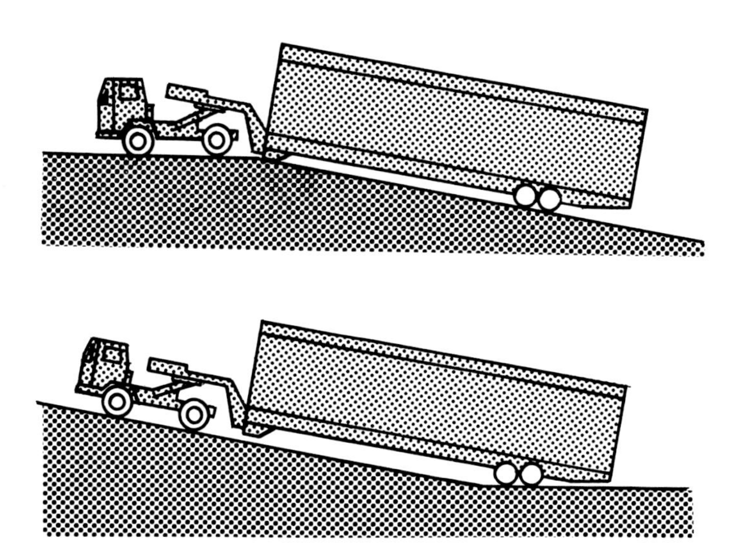

Excessive change of slope (e.g. between ship's ramp and quay) may cause grounding problems.

Where extreme tidal ranges may affect the gradient of the ramp, or even prohibit it from being landed on the wharf, Ships' Officers should make trim and draft calculations before arrival in the port of loading or discharging to ascertain the periods during which cargo may be worked, or in extreme cases periods when very heavy lifts may be handled over the minimum gradients. This information should be available to stevedores, planners, etc., so that maximum speed of operation may be achieved. Occasions have arisen when cargo has been presented to the ship, but, because of the angle of the ramp, the change of angle between ramp and quay, and the "vehicle envelope" requirement, the cargo has had to be turned away and the ship owner presented with the costs.

Stowage

Many Ro-Ro vessels have a predominance of the weight (e.g. ramp system) aft. This may require that tanks are used to maintain an acceptable trim during loading operations, and may also require that cargo is first in and last out to the forward lower decks.

Cargo may be taken on board the Ro-Ro vessel in one or more of the following ways:

(a) road vehicles with integral haulage power which will also remain with the vessel;
(b) road trailers which will remain with the ship throughout the sea transport leg;
(c) roll trailers which are not suitable for road haulage but which will remain with the ship during sea transport;
(d) cargo towed on board using roll trailers (or occasionally road trailers) and then cargo removed and stowed without its wheels.
(e) cargo secured on flats and carried on board either using roll trailers or by other mechanical handling equipment; both the flat and its cargo being stowed as a unit;
(f) pallets either singly or in groups (e.g. four at a time) carried on board using roll trailers or fork lift trucks;
(g) individual items of cargo brought on board by fork lift trucks.

Each of the systems described above have their own part to play in particular trades or environments. For instance short sea trades nearly always keep the cargo on its wheels for the short sea passage, the increase in cargo handling speed outweighing the loss of space.

Vehicles are usually close parked (and this includes trailers) in lanes of about 3 metres wide. This should allow access for lashing gangs to secure each vehicle properly. Australian waterside workers, for instance, require a minimum of 40 mm between vehicles when discharging car carriers.

Containers may be stowed fore and aft or athwartships, but care must be taken to ensure that suitably strengthened areas of the deck (usually with pads and locating cones) are in way of the corner castings.

As mentioned above trailers may be backed up the ramp and positioned so that at the port of discharge towing vehicles may have direct access to the coupling point of each trailer, and be able to tow straight off the vessel without the need to turn round.

It is important that different types of cargo, e.g. containers and pallets, are properly separated to prevent the one causing damage to the other. This separation, which in many cases also provides restraint, may be by means of timber dunnage, dunnage bags, sheets of plywood or hardboard, other cargo, e.g. tyres, etc.

Where containers with air cooled integral refrigeration units are stowed below decks, it must be ensured that adequate ventilation can reach these containers to allow the proper air cooling to take place, as well as sufficient space, c. 600 mm (2′), in way of the equipment end of the container, so that maintenance may be carried out and temperatures monitored.

Appropriate Dangerous Goods' regulations apply to all Dangerous Goods' cargo (see "Dangerous Goods").

Securing

Securing of vehicles on board Ro-Ro vessels must be in accordance with an approved system, making full use of trestles, pedestals, deck securing points, as are recommended by the builders. Securing points and appropriate trestles, etc., should be used to by-pass the springing system of vehicles, so that during the movement of the vessel at sea the vehicles remain rigidly secured (see also Part 1 — "Safety").

Containers should be lashed and secured in accordance with an approved system, preferably to locating cones and securing pins. Wire, chains, hooks and levers must be set up so as not to take undue strain and thereby rack or distort the container.

Equipment

When mechanical equipment, towing vehicles and road vehicles are operating on vehicle decks, proper and adequate ventilation must be used to remove excess exhaust fumes.

Using fork lift trucks to handle cargo direct from shore storage to ship stowage may be uneconomical in certain circumstances. The table below indicates suggested maximum operating distances for fork lift trucks:

Speed of FLT *km per hr*	*Max. Economical Distance* *metres*
7	64
8	74
9	83
10	92
11	102
12	111
13	121
14	131
15	140
16	150
17	159
18	169
19	179
20	188

Neobulk

A term used where large quantities of homogeneous or even similar cargo is loaded into one bottom, often with specialised facilities or specialised handling equipment to do so. While these types of cargo do not fall within the category of bulk cargo, they are of a sufficiently special nature to require techniques applied over and above those required for general cargo and bulk cargo. Neobulk carries no hard and fast definition with it and in fact the types of cargo loaded as Neobulk are classified and treated as general cargo in small parcels. Such cargo might include: steel, timber, forest products, motor cars.

Specialised ships may be employed for some of these items: specialised car carriers, specialised timber carriers and in some instances specialised steel carriers.

IRON AND STEEL PRODUCTS

Typical examples: ingots, billets, slabs; steel sheet in hot rolled coils and cold rolled coils; rolled sections and constructional steel; small section material, rods and wire; pipes and tubes of all dimensions; reinforcing bars; packs of sheet steel; individual steel sheets.

Characteristics

Nearly all steel — including some stainless steel — can suffer from rust development to some extent. This affects the products to a greater or lesser degree depending on (a) its state of production, and (b) its proposed end use. All steel can, to a greater or lesser extent, suffer from mechanical handling damage. This may occur on the quay, on the ship, at some inland point of transport, or while in the stow and subject to external forces of ships' movement and other cargo.

Steel may be packaged, that is protected with a cover of cardboard sheeting or crated or parcelled, e.g. packs of steel sheet wrapped in metal envelopes. Steel may be coated with oil or other protective against moisture. A great deal of steel is transported and stored without any protection at all, e.g. structural steel, reinforcing bars, pipes, etc. The degree of protection provided does not necessarily reflect the degree of care needed during transport.

Handling

Where steel is carried in large quantitities, specialised handling and slinging equipment is almost certain to be available, and this will vary from country to country and from berth to berth.

Slings and strops are normally constructed from wire rope, braided wire, and chains.

Slings, hooks, shackles, strops and chains must all be of sufficient robustness to provide adequate factors of safety when lifting the proposed loads. Chains are normally stronger than wire ropes or other slings (size for size) but may be unsuitable on account of possible damage to the cargo itself. When a load is lifted by a pair of legs or slings the angles subtended at the lifting hook should be small enough to ensure the cargo is not distorted or damaged by "pinching". This is particularly important when manufactured pieces are lifted which could be severely damaged if bent. A spreader or beam may need to be provided to prevent such damage. Sufficiently robust bars through the centre of steel coils, slung at the ends, may serve the same purpose. Steel should never be lifted by its own strapping unless it has been specially designed and approved.

Certain types of steel, e.g. pipes, coils, tubes, may have their own purpose built lifting equipment, e.g. pipe hooks, bars and spreaders.

Pallets, skids, bins, all may be utilised in the packaging of certain types of steel particularly coils of wire. Packages and units made up in this manner may be handled by mechanical equipment (see "Unit Loads"). "C hooks" may be used for lifting and handling of coils. This is a counter-balanced support arm which is put through the centre eye of a coil, and as the weight is taken the arm remains horizontal and supports the load. It is only suitable for handling and storing cargo that has plenty of access space — i.e. to put the arm (or bar) of the lower part of the "C hook" through the coil requires an equivalent amount of space adjacent to it so that the arm may enter.

It is important that the handling equipment used at the port of loading can be matched at the port of discharge. Instances have occurred of sheet metal being loaded by electro magnet, and the stow made so tight that it is impossible for conventional gear to gain access to the sheets at the port of discharge.

Damage

Steel products which are shipped in a partly processed condition, i.e. for further processing at place of destination, are less susceptible to handling and stowage damage than finished items. They may consist of hot-rolled bars, bar-sized shapes or coils, ingots, billets, slabs. Some of these can be seriously damaged by bad handling. Damage to coils may be caused by the edges being cut or dented. This can effectively reduce the width of the coiled sheet when it is offered for processing. Coils that become unstrapped may telescope or become kinked, and they may also be unsuitable for further processing.

Finished products which might sustain damage from careless handling or bad stowage include:

Rolled sections, i.e. constructional steel. These may be damaged if distorted or bent.

Steel sheets or packages may be subject to damage by mechanical handling or by rust. The edges of these packages may be particularly vulnerable, and unacceptable point loading may cause further damage.

Coiled finished carbon alloy and stainless steel bars, coils, etc., may be damaged by weather, and/or physical and/or chemical damage.

Pipes and tubes may suffer unacceptable damgage if the packaging or coating is damaged — i.e. to allow the ingress of moisture laden air.

STOWAGE AND SECURING

(A) Coils Steel Sheeting

Appearance

Coils can generally be divided into two categories namely hot rolled and cold rolled coils. Coils are either packed or non-packed and the type and quality of the packing material, whether metal envelopes or just paper wrappers are involved, appears in a wide range of types and colours. The packing components are secured in place by applying a number of circumferential and radial steel strapping bands.

Without going into detail about the various types of material within these two categories one can take it as a general rule that the carrier should at least care for the cargo in the same manner as the shipper does. Cargoes which are protected against humidity prior to shipment, either on trucks in barges or in warehouse, should never be exposed to rain during loading. In fact it is only non-packed hot rolled coils which may possibly be loaded in rain providing these are stored in the open and appropriate remarks as to its condition can be inserted in the relevant shipping documents. Further to this, consideration for cargo already loaded and/or to be loaded must be kept in mind.

Handling and Stowage

Coils should always be handled with braided wire slings or by means of special designed lifting gear in order to avoid lifting marks to the core packing and possible bending damage to the core windings. Forklifts should always be equipped with a round ram without any sharp edges.

Stowage of steel coils is to be performed with the utmost care and should comply with a number of rules which after many years of trial and error have eventually proved key elements in the safe carriage of this type of cargo.

Steel coils are to be loaded on a sufficient number of lines of flat dunnage, usually two or three lines depending on the width of the coil, with the centre cores placed in a fore and aft direction. The stowage is usually started in both ship's sides and is extended towards the middle of the hold taking care that at least two wooden wedges are properly placed against the base of the coil, always pointing towards the ship's side. The empty space which unavoidably remains in the centre of the stow should be provided by a so called locking — or key coil which is intended, by its own weight, to apply pressure to units in the sides in order to compensate for any slackness which might exist. This coil should be placed within the cantlines of the adjacent units and should sink down to at least one third of its diameter. In case the space is not sufficiently large two locking coils are to be provided as shown in **drawing X**.

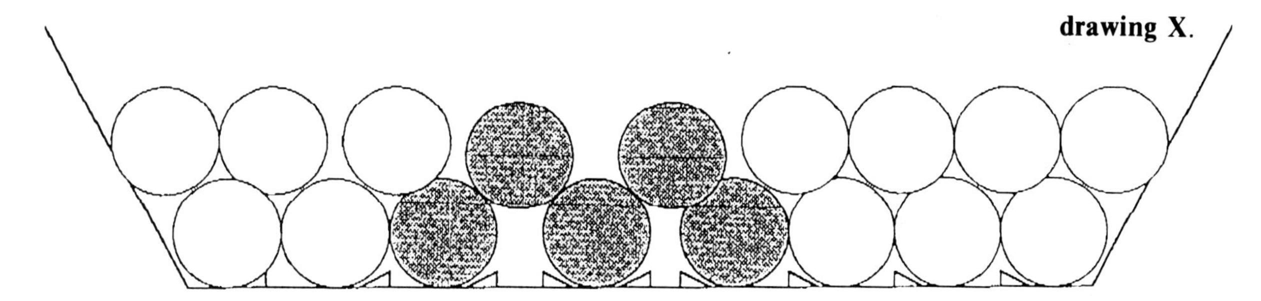

Coils should, for reasons of safety, if possible, always be stowed at least two units high as it is obvious that in this case any of the coils in the second layer are serving as locking coils and sink down whenever slackness would occur in the stow. A two or three high stowage nevertheless does not overrule the key coil system as described above.

In case a one high stowage (single tier of coils) is performed, as for instance in smaller sized vessels where the tanktop strength does no allow a higher stowage, additional securing is to be performed as described under the following heading.

In a multi layer stowage the second and possibly further layers of coils are loaded in the cantlines of the previous loaded coils. Care is to be taken, in case of different diameters and sizes of coils, that all coils are properly resting on the cantlines of the understowed units which involves a selection of the coils prior to shipment.

In connection with the stowage of coils on vessels equipped with side tanks it is common practise to stow coils by means of forklifts, equipped with a special designed hook on the ram, on the side tanks. It is nevertheless not essential for coils to be stowed in this manner and a pyramid stowage is perfectly acceptable providing additional precautions are taken when lashing the goods.

Lashing and Securing

(1) Coils in one high stowage

Coils in a one high stowage should be avoided but is sometimes necessary in smaller vessels for reasons of tanktop strength, or in larger vessels where individual coil weights are in excess of 15 tons. With this type of stowage it is advisable to perform a double lashing of the locking coil by means of steel wires or metal strapping bands. It is preferable to perform a short lashing through the eye of the locking coil towards the eye of the understowed coil rather than to lash the coil over its circumference. In any case, the method of securing should involve lashing the locking or key coil to adjacent coils below through the cores of the individual coils.

(2) Two or more high stowage

Two systems are generally used for lashing of coils in a stowage of two or more units high. The securing which is performed by means of chocking all empty spaces in the toplayer by means of square timber is less and less used and is in many instances replaced by the "metal strapping band system".

In the first case large quantities of square timber are used to fill up all empty spaces in the toplayer of coils and additional wire lashing is performed to those units which are vulnerable to shifting, especially the coils stowed in the ship's side in case a pyramid stowage is performed. With the "metal strapping band system" one and every coil in the toplayer is lashed by means of pneumatically driven tools. With the latter system strapping bands are abundantly used and therefore the usual chocking with timber can be dispensed with. Every coil in the top tier is lashed through the eye to the understowed units preferably with a short lashing. Every coil is thus lashed by means of two strapping bands each of which are secured by means of two steel clips. The short lashing system demands that a lashing gang has to standby during the loading operations and therefore, for economical reasons, the system has been somewhat altered within the years. It is now most common practice that lashing starts only on completion of loading operations in a hold. In this case the strapping band is lead through a coil stowed in the second tier from the top, is crossed over the coil to be lashed and is passed through the eye of another coil in the second tier on the other side of the coil to be lashed. The loop is further

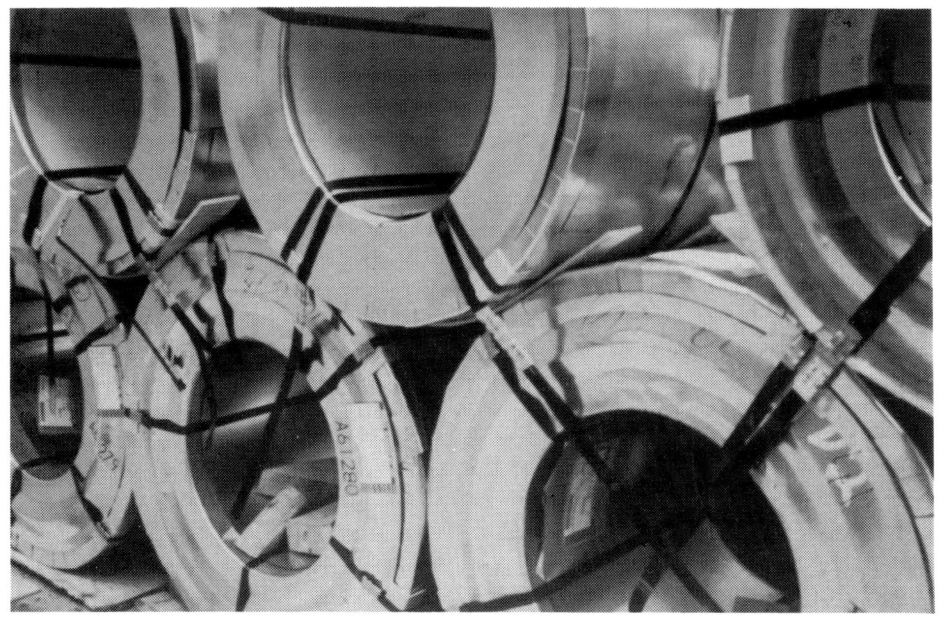

View on method of the lashing and securing of steel coils by means of flat metal strapping bands.

Steel coils in stow.

General view on the method of stowing steel coils in a square shaped cargo hold.

General view on the stowage of steel coils.

extended and runs via the eye of the coil which is to be lashed, the latter in such way being kept in position. Additional double lashing is usually performed for those coils which are sitting in a dangerous position such as the external coils in a pyramid stowage.

Where appreciable spaces exist either forward or aft of the face of a coil stowage, the top coils in a multi tier stowage in the area of the brow (face) of the stow, and the coils in the same area in a single tier stow, must be block lashed to prevent the end coils moving out of stow when the vessel is labouring in a seaway. Usually the lashings extend three coils deep into the stow and in the olympic block lashing system there are nine coils per block with adjacent lashings overlapping. For such an arrangement to be effective spaces between coils within the block lashing must be wedged before the lashings are tightened.

(B) Coils Wire Rods

Appearance

Steel wire rod which is a semi-finished hot drawn product is usually wound into coils with a weight ranging from 0.5 to 2.5 tons average. The coils, or sometimes bundles which are consisting of two or more smaller units, are usually secured by means of a minimum of four steel wires or strapping bands. In case of moderately compressed bundles being secured by strapping bands a loping wire is sometimes fitted over the circumference in order to secure the straps against shifting together during handling. Great care should be taken in checking the securing of the bundles as a single missing or broken securing band may not only cause the bundle to fall apart during further handling but might cause the stowage to become unstable.

In connection with physical damage it is so that in further processing the wire is lead through a number of dies which eventually results in its gauge being reduced to a required diameter. It is not as is sometimes argued, that imperfections such as gouges and score marks to the winding surface will eventually reduce or disappear during the manufacturing process. Any defect will follow the wire through the process, will increase in importance compared to the reduced wire diameter, and will eventually cause the wire to break. Care should be taken not to drag coils over the quay during loading, to use suitable lifting gear, and to avoid steel to steel contact in the stow.

If wire is wrapped one should interpret this wire to be of a special quality which should not become wet and for which physical damage might have a detrimental effect.

Handling and Stowage

This material is either lifted by means of "cobra wires", which is a steel wire with rope protection woven within the steel strands, or by means of "C" hooks. In case cobra wires are used every sling should only lift one bundle at the time but more slings can be used to lift more bundles.

Adequate sized coils of steel wire and unitised bundles are generally stowed with their cores in a fore and aft direction in the ship, but there is no objection to perform an alternative stowage. When the full tanktop space is used, it is possible to build up the stow in a fore and aft direction under the hatchsquare and to fill the ship's sides with bundles stowed in an athwartships direction. This system avoids coils from being dropped into the ship's sides but implements a perfectly tight and stable stow so that steel driving plates can be laid on the bundles and forklifts can be landed on top for stowing the bundles in the wings.

In general dunnage wood is used on the tanktop plating not in order to spread the weight but in order to avoid steel to steel contact and for the protection of the wire against contact with any moisture collecting on the tanktop.

In those areas where the wire could come in contact with structural parts of the vessel dunnage wood or plywood is equally used in order to avoid scoring of the windings.

When stowing the bundles of wire rods, which stow should extend from side to side of the compartment, care has to be taken to arrange it in such a way that the coils are tightly and compactly stowed.

Structural steel in the wings loaded in a fore and aft direction to protect the ship's sides against possible movement of athwartships stowed cargo.

Slabs and long bars in stow.

View of stow of long steel plates athwartships and fore and aft stow of structural steel.

View of Californian method of stowage of steel slabs in a box shaped hold.

Lashing and Securing

In case the stow does not occupy the whole of the cargo space and terminates leaving an empty space, it is essential that it is secured by pulling the outer bundles to the bulkhead. This is generally performed by fitting a wire in front of the rows in a athwartships direction and by pulling this wire to the bulkhead, through the eyes of the coils, in several positions.

Further securing, unless for those coils which are in an unstable position, is not required as wire rod stowages settle considerably during the voyage.

(C) Tin Plate

Appearance

Thin steel sheeting covered with a fine layer of tin, used in the canning industry for food conservation, presents itself for shipment in form of coils and packages.

The coils are usually placed eye to the sky on a strong wooden or steel skid, externally protected by a metal envelope, being secured by means of a number of steel strapping bands. The sheet form tinplate, shipped in rather small square bundles, is contained in steel boxes which are fitted with bearers beneath to facilitate handling.

Handling and Stowage

Tinplate is very sensitive cargo and should always be kept dry whilst physical damage should at all times be avoided. The slightest damage or distortion of boxes of tin plate may result in expensive delays to the fully automated machinery used in further processing. One should also consider, and this applies to the boxes as well as to the coils, that any excessive pressure applied in the stow to the external packing results in deformations and imprints on the plate surface which eventually causes a considerable amount of material in the package to be rejected. Instances where the lashing gang had left bulldog clips lying around on the packages whilst stowing have caused not only deformation to the sheets but have punctured the metal envelopes giving sweat water free access to the contents.

Coils as well as boxes are always to be lifted by means of braided wire slings unless the pallet is made in such a manner that chains or wire slings can be used without coming into contact with the packing material. For this reason, in case more than one box of tinplates is lifted, a spreading device is required in order to avoid squeezing of the top package.

Coils

The coils are usually stowed two or three units high from side to side of the hold when care should be taken that, for the two high stowage, sufficient dunnage is placed in an athwartships direction in order to lock the coils together. The empty space left in the ship's side of non-box type vessels should be chocked off whilst the coils are to be blocklashed by means of steel wires or preferably by means of mental strapping bands. The metal strapping band is easier to apply and due to its nature avoids imprints of the lashing material which might be the case if steel wires are used.

It is common practice for stevedores, providing they have the equipment to do so, to the fill the remaining space in the ship's side by means of other suitable cargo.

In case the wooden skid extends beyond the diameter of the coil it is obvious that a non-tight stowage is performed in which case the empty spaces existing between the coils themselves should be chocked by means of wooden wedges and/or timber. The timber stanchions or the wooden wedges, in case the remaining space is smaller, should be linked by nailing a wooden board over the top in order to prevent them from dropping down in case slackness in the stow develops.

Boxes

Boxes of tinplates are stowed in layers from side to side in the hold when care is taken, in spite of the wooden skids provided, to place layers of dunnage athwarthships at least in between every second layer or preferably in between every layer. Failure to do so creates a very unstable stowage as the packages are stowed one on top of the other to form unsecured columns of boxes in the stowage in which case they will chafe and move during heavy weather and cause the packing and/or contents to become damaged as a result of friction. Providing the packages are not too high it is usually possible to extend the dunnage out to the lower wing tanks so that the package in the next layer can rest partly on the adjacently stowed package and partly on the lower wing tank slope plating. This prevents empty spaces in the ship's side which for this type of cargo, in view of the dimensions of the packages and the height to which this cargo is sometimes stowed, should be avoided.

Any empty space in one of the layers should either be bridged by means of dunnage or if the space is too big it should be filled by means of square timber. The toplayer should preferably be complete and the remaining empty space, providing it is reasonably small, should be tommed off whilst all smaller empty spaces between the packages are to be wedged. The wedges are to be linked by a nailed wooden board placed over the top in order to prevent them from dropping down in case slackness develops.

In case a large empty space would unavoidably remain in one or more layers of the top stow it is preferably to start off with a new layer or to fill the space with other cargo rather than to build large timber stools.

(D) Steel Slabs

Appearance

Steel ingots are held at a high temperature until they are transferred to the rolling mill where they are rolled into a semi-finished block of metal referred to as a steel slab. These slabs will eventually be rolled into a steel plate and ultimately form hot rolled steel coils.

The finished slab has dimensions of about 250 mm thick and 2,500 mm wide and is produced in various lengths. Slabs are unwrapped and unprotected against rust. The nature of this material is such that it is impervious to damage.

Stowage, Lashing and Securing

A relatively new system which is now in use for loading steel slabs is the so-called "California block stow" system.

This system consists of loading slabs straight upwards, one slab on top of the other, the different tiers being interlinked with square timber. Olympic lashings by means of metal bands are applied to the top tiers only.

This manner of stowage should only be employed in vessels with box shaped holds. Such vessels have wing tanks which extend for the entire depth of the hold in which case the hold side plating is vertical and terminates at a right angle with the tank top plating — there are no slopes to port and starboard.

Such stowages, in the normal type of bulk carrier, which are not winged out to the ship's sides are unacceptable because they are a potential danger to the ship and her crew. Such heavy weights, as form a slab cargo, are less liable to shift than relatively lighter items of cargo. However, once they commence to move they usually do not stop moving until they have caused considerable and serious damage.

The stowage of slabs in the normal type of bulk carrier should be winged out, over the lower wing tanks, to the ship's sides. Being overlapping and fully dunnaged in every tier.

(E) Plates

Appearance

Mild hot rolled steel plates, either coated or non-coated, are shipped as loose pieces or in bundles being secured by a number of steel strapping bands. It is usually the smaller sized plates which are made up into bundles, some of which are occasionally tarpaper wrapped. The larger sizes, which might present themselves in lengths of up to 12 metres, are generally used in shipbuilding and tank construction whilst the smaller plates will be used as general purpose plates in the fabrication industry.

Handling and Stowage

Large sized plates are usually lifted by means of two sets of steel hooks attached to a spreader bar. For small gauge plates in lengths of over 6 metres the quantity of hook sets is increased and possibly assisted by an end hook at each short side of the plate in order to avoid the lifting gear to shift during manipulation. Care should be taken not to lift plates with different width in the same load as otherwise, apart from the dangerous nature of this operation, the protruding plates will unavoidably become bent at the edge. Closely check whether the weight of the bundle does not cause the plate edges to become bent and should reduce the quantity of plates lifted if this is the case.

In case plates can not directly be lifted with shore cranes, they are brought alongside the vessel by means of forklift. In general a single forklift is sufficient for safe handling of most plates unless we are dealing with small gauge plates or plates in lengths above 6 metres in which case two forklifts should be used. This especially applies to long plates where handling with a single forklift has in many cases led to the plates becoming permanently deformed. Another damage which is not readily visible is score marks to the surface resulting from handling with forklifts which, in case of tankbuilding plates, results in very high re-conditioning costs.

Stowage of steel plates on board a vessel is somewhat contradictory to the ideal stowage of steel plates. Loose steel plates are in many cases stored on the quay without the use of any dunnage resulting in a perfectly level stow. It is obvious that this stowage can not be performed on board vessels in view of the danger of shifting resulting from steel to steel contact and because lifting of the plates prior to slinging would drastically increase the time necessary for discharging. Dunnage is therefore used and dunnaging should be performed with the utmost care. Dunnage wood should be of a uniform thickness. It should be kept in line vertically throughout the stow in order to avoid permanent deformations to the plates and the stow should be kept level. In case longer plates overstow smaller plates, any gaps in the stow must be properly filled by means of either dunnage or square timber so that the next layer lies flat after landing.

Contact with structural parts of the vessel must be avoided as this might result in damage to both vessel and plate. Special care should be taken for plates not to rest on the vessel's side tanks as settling of the stow during the voyage might result in the plate edge being bent upwards.

As a general rule one might say that the use of dunnage should be restricted to the minimum and should only serve the following two goals.:

— to keep the stow level at all times
— to facilitate lifting at the discharging port

Lashing and Securing

Providing the stow is level and no cargo is protruding above the stow securing can be restricted by choking all empty spaces in the top layers of plates by means of wooden wedges and square timber. Any protruding plates or bundles are to be secured against shifting either by means of choking or wire lashing.

(F) Structural Steel

Appearance

Structural steel is shipped in lengths of up to max, 18 metres (60 ft.). Structural steel covers all types of steel sections, including channels, angles, beams, flats, rounds and reinforcing bars, which are used in structural work. They are usually shipped as unwrapped single units or in the form of unwrapped bundles.

Stowage

Structural steel would normally be expected to occupy the lower stowage position within a ship's hold. The stowage of long structural steel commences by placing dunnage on the vessel's tanktop to hold the goods clear of the ship's metal work. This assists in spreading the weight, protects the ship's structural parts and guards against the goods being contacted by any moisture which might collect beneath the cargo. Flat pieces of timber are used to hold the steel clear of the ship's side frames and tank sides, and under no condition it is admissible to permit the goods to be in contact with any component parts of the vessel's structure. When stowing beams and similar constructional steel, it is important that the webs of beams are kept vertical and that the flanges overlap in an over and under manner, otherwise there is risk of the steel becoming severely distorted. Mixed sizes should be avoided where possible as this can cause gaps in the stow which later may lead to the whole stow collapsing. Furthermore, sufficient dunnage is to be used throughout the stow. The purpose is twofold. The first consideration is to bind the mass of steel into a solid block within the cargo space. The other purpose is to facilitate slinging at the port of discharge. In order to avoid warping or bending of the long lengths of steel the dunnage must be kept vertically in line throughout the stow.

Athwartships stowage of steel beams is best avoided. In particular one should avoid the ends of beams being stowed in the wings of the compartments. If during the voyage the dunnage compresses, the beams may settle, leaving the ends resting against the lower wing tank sides and the middle of the beams unsupported. As a result the beams may become permanently bowed, as well as risking damage to the tanks themselves. Structural steel should be stowed in a fore and aft direction. Notwithstanding this, situations do arise when in order to prevent part of a cargo being shut out from shipment, some structural steel does have to be stowed athwartships. Should this type of stowage be unavoidable, it is recommended some lengths of steel are placed fore and aft in the wings, adjacent to the ship's side frames, before the layer of athwartships steel is loaded.

Lashing and Securing

Stuctural steel, whether bundled or not, can be secured against movement by blocklashing the uppermost tiers of the cargo with strong steel wires. Wedges or square timber can be driven between any gaps. This will ensure that there is no pile up of the steel through rolling of the ship or movement in a fore and aft direction.

(G) Sheet Piling

Appearance

Sheet piles are hot rolled "Z" shaped profiles in lengths of up to 60 feet. The sheet piles are either shipped loose as single pieces or connected one to another in which they will be described as double sheet piling. The keying arrangement with which one piece is connected to the other is called a finger edge. The sheet piles are driven in the ground and are generally used in instances where a building pit, whether on land or at sea, is to be made and a water and/or earth barrier is to be formed.

Handling and Stowage

Sheet piles, in view of their construction, provide a sufficient lengthwise strength so that they can be lifted at two points. The handling of longer lengths requires the use of a spreader bar not only to avoid unnecessary stress in the piles but also for safety reasons. The finger edge is the most critical part of the sheet pile and damage in this area should be avoided. Braided wire slings have been used to

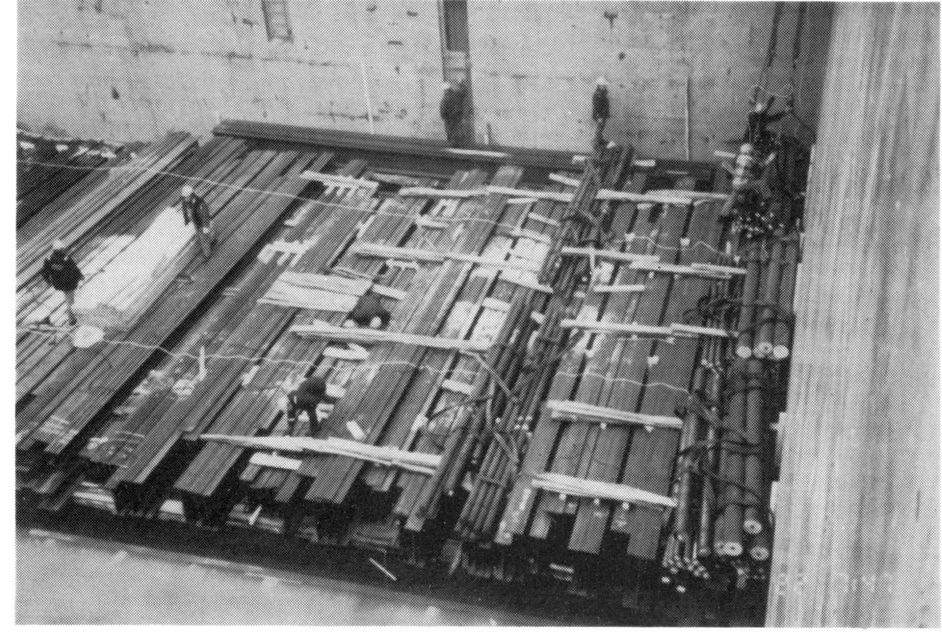

General view on the method of stowage of long constructional steel — stowed in a fore and aft direction.

Steel and pipes lashed in stow.

View on stowage of large pipes in the hold of a ship. Method of lashing applied to the top layer of pipes to avoid shifting.

handle this material for a number of years but experience has proved that a doubled chain sling at each end of the pile(s) does not result in damage and provides a safer way of handling.

Single sheet piles are difficult to stow and require large quantities of timber to be used in order to level the stow before a next layer can be stowed. Therefore piles are usually shipped as double sheet piling in which case the stowage, although requiring care, is much facilitated.

The double sheet piles are usually loaded one on top of the other without the use of any dunnage so to form an unsecured hood shaped bundle of 7 to 10 pieces high. Higher stowage would result in the piles sustaining damage through inside bending or cracking. At this stage square hardwood timber is laid all over the stow in an athwarthships direction which is supported by means of stools built up crosswise laid timber in the gap between every load. The quantity of lines of square timber and stools to be used should be guided by the length of the sheet piling.

The next layer of piles is thus stowed on the timber flooring performed.

Lashing and Securing

In view of the considerable work involved in lashing of this type of material, as a result of the large gaps in between the different stack, sheets pilings are usually overstowed by other steel cargoes. In case no other cargo is available one should fill all empty spaces in between the topstow by means of nailed square timber. Preferably block lashing should be performed by means of steel wires which are to be passed under the cargo prior to commencement of the stow.

(H) Pipes and Tubes

(1) Large Diameter Pipes

Appearance

Large diameter pipes are shipped as single units and are used for the transport of gas, fuel or water. Large diameter pipes can be either coated, when the coating may consist of bituminous materials — carbolac — polythene — epoxy paint etc., or non-coated. The ends of the pipes may be plain, provided by bevelled ends for welding purposes, or with collar ends.

Handling

Every type of pipe listed above, will be handled differently and requires in many occasions special loading gear.

Any type of damage either to the outer or inner circumference, whether we are dealing with coated or non-coated pipes, or to the ends is to be avoided. Usually, especially when we are dealing with full shipments, the pipes are stored on the open quay or on wagons and will be directly lifted by means of shore cranes. The use of forklifts should be avoided but in case these should be used utmost care is to be taken and the necessary precautions followed in preparing the forklift as well as the transport so as not to cause any damage. Pipes are lifted either by means of copper or Teflon coated steel hooks attached to each end or by means of adequate belts but never by means of steel wires or chains.

Special care should be taken in case pipes with bevelled ends are shipped, as any appreciable dent — to the bevel will require a piece of pipe to be cut off, in order to perform a new bevel, with high handling and transport costs being involved.

Stowage

Large diameter pipes are always to be stowed in a fore and aft direction in order to avoid shifting during the voyage. The stow of large diameter pipes should always be kept level which involves specialised stevedores and a considerable number of labourers in order to fit the necessary timber stools. Loading starts off from the tanktop on which sufficient layers of flat dunnage or square timber are laid, usually about 3 metres apart. In the case of collar type pipes large sized square timber is to be used in order to keep the collars free from the tanktop. Collar type pipes are kept apart in an athwarthships' direction by fitting wooden stanchions in between the bottomlayer of pipes. In this case the bottomstowed pipes are also properly secured by means of nailed wedges.

The stowage starts at one side of the hold and the remaining empty space at the other side of the hold is filled by means of wooden stools being composed of crosswise laid and nailed pieces of square timber. A piece of plywood cut out in the diameter of the pipe will assist in raising the wooden stool to the level required by the pipe, which will be loaded on top of the stool, will eventually be at the same level of the pipes, which will be loaded in the cantlines. This procedure should be continued throughout the loading operations hence the reason why a number of labourers need to be on standby. Pipes are usually landed on timber being placed on the previous layer and are rolled by manual force into the ship's side.

In case of collar type pipes it is obvious that every layer, in view of the collar, is loaded with the collar alternatively in a fore and aft direction.

In case of longitudinally welded steel pipes it is preferably, if possible, to keep the welding seam free of cantline contact with over- or understowed pipes.

The stacking limitation is regulated in the A.P.I. rules and it is preferable to inquire with and get a written confirmation from the shippers.

Polythene or epoxy coated steel pipes will require padding to be laid in between the pipes, such as straw filled plastic strips, this according to the instructions of the shippers.

In case **deck stowage** is performed, the most common method is to stow the pipes in pyramid fashion over the hatchcovers from side to side of the deck. The key points in this stowage are the bottomstowed pipes on both sides of the deck which should be adequately supported either by a strong bulwark or by pieces of steel beams welded on deck in case of an open railing at the ship's side. In order to keep the stow on deck and over the hatches level, this type of stowage requires large quantities of square timber to be used. In view of the rather complex procedure to be followed for this type of stowage it is required to call in expert advice.

Lashing and Securing

The lashing of under deck stowed pipes is usually performed by chocking all empty spaces in the topstow and by pulling a number of steel wires over the top. These wires are attached to available padeyes in the ship's side deeper in the stow. The lashing wire should be provided with two turnbuckles as the wire, over this length, is bound to stretch much more than the compensating capacity of a single spanscrew.

At the open end of the stow is advisable to secure the topstowed pipes to the bulkhead by fitting a wire athwarthships in front of the layer and pulling it at regular intervals to the bulkhead by a pass through the inside of the pipe. An alternative system is to hook shackles to the pipes ends and to secure them to the bulkhead by leading a small sized wire through the available eye of the shackle.

Lashing and securing of **deckstowed** pipes consists of chocking all empty spaces in the toplayer and by passing a sufficient number of adequately sized steel wires (usually 18 or 22 mm.) over the stow from side to side. The steel wires, attached to an available point in way of hatchcoaming, are passed under the stow of pipes on deck, and are eventually secured together on top of the stow by means of at least two adequate sized spanscrews. The steel wires should be tightened by means of a mechanical or hydraulic jacking system. Chain lashing can be used taking into account that, as for wire lashing, proper protection should be provided in order to avoid damage to the body of coated pipes.

(2) Smaller Diameter Pipes

Appearance

Small diameter pipes, whether loose pieces or in bundles, usually have a protective coating. They may be oiled, greased, varnished, painted or galvanised and are usually 6 metres in length. The ends may be plain, bevelled, or provided by sockets or couplings. In most cases the pipes are shipped in bundles either round or hexagonal shaped and are secured by means of a number of flat metal steel strapping bands. Occasionally pipes will be seen to be stowed in wooden or protected steel racks. In this case we are dealing with drilling pipes or joints for the oil industry and special care is to be taken in handling. Special types of pipes such as stainless steel pipes or pipes with very small diameters are usually shipped in cases.

Handling and Stowage

Smaller sized pipes and bundles are usually brought alongside the vessel by means of forklifts and are very susceptible to physical damage such as bending if inadequately transported or lifted. Lifting is preferably performed by means of protected steel wires or by belts in order to avoid score marks to the periphery of the pipes.

In connection with the pipes destined for the oil industry, commonly known as "joints", stowage and handling requirements should be obtained from the shipper. Some of these pipes, namely those containing a high chrome percentage, should not come in contact with carbon steel in which case non-protected forklifts and crowbars can not be used. Also the lashing wires should be protected as some of these pipes can be bacteriologically contaminated even through contact with dunnage wood.

Small sized pipes, in view of their frail nature, are usually topstowed and should preferably not be overstowed by any other cargo unless it is very light.

A perfect stowage, with the exception of the hexagonal shaped bundles, is usually not possible in view of the round shape of the majority of the bundles offered for shipment the aim should be to perform a stowage which is as level as possible. A sufficient quantity of dunnage should be fitted between in the layers in order to facilitate the discharging operations and also in order to avoid score marks whilst pulling the lifting wire away after landing.

Lashing and Securing

Securing is performed by filling all empty spaces in the topstow by means of wooden wedges or square timber. Any incomplete layer (always to be avoided if possible) should be wire lashed to the ship's side. Securing with "flat metal" strapping bands is very convenient for this type of material.

TIMBER

Timber measurements are the most complicated and laborious of all the various measurements in use for shipping purposes. The unit of measurement in use in the U.K., North European countries, etc., is "a standard" of which there are many varieties, bearing no relation one to the other as shown below. In North America the unit of measurement is the 1,000 board feet; in France, Italy, Belgium, etc., the unit is the "Stere", equivalent to the cubic metre of the metric system, while "Petrograd standard" is almost exclusively used in the U.K. — wholesale transactions in battens, boards, deals, planks, etc., being on that basis.

Timber cargo may consist of a "composition" of logs, deals, battens, small batten boards, small boards, scantlings, slatings, and laths — the latter two being usually in bundles. A good composition for cargoes is two-thirds deals and battens and one-third boards, and the vessel with clear holds should stow this at about 225 cubic feet per standard.

If the cargo consists of more boards and contains quantities of small boards, slatings and laths, the stowage factor would be subsequently higher. For example bundles of laths stow at about 320 cubic feet to the standard.

By a custom prevailing in the Baltic wood shipping countries a shipper is supposed to have a margin of one-sixteenth of an inch in cutting his planks. This margin is very often exceeded and may be found to be as much as one-eighth which the ship carries free.

Newly cut timber, being full of sap, is naturally much heavier than timber cut the previous season and the vessel is not able to carry so high a deck load as with old cut timber.

If the vessel is to load a cargo of new cut timber, with the excessive marginal cut, and the "composition" is a poor one the intake would be more than 5 per cent less than her ordinary capacity, and it is impossible for a ship owner to estimate what his ship will load. Experience in a 1,200 standard ship has shown the intake to be as much as 70 standards below capacity under these circumstances.

The average weight of Baltic sawn wood is about 2½ tons to 3 tons to the standard.

The best type of timber carrying vessel is that which has a large beam in proportion to draft with a minimum number of obstructions in the hold, such as stanchions and web frames.

The vessel will carry a "good cargo" or otherwise according to whether the Charter provides for a proper proportion of short lengths, laths, pickets, etc., to be provided for filling broken stowage and that such be delivered alongside when the loading commences. Full stowage below decks adds to the amount of deck load which can be carried.

Receiving

In some timber loading ports, the shippers demand receipts for cargo when brought alongside by raft of lighter and before it is received on board.

As part of all such cargo may be lost through drifting away, sinking, fire, or even by theft, the risk involved in the granting of receipts for cargo before it is actually shipped is substantial and, in so far as the receiver of the cargo at destination is concerned, the customary noting of protest, following loss of cargo from alongside avails nothing — at the same time, the noting of protest where cargo is so lost is necessary in the interests of the owner.

Unless the charter-party specifically provides for the issuing of Bills of Lading so delivered, no documents should be signed until the cargo is actually on board.

Lumber Bills of Lading not infrequently are presented with a clause to the effect that the lumber is "free from splits and shakes" which term is often taken to mean that the goods have not been split or damaged by rough handling before shipment. "Splits and shakes" when present are in the main, due to latent conditions, these defects developing as the wood dries.

From this it is evident that, with such a clause in the Bill of Lading even if the cargo be quite free of such defects when shipped, the ship may be liable for claims for deterioration which has developed along natural lines during the voyage.

Bills of Lading should therefore be endorsed "not responsible for condition or quality".

Slings

While the use of chain slings is permissible for handling deals and battens, rope or webbing slings should be used for boards, box boards, slats, laths and similar classes of timber as well, of course, as with prime woods. Slings with spreaders will be required for packaged timber.

Damage to Account of Charterers

Careful note should be made of any damage sustained by the ship or her fittings during the loading and or discharging. All such should be reported by letter to the charterers or their local agents. When loading heavy logs careless lowering or rough dragging of logs when discharging, frequently results in the buckling of frames, damaged pipes, cement chocks, etc.

Packaged Timber

Sawn timber is almost invariably packaged or unitised. These packages may vary in length and size, depending on the handling equipment, consignee, etc. Specialised handling equipment, such as straddle carriers, may be used, and special terminals provided for the handling and stowage of this type of cargo. Where packages consist of sawn lengths of varying length, it is normal practice to square off one end to give a flush face. Where such packages of varying lengths are made up, and where the holds of the vessel are not suited to cargo made up in this way, there may be a great deal of wasted space.

Timber stowed on deck. Note broken bundles on wharf.

Deck Loads

When deck loads are carried, which is more frequent than not, the upper deck should be assisted to carry its load by hard wood wedges driven between the deck beams and a plank placed athwart on top of the timber cargo below, and adequate compensation should be made in the same means for any stanchions which have to be removed — the practice of removing stanchions should be restricted as much as possible.

When dunnaging under a deck cargo of timber, use rough 25 mm boards placed diagonally 750 mm to 1,000 mm apart, so as to distribute the weight evenly over the beams, etc., and avoiding buckling deck plates.

Stanchions to support deck loads at the side should be of sufficient length as to extend not less than 1,200 mm above the finished level of deck cargo, to permit manropes being fitted to same for the protection of the crew. Stanchions, which usually are supplied by shippers as part of the cargo, should be spaced about 2,500 mm to 3,000 mm apart and be inclined inboard.

Chain lashings are probably the most effective lashing method for deck cargoes of timber. Certain systems are designed around webbing made of man-made fibre.

Lashings are usually shackled to eye plates or wing bolts riveted to the part of the sheer strake, extending above the deck stringer bar, or to the deck, spaced 2,500 mm to 3,000 mm apart; each length of chain long enough to meet its mate in the middle line, where the two free ends are connected to a heavy turn buckle and slip hook by which the chains are set up.

Regulations and Recommendations

Various countries have regulations, recommendations and codes of practice for the carriage of deck cargo of timber. These should be carefully examined and any appropriate regulations carried out. Typical of these include the British Merchant Shipping (Loadlines) (Deck Cargo) Regulations 1968; IMO Code of Safe Practice for ships carrying timber deck cargoes, 1991. This publication is an update since the first publication in the early 1970s and incorporates new systems and techniques learnt from casualties during the interim period and allows for the larger and more sophisticated ships that have come into service.

Pit Props

Pit props are short, straight lengths of timber, mostly fir, debarked, and are exploited in large quantities from such countries as Scandinavia, Baltic Russia, etc.

The props vary from 75 mm to 250 mm in diameter and are shipped in standard lengths, 3.5 ft, 4.0 ft, 4.5 ft, 5.0 ft, etc., up to 11.0 ft, which is about the maximum. Crooked or split props should be rejected. The unit of measurement in the pit prop trade is the English (cubic or pile) "fathom", i.e. 6 ft × 6 ft × 6 ft = 216 cu. ft, freight being payable on that basis.

A fathom of props varies considerably in weight, from 2.5 tonnes or less to on occasions, as high as 4 tonnes, such depending partly on the class of timber but more particularly on its moisture content. When receiving from wharfs or railway trucks, the timber being relatively dry, approximates for less weight per fathom, when the props are brought alongside in leaky barges, or are rafted alongside, the weight per fathom is consequently high.

Pulp and Paper Products

Paper

Paper is made from vegetable matter reduced to pulp, such as wood, esparto grass, flax refuse, straw, jute, also rags. Spruce, Balsam, Hemlock, Cottonwood and other timber are used.

Paper is usually in shipped rolls, the ends of which are, in some cases, protected by circular discs of wood; in other cases the rolls are simply wrapped with thick paper with extra layers over the ends.

Rolls of paper vary from 500 mm to 2,135 mm in length, i.e. width of paper, while the diameters vary considerably averaging in mixed shipments to about 900 mm.

The main categories of paper consist of newsprint, printing paper and kraft. All are extremely sensitive to mechanical damage particularly at the flat ends of the rolls. Any such damage will reduce the effective width of the paper, and may make it totally unsuitable for use in the printing process. Flattening of the reels and distortion of the core may make them useless for modern high-speed printing presses, and result in heavy claims for damage.

Holds must be properly prepared before loading. They must be clean and any sharp protrusions should be cushioned in the best possible manner to avoid damage to the paper. Ventilator or fan openings must however be left free as air circulation is very necessary to avoid condensation damage to the paper.

Paper reels must be stowed in several ways.

on end;
on the bilge or roll, athwartships;
on the bilge or roll, fore and aft;

or any of these in combination.

The chosen method of stowing may be determined by several factors, namely the nature or type of vessel and its equipment, the sizes of the reels to be loaded, the facilities available at the loading or discharging ports, and possibly any special requirements of the consignees.

Stowage of reels on end in the vertical position enables the use of special handling equipment which has been developed for the purpose of loading and discharging this cargo, such as:

The Jensen Sling, available in different versions for lifting between two and eight reels. The Jensen Sling is a semi-automatic lifting device developed for the purpose of lifting paper reels in the vertical position. Operation is reasonably simple and labour is required only to guide the device into the hoisting position. On landing the load is automatically released.

The vacuum lifting device which is available for lifting reels in pairs of six, eight, ten or more simultaneously. Vacuum handling, however, makes certain demands on the package, and the person in charge must be familiar with these and satisfied that all requirements are met. Furthermore, for safety reasons, personnel should not be allowed to remain in the hold or on the sector of the quayside where handling with vacuum attachments is taking place. It may be necessary to fence off the working area.

The Core Probe is another semi-automatic device, which is applied in the reel core and expands to grip the walls of the core when it is lifted. Core Probes may be used with frames or spreaders to handle anything up to 20 reels in one lift, depending on the weight of the reels and the handling capacity of the crane available.

If paper reels are to be discharged conventionally by putting machines into the ship's holds, it is important that a "break in" area be provided in the square of the hatch to facilitate discharge. This may best be done by stowing preslung reels on the bilge or roll in a sufficiently large area to enable a clamp truck to manoeuvre.

Rolls of paper should be stowed solid and well chocked off to avoid movement when the vessel is at sea. If consideration of space renders desirable the stowing of the top tier on its side or bilge over paper stowed on end, it is essential that every precaution be taken to ensure that this top tier is adequately chocked off. If the vessel is stiff or normally has a violent motion in a seaway, providing other suitable cargo is available, this form of top tier stowage should be avoided. It is extremely difficult to chock the top tier on its roll over paper stowed on end, and serious claims have resulted in such stowage breaking loose.

In end holds, the greatest care should be exercised to ensure that the platform on which the ground tier of rolls is stowed is both level and firm. When a full cargo of paper is carried, the most satisfactory way of doing this is by building a series of platforms (or bridges) of suitable width to take the diameter of the larger rolls, the platforms themselves resting on substantial bearers.

All stanchions, ladders, etc., should be well covered with burlap or other protecting material to avoid chafage of rolls; and dunnage should be thorough throughout, to prevent movement of the rolls, particularly so if the cargo does not entirely fill the hold or compartment.

Slings made of webbing (man-made fibre) or soft rope should be used, and utmost care taken to avoid chafe and damage while handling and lifting. Loading or discharging of paper rolls by swinging derricks should not be resorted to owing to the difficulty of keeping rolls from banging against hatch coamings, the ship's side, etc., which tends to destroy their shape and inflict other mechanical damage.

The dragging of rolls from wings or ends of compartments to the square of the hatch should be prohibited. The use of cargo hooks, crow or pinch bars should never be permitted when handling paper.

The amount of broken stowage with a cargo of paper is very considerable. The smaller rolls may, with reasonable care, safely be utilised for "fillings" in the wings etc.

Rolls of paper vary considerably in their moisture content. In a totally enclosed space therefore (e.g. a closed box container) the migration of moisture might take place, i.e. sweat.

When rolls of paper are stowed on their ends, adequate space between tiers may be needed to access the mechanical handling equipment. In this case a well proven method of restraint may be provided by inserting pneumatic dunnage bags.

Paper pulp.

Pulp

Wood and paper pulp is shipped in compressed bales both as "dry" and "wet" pulp. It is made from various kinds of timbers, and is shipped in large quantities from the timber areas of the world, e.g. Scandinavia, Canada, British Columbia, U.S.A., etc.

Wood pulp is very liable to damage and contamination by dirt (particularly fibres) or the remnants of previous commodities in a ship's hold. During handling it must be kept clear of any contact with ropes, etc., and should be loaded and discharged with wire or chain slings. Remnants of the previous cargo (and particular grain) should be carefully removed and the space cleaned before the commencement of loading pulp.

Bales of pulp may be unitised with wire banding securing the bales. It is very often permissible and acceptable to lift by this banding, and some proprietary equipment is designed so to do.

It is advisable, however, that certificates of strength be provided for the banding.

Timber: Terms and Definitions, etc.

Battens

A sawn piece of timber from 150 mm to 180 mm wide and not less than 100 mm thick. Stow at 220/225 cu. ft per Standard.

Battens, Small

A sawn piece of timber under 150 mm wide and less than 50 mm thick. Stow at 230/240 cu. ft per Standard.

Battens, Ends

A sawn piece of timber under 2,500 mm in length.

Baulks

A large heavy beam of timber — hewn or sawn.

Boards

A sawn timber 50 mm thick and under, any width. Above 150 mm stow at 230/240 cu. ft per Standard and 250/260 cu. ft below 150 mm thick.

Deals

Sawn timber no less than 50 mm thick and 230 mm or 250 mm wide. A Petrograd "Standard Deal" is 1 piece, 75 mm × 280 mm × 1,830 mm. A "Slit-deal" is 30 mm thick. A "Whole deal" is 15 mm thick. A "Hundred deals" = 120 pieces, 150 mm × 75 mm × 280 mm × 165 cu. ft.

Flooring

White or yellow boards, chiefly 19 mm to 38 mm thick, planed, either square edged or tongued and grooved.

Laths

Thin narrow strips of wood for the building trade (plasterers' laths). The usual dimensions for sawn laths are 25 mm to 32 mm to 8 mm. American laths vary from 6 mm to 12 mm thick, different sizes being known as "Lath", "Lath and half", and "Double lath". Laths stow about 320 cu. ft to the Standard.

Lumber

The term usually applied in the U.S.A. to timber.

Log

A heavy piece of timber either round, hewn or sawn.

Pickets

Sharpened stakes — shipped in bundles.

Pit-props

Are short, straight lengths of timber mostly fir, denuded of the bark (see "Pit-props).

Plank

Any substantial piece of sawn wood of substantial thickness.

Prime-wood

This expression applies to the more valuable classes of timber, such as oak, mahogany, beech, whitewood, etc., used in the furniture and allied trades.

Railway Ties

See Sleepers, below.

Rickers

Light poles varying from 6,000 mm to 16,750 mm in length used for scaffolding, making builders' ladders, etc.

Scaffolding

Light poles of varying length (see above).

Shingles

Thin slats of wood used for roofing — one end being thinner than the other. Shipped in bundles.

Shooks

A complete set of staves (body and head) for making casks or cases ready for assembling. Shipped in bundles usually but occasionally loose.

Slats

Strips of wood used in the manufacture of light cases, usually of deal — shipped in bundles.

Sleepers

English sleepers usually are of Baltic fir, in other countries they are composed of the cheaper hard-woods such as oak and jarrah, the latter being very suitable for the tropics.

Staves

See Shooks, above.

Waney Timber

Pieces of timber which have round edges — either hewn or sawn square leaving part of the round on edges.

Wood-goods, Heavy

"Any square, round, waney or other timber, or any pitch pine, mahogany, oak, teak or other heavy wood goods, whatever . . ."

Wood-goods, Light

"Means any deals, battens or other light wood goods of any description"; and for deck cargo purposes the following is added: "Each unit of goods must be of a cubic capacity not greater than 0.42 cubic metres".

Cars

At time of shipment cars should have their petrol tanks drained. Cars carried with fuel in their tanks may be subject to IMDG regulations.

When cars are being manoeuvred on to vehicle decks, full mechanical ventilation must be utilised to remove fumes.

Only properly authorised and trained drivers should be employed. They should be issued with clean overalls to avoid soiling the upholstery.

Loading ramps should not be so steep as to either cause risk of damage to vehicles with small ground clearance or to cause engines to labour unduly when cars are ascending the ramps. Speed of loading/discharging will be adversely affected if the angle of the ship-to-shore ramp is too severe. A better than one-in-ten slope should be achieved if possible. Change of slope (from ramp to deck, for example) should not exceed 7 degrees. Decks should be properly treated with a good non-slip compound on vessels where the cars are not firmly lashed down.

Cars arriving for shipment under their own power should be checked for moisture, snow, mud or other debris that may be adhering to the vehicle, and affect it in subsequent transport. Any such dirt might cause the car to be subject to quarantine regulations of the country of destination.

REFRIGERATED CARGOES

Preamble

Today refrigerated cargoes move in very substantial quantities. Many reefer vessels are specialist carriers almost wholly dedicated to the carriage of refrigerated cargoes under a variety of temperatures.

In recent years the tendency has been for the carriage of cargoes under hard frozen conditions to be at significantly lower temperatures than hitherto, and this is particularly so in the carriage of fish, the movement of this commodity now being a significant feature of refrigerated cargoes.

The introduction of ISO containers has proved to be of outstanding value in ensuring good out-turn of cargo under refrigerated conditions. Their use helps protect the cold chain from producer to consumer. It significantly reduces the physical handling requirements, with the attendant possibility of dirt and contamination. Conventional break bulk shipments may be handled 15 times between leaving the freezing works and arriving in the market place. The container also reduces opportunities for pilferage.

However, notwithstanding the above, for the foreseeable future refrigerated cargo will continue to be carried extensively in conventional break bulk vessels. The size of port, development of trade, the seasonal nature of many refrigerated commodities (e.g. fruit), all contribute to the choice of transport mode to be employed.

Introduction

Organic matter is under continuous attack by microorganism, bacteria, yeasts, mould, etc., which require warmth and moisture to exist, multiply and go about their destructive business. This has long been recognised, particularly with regard to foodstuffs and the practice of drying and smoke curing as a means of preservation has long been practised. It is also recognised that cold weather increases storage life, again by inhibiting the growth and activity of these microorganisms. However it was not until the development of techniques to artificially control temperature by mechanical means that the prolonged and controlled preservation of foodstuffs was possible. Refrigeration is now used extensively to produce atmospheres suitable for the storage and transport of a wide range of products both organic and inorganic, living and inert, all of which benefit from control conditions of temperature and humidity including in some cases, the extraction of harmful vapours and the introduction of beneficial gases.

Storage in a Controlled Environment

Although various different commodities require widely varying conditions for prolonged storage, if these specialised conditions are to be obtained then the storage space must be insulated from the ambient conditions, both as regards temperature and humidity. In other words these spaces require to be basically airtight and insulated. This may be realised by specially constructed buildings on land, suitably constructed cargo spaces in ships and properly constructed portable containers with the necessary machinery for the creation and maintenance of the required environment. Thermal insulation is provided by a variety of substances, all of which may be easily damaged and thus need to be sheathed to prevent mechanical damage from the cargo and contamination from incidental spillage or odours emanating from the cargo.

However though this insulation will delay the transfer of heat either way, it will not prevent it. So it is necessary to remove such heat that will penetrate into the cold storage chamber — threatening to raise the temperature of cargo above the required storage temperature — or to replace the heat lost from cargo requiring a warmer environment. There are two basic ways of doing this. One is to arrange a series or grid of pipes around the space, sides, roof and in some cases floor through which hot or cold brine is circulated to deal with the heat leakage and maintain the desired temperature. This method has now been largely superseded by a similar system of air ducts through which fans force air — after it has been passed through nests or batteries of pipes containing hot or cold fluid (brine), thus counteracting any heat leakage — as mentioned above.

Having dealt with the heat leakage problem of the chamber, the cargo itself must be considered. If the cargo is inert and precooled to its correct storage temperature, no further problems should arise in so far as temperature is concerned. The living cargoes, however, will generate heat and this must be removed if the correct storage temperatures are to be maintained. To achieve this it is necessary to circulate cool air throughout the cargo, which means that the stow must not be a tight block but have air channels and ducts within itself. In some cases the shape of the cargo itself (e.g. frozen carcasses, bulging crates, etc.) will prevent close block stowage and provide sufficient free space for air circulation. Where regular flat-sided packages (e.g. boxes and cartons) are concerned, if the packaging itself does not make provision for the through movement of air, dunnage or battens may have to be inserted at regular intervals (e.g. every second row; every sixth tier) to allow the controlled temperature air to contact one side of each box or carton. Packaging that has been specially designed to allow air entry — e.g. cartons with specially positioned ventilation holes — may have to be stowed "in register"; that is with one carton directly on the next so that the ventilation holes match up vertically and/or horizontally.

In ships' holds and 'tween decks it is common practice for the air to be introduced through ports cut in air trunks running along the bottom sides of the compartment, and dunnage will be required to allow it to reach the centre of the stow. The ground layer of this dunnage must be laid in line with the air flow, care being taken to see that cargo, particularly bagged cargo, is not allowed to sag between the dunnage and block the air flow which must be able to rise up through or around the cargo and be extracted by similar air trunks via the roof or deckhead. This air circulation through the cargo may be from the bottom to top, side to side or end to end, and it is essential that the direction is known and understood so that the dunnage can be laid correctly. In some cases permanent dunnage may be a feature of the space, e.g. ISO Refrigerated Containers.

Storage temperature, however, is not the only consideration. Controlled humidity plays a vital part in inhibiting the growth of mould, bacteria, etc., always present in the air to varying degrees and inevitable in living cargoes. Inert cargoes also benefit from controlled humidity conditions. Cartons can absorb a certain degree of moisture to their detriment — wrappings and labels can be shed, steel rusted in a humid atmosphere. In an airtight air circulation space the humidity can be controlled and excess moisture extracted and condensed on the cooler coils. This may appear as snow and requires defrosting from time to time (to maintain efficiency) and the water run off to the bilge. (N.B. Bilges may have to be "brined" prior to reducing the temperature in the compartment, so that scuppers do not ice up). Care should be taken not to over-dry certain commodities in storage which can result in loss of weight, shrivelling and even brittleness (sometimes known as "freezer burn"). This may be eliminated or reduced by either treating the commodity itself — e.g. glazing — or provision being made for introducing a degree of moisture into the circulating air — normally by venting dry air and introducing a proportion of humid ambient air.

Mention has been made of the presence of gases in cargo chambers. These fall into two main groups:

those introduced to aid storage;
those produced by the cargo — usually to its detriment.

CO_2 is the most common gas, a controlled percentage of which will inhibit mould growth. It is also used as a direct method of refrigeration in the case of some insulated containers though many commodities will not survive this treatment. Similarly nitrogen is also used in this manner. Of the gases produced by living cargoes, principally fruit, CO_2 dominates and will build up to harmful concentrations of not removed. Likewise ripening fruit will produce ethylene which will speed up the ripening process — which is not desirable during storage, and these gases must be removed. Comparatively easy with forced air circulation conditions, as a proportion of the vitiated air drawn from the cargo space, can be blown free to atmosphere, and replaced by fresh air suitably cooled and dried before being introduced into the space. It is important to make sure that this incoming air is fresh and not contaminated by exhaust fumes either from adjacent cargo spaces being vented or by nearby engine exhausts, funnel fumes, etc.

Commodity Groups

There is presumably an optimum environment for the storage of every commodity, although it is difficult to imagine, say, a block of granite coming to much harm whatever the conditions that surround it. On the other hand many articles are very sensitive to environment, especially living organisms such as soft fruit. Between these extremes most products will fall into one of several groups, though it is usually only where valuable and sensitive commodities are concerned that refrigeration becomes economic.

Living cargoes, such as fresh fruit, vegetables, eggs, flour, bulbs, cheese, continue their ripening process while in storage and require oxygen for this purpose, during which they also give off carbon dioxide. This process can be slowed down by reducing the temperature thus prolonging the storage life, but it is essential that the temperature never reaches the product's actual freezing temperature or it will die and quickly rot. Further, since they produce CO_2, without adequate ventilation this would soon build up harmful concentrations although up to 2 per cent may assist in prolonging storage life and inhibiting moulds. To achieve these conditions calls for very careful attention and control. Cold air should be delivered as near as practical to the freezing temperature of the product but never reach it. The temperature of the returned air will indicate the effectiveness of the cooling process and remove self generated heat. If too high it may be necessary to increase fan speeds and air circulation as the incoming air temperature must not be lowered. In any case fans should be regularly reversed to circulate the chilled air through the cargo in the opposite direction and promote a more concentrated temperature throughout. A constant temperature is required for good storage, fluctuations being harmful to the product, so return temperatures should be carefully monitored and prevented from grossly exceeding the delivery temperature. While high fan speeds aid temperature control, they also may result in over-drying of the cargo. On the other hand a dry atmosphere will prevent mould growth. So although humidity is an important factor in good storage, it tends to become overridden by the more vital considerations of temperature and CO_2 control CO_2 meters will indicate the concentration in the return air which should be kept to about 2 per cent by ventilating to the outside atmosphere. This action will also assist in clearing the space of enzymes produced by the fruit in its ripening process. The expelled air is replaced with fresh air from outside. It is important that this is fresh and not contaminated by odours which would adversely affect sensitive cargoes. Passing over the coolers the air will then enter the chamber at the required temperature and be dried to its corresponding dew point. The temperature range for this class of cargo is from 0 degrees C (32 degrees F) up to 13 degrees C (55 degrees F) — as in the case of bananas where a degree of control ripening may be required. More specific details are given in the alphabetical section. Usually the shipper will issue instructions for the carriage of his cargo, but if these are not forthcoming, expert advice should be sought.

Non-Living Organic Cargoes — These are principally meat, fish and poultry, etc. Since these are not living commodities they do not of themselves generate heat, neither do they require oxygen nor produce carbon dioxide. So the main consideration in prolonging their storage life is to reduce the temperature sufficiently low to inhibit the growth and activity of the microorganisms. This means storage in a hard frozen condition. A comparatively easy cargo to handle if pre-cooled, the main exceptions being chilled beef or horse meat which are explained under those headings, and carried under conditions approximating that of living cargo. The general carrying temperatures of hard frozen cargoes range from – 15 degrees C to – 10 degrees C (5 to 15 degrees F), generally considered to be the most economical degree of refrigeration necessary.

Where the Codex Alimentarius applies, even lower temperatures are required in order to allow for a permissible rise in temperature to occur during handling and transportation after discharge from the vessel either in container or as break bulk. This consideration should be fully understood before loading as it may be beyond the ship's capability to reduce the temperature on passage.

However an ever increasing amount of cargo is now carried and stored under deep freeze conditions though the fundamenal consideration here is quick freezing. At these very low temperatures down to −20 degrees C or even lower microorganism activity is stopped and the storage life considerably lengthened and a wide range of meat, fish, poultry, fruit and vegetables can be successfully stored for considerable periods. The essence of deep freezing is very quick freezing so that the ice crystal formation in the cells within the commodities does not have time to grow large and rupture the cell walls, which would result in the loss of essential juices upon thawing. This rapid blast freezing is only possible on comparatively small parcels so that the very centre will acquire this quick freezing treatment, this it is not used for sides of beef which are best carried under chilled conditions. Pre-butchered small joints, steaks, chops, divided into ready-for-use portions may be rapidly frozen, individually sealed or polythene wrapped to prevent the dehydration that would otherwise occur in storage at these low temperatures, then packed in boxes or cartons of a suitable size for close stow and transportation and present a comparatively easy cargo to carry — provided that a constant low temperature is maintained.

Inert Commodities — Specialised loads which may be presented for carriage under refrigeration under certain circumstances or in particular conditions could include confectionery, pharmaceuticals, X-ray film. They require closely controlled conditions and the carriage instructions should be carefully considered with regard to their storage and their effect upon other cargo — especially if of an obnoxious, inflammable or otherwise dangerous nature. Usually quite stable materials with long storage lives under normal conditions, although they may be subject to chemical changes if exposed to extremes of temperature and humidity. Some cooling may therefore be necessary when in tropical conditions. Likewise it may be necessary to warm the space in arctic conditions. Also a low degree of refrigeration may be used to advantage to dry the air in certain cases. The temperature range will vary according to the commodity but will normally be within the chilled range. Advice should accompany any special chemical products. Special attention must be made to the range of easily vapourising low flash point products which must receive constant refrigeration to prevent the production of fumes which may well taint other cargoes and indeed the insulating material itself — or worse, be of an inflammable nature presenting a grave risk of fire and explosion; in this case spark-proof machinery only may be used both for handling and refrigeration, and appropriate precautions taken as laid out in the IMDG code (see "Dangerous Goods").

Receiving Cargo

Ideally cargo presented for carriage under refrigeration should be pre-cooled to the carrying temperature since normally the vessel or container is only provided with sufficient power to deal with heat leakage and the modest amount of heat generated by living cargoes. Even with full use of reserve power the actual freezing of say a large tonnage of meat at killing temperature would involve a completely unacceptable delay when loading. Thus perhaps the prime consideration when receiving cargo for refrigerated carriage is to see that it is at the correct temperature. There is always liable to be a slight rise in the temperature of the surface occurring during trans-shipment which the ship can well take care of provided the internal bulk of the packages is at the correct temperature. Spear thermometers are available to determine this. It may even be necessary to drill holes into frozen meat to ascertain the actual bone temperature. Where the thin flanks of carcass meat are soft they must be re-frozen before storage or badly distorted carcasses will result on discharge. Care should be exercised in such cases to see that the soft carcass has not just received a cold blast to freeze the outside flank which will now appear in a good hard condition whilst the inside is still warm hence drilling for the bone temperature. Blood-stained shirts will indicate that a carcass has at least partially thawed since initial freezing and should be considered with suspicion. Again if soft or wet carcasses are stowed in this condition they may well distort, nesting one into another and blocking the necessary air flow. If wet from rain these may freeze together, resulting in considerable damage when prised apart on discharge. Similar conditions can arise with frozen cartons, which must be clean, dry and free from frost at time of loading. Chilled cargoes are of less concern in this respect, but owing to their more sensitive nature great care and inspection is called for during receiving and stowage. Most packaged fruits, and vegetable cargo will be presented pre-cooled but must be carefully examined for the odd warm or over-ripe lot which should be rejected. The general condition noted and the space sealed and the carrying environment reached as soon as possible. If as sometimes occurs a fruit cargo is presented for loading at orchard temperatures, the ship may have to undertake the cooling down to the correct carrying temperature. This facility must be requested by the shipper in writing, or the Bill of Lading claused accordingly. Once agreed, the cooling down must be carried out as quickly as possible. However, very rapid cooling cannot be achieved since the incoming cool air must not be at or below the freezing temperature of fruit. Maximum air circulation must be possible throughout the stow, with adequate intermediate dunnage and battens to assist the cooling process. If the stow does not cover the whole deck encompassing all the air delivery outlets then those not covered must be temporarily blocked off to prevent short cycling of the air. Similar considerations concern deep frozen cargoes. When close stowing regular shaped cartons it is important that none are significantly above the carrying temperature, since in a close stow the cool circulating air will not be able to penetrate and reduce the temperature of any one package or group of packages.

Preparation of Spaces to Receive Cargo

The generally sensitive nature of refrigerated cargo requires very careful preparation of the space to receive it. Cleanliness is obviously of great importance particularly with foodstuffs. The space must be free of odours and microorganisms and may well require fumigation. The introduction of ozone will deal with remaining airborne smells but lingering ones which have been absorbed by the insulation or dunnage may well require the removal and renewal of the affected portions. Fans should be run in both directions to clear smells and dust, etc., from the air trunking. Bilges and scuppers must be clean, tested and U-bend vapour traps sealed with brine to prevent cross taint between compartments. Adequate and clean dunnage must be provided or, if already in place, inspected to ensure an adequate air circulation under, around and through the cargo as required. Old dunnage may well get moved during discharge with the possibility of blocking air circulations of some sections. This must be then repositioned. Extra dunnage may be required for adjacent compartments which are of widely varying temperatures, e.g. over double bottom oil tanks liable to steam heating; chilled cargo compartments over deep-frozen holds. Thermometers, gas sampling points, fire detection and extinguishing equipment should be carefully checked and inspected. Inspection should also be made of any pipes passing through the space, particularly their joints, for any sign of leakage and tested if in any doubt. Closing arrangements, locker doors and hatch plugs should also receive careful examination. The space can then be cooled down to slightly below the carrying temperature and held there for at least 24 hours to ensure that all the residual heat is removed from insulation dunnage and other fittings within the space. The air temperature will quickly rise when the compartment is opened for loading and every opportunity should be taken to run the fans during breaks in the loading operation.

Metals with Refrigerated Cargoes

Quite apart from freight-earning consideration, the stowage of pig copper, lead, tin, etc., under frozen cargo in lower holds has its advantages in that it gives increased stability which, in the case of most vessels carrying full cargoes of refrigerated goods, is desirable and also tends to stabilise the temperature of the compartment throughout the voyage.

The metal should be spread over the entire compartment to facilitate cooling down; the operation of cooling a deep compact block of metal being a slow one, a greater depth than 1,200 mm of metal should, if possible, be avoided.

The lower or ground tier of pigs or ingots should be spaced apart in such a manner as will ensure the requisite air courses.

Battens, as already referred to, should be laid on top of the metal to receive the frozen cargo, and the whole mass brought to the loading temperature. An additional six to eight hours more than the cooling down time should be allowed for when carrying metals in a frozen compartment.

Dunnage should on no account be nailed to any metal. Serious claims have had to be met from damage arising from same.

If chilled meat is to be carried, the temperature should be brought down to the carrying temperature and maintained there for 12 or 18 hours before commencing to receive cargo.

Taint

Odour, either pleasant or obnoxious, is closely associated with taste, and it is undesirable that even a pleasant smell should intrude into the one expected from a favourite food or perhaps completely override its own delicate flavour. It is in this connection with foodstuffs that taint is principally a problem. Some products produce strong odours in themselves — others may be particularly susceptible and readily absorb foreign smells. These properties of the various products are noted in the alphabetical section and they should not be stowed together, even though they may require the same considerations of temperature and humidity. The separation of odious and sensitive cargoes into separate airtight compartments would in most instances solve this problem (see Part 2, "Obnoxious Cargoes"). Badly fitting locker doors, container doors, hatch plugs, or fan spaces, might allow for a certain degree of cross tainting as indeed would common scupper systems not sealed with U-bends or traps filled with brine. Pipes passing through a cargo space present a particular hazard as a leaking oil pipe or oil tank sounding pipe may result in obnoxious fumes entering the space. Special care is required with living cargo when air changes are necessary to ensure that fresh air introduced is completely free of taint, similarly that the vented air expelled from a space is not drawn into a space where it might cause contamination. An odious cargo may well leave a well tainted atmosphere behind it after discharge, and a sensitive cargo should not be worked through this space either loading or discharging until it has been well ventilated and the odours removed.

Ozone

An oxygen enriched gas may be introduced to eliminate the odours by oxidising the offending molecules. This gas may be produced by electrically operated apparatus placed in the fan spaces and the gas circulated throughout the affected area. It should be noted that the ozone will only deal with airborne odours. Those which emanate from spills will continue until the spill has been cleared up and any damage to insulation or dunnage which has absorbed the spill quickly removed or renewed. Furthermore all traces of ozone should be removed by fresh air ventilation before the space is entered, or sensitive cargo worked through and loaded into the space. Particular care should be exercised before entering into the fan space to attend to or remove an ozone machine where there is liable to be a concentration of gas.

Containers

Unlike permanent cold stores or refrigerated ships, where robust equipment is under constant care by qualified personnel, the ISO refrigerated container may travel by many different modes and be in the care of many and varied people. Prior to being despatched to load refrigerated cargo (usually at shippers' premises), the container and its machinery should be subjected to a rigorous examination. External damage received during previous handling must be noted and if necessary repaired — and in cases where the external sheathing is pierced, insulation must be examined with particular reference to the ingress of water. Internally the successful carriage of refrigerated cargo depends on air circulation. All airways and battens should be inspected for damage particularly floor extrusions where fitted and fans tested. Cleanliness is of paramount importance and should be dealt with as previously outlined. Doors and their fastenings including hinges should receive special attention and an airtight seal ensured when closed.

Stowage position on board the vessel will be governed by: the commodity carried, its temperature and other requirements; the type of container — port-hole or integral, air or water cooled, possibly with Clip-on unit.

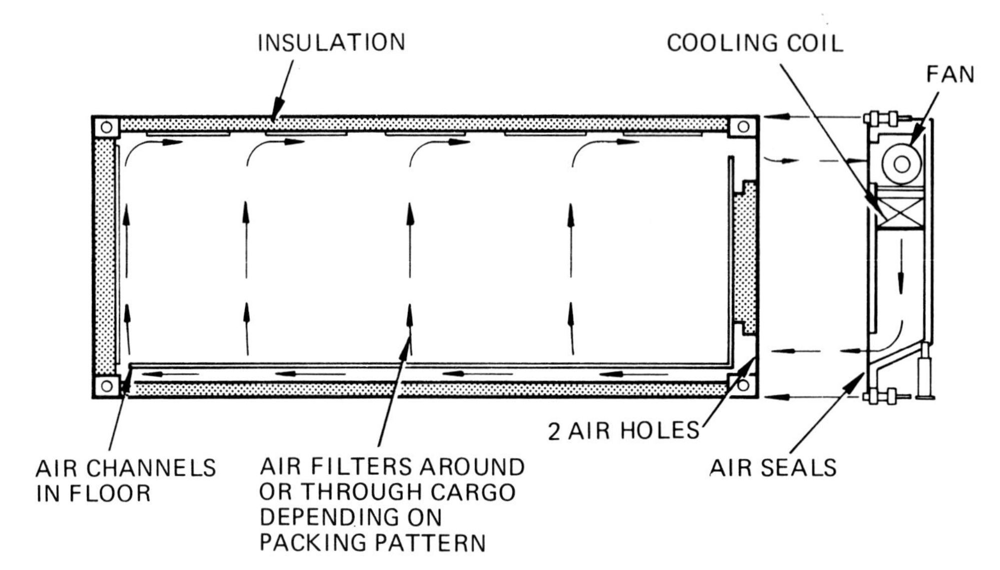

INSULATED CONTAINER REQUIRING EXTERNAL REFRIGERATION SOURCE. BOTTOM AIR DELIVERY.

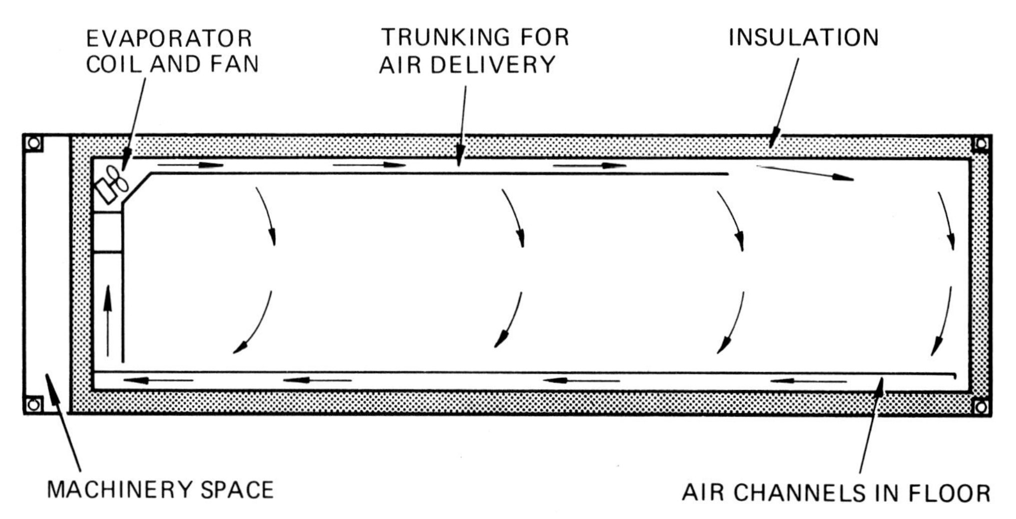

INTEGRAL REFRIGERATED CONTAINER. TOP AIR DELIVERY.

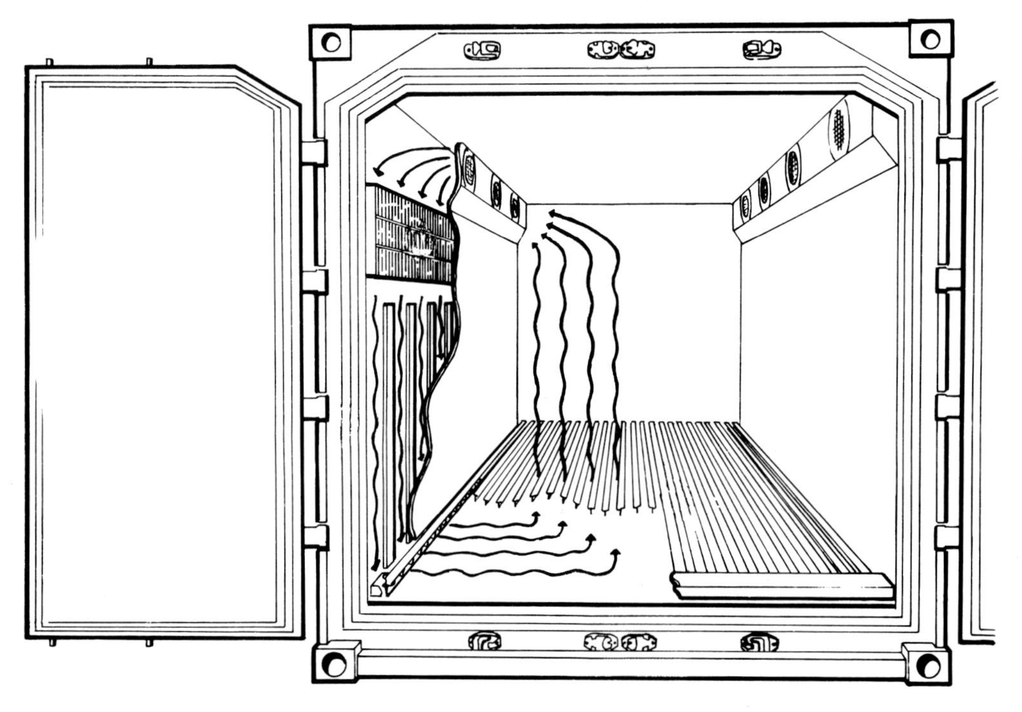

INTEGRAL REFRIGERATED CONTAINER. SIDE AIR DELIVERY.

(vii) Insulated ISO container with supplementary cryogenic refrigeration.

(viii) Insulated ISO container attached to shore side terminal holding unit.

(ix) Insulated ISO container with mechanical clip on refrigerating unit.

Photo: Associated Container Transport (Australia) Ltd.

The insulated container, cooled using the ship's refrigeration by way of port-hole apertures in the front, is usually stowed in cell guides, care being taken to ensure that the port-holes are facing the correct way, and that 8 ft and 8 ft 6 in containers are neither mixed nor placed in slots designed for a different size.

Air-cooled integral containers, and containers fitted with Clip-on units are usually stowed on deck, though vehicle decks of Ro-Ro vessels, with appropriate ventilation facilities, may be suitable and acceptable.

Water-cooled integral containers must be stowed so that the hose connections are facing the correct way, and the comments (above) regarding 8 ft and 8 ft 6 in container problems also apply.

The vent closures of port-hole containers must be open prior to loading, though any empty insulated containers in the system should have their vents closed to prevent short-cycling of cold air.

Inspecting and testing should be carried out by qualified personnel and a certificate issued. Containers may require to be pre-cooled before loading (depending on the type of cargo and local regulations). When a Clip-on unit is fitted, or in the case of integral unit containers, the correct temperature for the cargo to be carried must be set, and the recording chart fitted after first winding up the clockwork driving mechanism. Information recorded on the chart should include:

Date and place of stuffing.
Carriage temperature required, plus any additional information, e.g. the requirement to change delivery air temperatures at any stage of the voyage, etc.
Commodity, type and packaging.
Container number.

When appropriate (e.g. for live fruit cargoes) the air intake/outlet vents should be set open to allow CO_2 and other gases to exhaust from the container. When loaded, sealed and delivered to the ship in the case of port-hole varieties there is little that can be done other than external examination including seals and port-hole fittings and inspection of the accompanying documents. If a Clip-on unit is fitted thermometers and thermostat settings should be inspected and the air temperatures recorded along with its general condition and if a temperature recording chart, this should be studied and removed for future reference. Similarly an integral unit, if running on its own diesel generator, fuel and lubricating oil should be checked and the unit should be stowed so that access is available to machinery. If the ship is to supply the electric power it must be ascertained that the required cable is available with suitable connections. Although largely standardised, differences are sometimes found, and converting links should be available. The voltages and transformer settings should be carefully checked before connecting up. Thermostat settings should be checked correct for the contents and the plant operating smoothly. Though major machinery breakdowns are a specialist's concern and unlikely, regular checks for the temperature must be made and variations investigated. The causes may be simple and easily rectified. A blown fuse, slipping or broken fan belts, thermostat settings altered by vibration. Simple tools with a few spares should be carried. Attention to minor defects can well save a valuable cargo.

Get-you-home Procedures

In the event of container machinery failure beyond the capability of local resources to rectify, it may well be possible to yet save some of the cargo by adopting certain emergency procedures. If the cargo is hard or deep-frozen in a well insulated container with good door seals it may survive for several days depending on the ambient temperature. Shielding from direct sunlight will help. Even then the probability is that only the outer portions of the cargo will suffer, but undoubtedly the best way of dealing with such a situation is to introduce coldness to the cargo. This can be done by dumping liquid nitrogen or carbon dioxide into the container and allowing it to vapourise within the space, when any heat that has leaked in will be consumed. Points to bear in mind: ensure that the gas used will not harm the cargo (not normally a problem with hard frozen products); also that it will not adversely be affected by the very low temperatures produced, particularly locally. It is not possible to regulate the temperature when resorting to this method of refrigeration, and it is quite unsuitable for chilled cargoes. Solid CO_2 may be a suitable alternative. Normally supplied wrapped for protection during handling, it is necessary to open the wrapping to permit evaporation. The more exposed are the blocks from their wrappings, the more rapid the evaporation the greater cooling effect — and the shorter the life of the blocks. See graph, Appendix 14, for approximate quantities required, although there is little point in economising on CO_2 to the possible detriment of the cargo. Care should be taken when handling blocks of dry ice; the extreme temperature may cause severe burns on contact with the bare skin.

Neither of these gases, nitrogen or CO_2, will support life so apart from the damage they would do to living fruit cargoes, personnel when opening a container which has received this treatment should ensure that adequate time has been allowed for ventilation before entering. Being heavier than air these gases should flow out readily like water, but beware of pockets trapped in a tight stow. Dumping of CO_2 or nitrogen, or the use of dry ice, are not suitable for chilled living cargoes. The rapid introduction of nitrogen or carbon dioxide would kill the product both on account of the very low local temperature and the effect of the gas itself on fruit's respiration (although it is possible for these gases to be introduced in small quantities using sophisticated control equipment).

Where chilled live cargoes are involved, shading from the sun, even hosing down of the container will help, but inevitably the temperature will rise and the ripening process increase. Ventilation at this time is essential to remove harmful vapours produced by the cargo and replace the oxygen necessary for the health of the fruit, though this may give rise to an accelerated rise in temperature. It is only by speedy transfer of the cargo to a more acceptable environment that it will continue to prolong its storage life.

Temperature Measurement

There are many instruments capable of doing this; perhaps the commonest is the mercury or alcohol glass thermometer.

When using these instruments (apart from their fragile nature as with all types of thermometers) it is important that the sensing head (in this case the bulb) is kept in the environment where the temperature is required to be known long enough for all the liquid in the bulb to settle to that temperature and ensure the correct degree of expansion or contraction. It must also be shielded from any extraneous source of heat or cold. In liquid cargoes it is simple enough to immerse the bulb for a few minutes. In the case of solids, if not too dense, it may be possible to carefully insert the thermometer. Special metal sheathed cases are available with pointed ends to enable proper penetration to be obtained, but due to the presence of the metal case, will require a greater time for the whole to achieve the true temperature of the commodity being measured. In other instances it may be necessary to drill a hole (particularly in hard-frozen commodities) to ensure penetration to the heart of the article to obtain its true temperature. In the case of sealed packages such penetration may well destroy or seriously reduce their value, and a close approximation of the temperature can be obtained by sandwiching the thermometer between two packages and wrapping them together for a while to exclude draughts and other outside influences. It may well be considered advisable, on receiving a cargo of sealed packages, the temperatures of which have to be ascertained, to penetrate a selected sample and these should be conspicuously marked so that the same package can be used for measuring the internal temperature if required on discharge, thus reducing the number so damaged.

A bi-metal instrument consists of two strips of dissimilar metals with different coefficients of expansion joined together at one end, such that the varying temperature causes them to expand or contract at a different rate, this variation being accurately measured as a temperature reading, and connected to a calibrated dial. More robust than the glass thermometers, they are easier to read, but require appreciably more time to settle and for the temperature to be measured.

Electrical resistance thermometers depend on the fact that the resistance of a wire or filament will vary in ratio to its temperature. Here it is necessary to have a suitable instrument to measure the resistance, calibrated in degrees of temperature, but the sensing probe may be at a remote distance from the instrument. These distant reading thermometers are of particular value during the actual carriage of the cargo, as the sensing probes can be distributed amongst the stow and then the temperature read off externally during passage. Such instruments can also be fitted with a recording chart to give a continuous record of the temperature of the cargo during passage.

In certain cases the importing country will require proof that the requisite temperatures have been maintained for the required time. This may be to ensure cold sterilisation of the commodity with regard to parasites in living cargoes, or for other temperature monitoring requirements. A temperature recording chart fitted to integral refrigerated containers may serve a similar function.

CONTROLLED ATMOSPHERE (C.A.) CARGOES

Controlled atmosphere storage or transportation provides a means to augmenting (but not in any way replacing) high quality refrigeration in order to enhance the shelf life of fresh fruits and vegetables. It is being adopted for use in the hold spaces of refrigerated ships.

The respiration rate of produce is closely related to its rate of maturation and subsequent deterioration. Respiration rate is reduced by holding at an optimum temperature as low as possible without freezing or chilling the produce. It may be reduced further by reducing the oxygen content of the surrounding air. Atmospheres containing below 2 per cent oxygen are found to be most beneficial, though the exact composition of an optimum atmosphere is species and variety dependent. The degree of benefit obtained from c.a. varies widely between different types of produce. It may be advantageous for some produce to use additional controls on carbon dioxide, humidity and ethylene level.

Low oxygen and high carbon dioxide atmospheres have also been used to slow microbiological deterioration of fresh meat, though this is more often done nowadays using vacuum packaging or similar techniques.

A common means of providing c.a. in a ship's hold involves the use of an air compressor and a membrane separator. The membrane acts rather like a sieve and effectively separates nitrogen and oxygen from the air. A nitrogen enriched stream (perhaps 98%-99% nitrogen) is then injected into the hold space under appropriate controls in order to maintain the desired atmospheric composition. For the system to be effective, good sealing and good control are needed.

It should be noted that the shelf life of produce, previously subjected to c.a. and then stored in air, has not been fully investigated for many varieties. It should also be noted that whilst some produce (e.g. apples, bananas) may be transported without c.a. and could therefore possible be carried without major loss in the event of c.a. failure, much subtropical produce may not survive the length of a sea voyage without c.a.

Of necessity, c.a. involves flooding hold spaces with an atmosphere which will not support human life. Safety is therefore a paramount issue. Systems of controls on access, warnings, and alarms are needed. Specific recommendations are provided by classification societies such as Lloyd's Register of Shipping.

BULK CARGOES

Introduction

Cargoes shipped in bulk are many and varied as may be seen from the Appendix, and stowage factors can vary between as little as 0.31 M3/tonne (11 cu. ft to the ton) to something in excess of 2.81 M3/tonne (100 cu. ft to the ton).

However the same considerations must be borne in mind as with any other cargo, namely weight distribution, stability of the vessel, the nature of the cargo and whether it may have any tendency to shift, as well as the safe carriage of the cargo to ensure delivery in a satisfactory condition.

In addition, possible chemical reactions should be taken into account, namely whether the cargo is likely to emit toxic or explosive gases, liable to spontaneous combustion, or possible corrosive effects.

In this case the IMO Code of Safe Practice for Solid Bulk Cargoes stipulates that it is essential for shippers to provide adequate information to all personnel involved regarding the physical and chemical properties of the material presented for shipment so that adequate precautions may be taken to ensure safe shipment. The IMO Code should be referred to for full information, but Appendix 13 of this book contains the following information from the Code:

1. List of cargoes which may liquefy.
2. List of bulk materials possessing chemical hazards.

It is, however, stressed that these lists are not exhaustive, and in each case when the loading of bulk cargo is contemplated currently valid information should be obtained from the shipper regarding its physical and chemical properties prior to loading.

The methods of loading bulk cargo are almost as varied as the cargoes concerned and may range from antiquated and obsolescent systems giving load rates of as little as 50 tonnes per hour or less to modern loading facilities able to pour the commodity into the ship at rates of up to 7,000 or 8,000 tonnes per hour.

Whatever the method, close liaison with the shore at all times during loading is essential. It must be remembered the equipment in motion cannot always be stopped immediately, and there may be a considerable quantity of cargo which of necessity has to be run off the belts even after the decision to cease loading has been made. Many loading terminals are in isolated areas where there may be no facilities at all for discharge and in the event of any overloading taking place it may well take several days at high cost to discharge the excess cargo which took only minutes to load.

Panamax bulk carrier.

Proper distribution and disposition of bulk cargo is essential. Ships have broken their backs both in loading and discharge through insufficient attention to this factor, although it might be considered obvious that undue stresses and strains on the hull must be avoided (see also Part 1, "Safety").

Concentrates

Concentrates are partially washed or concentrated ores. Copper, lead and zinc are shipped as bulk in concentrate form. Concentrates are powdery or lumpy in character and may carry a considerable moisture content, however this is normally regulated by the exporting country. Prior to loading the shipper should advise of any known characteristic or hazard in particular whether it is liable to moisture migration or becomes flammable or toxic under certain conditions. It is particularly important that the question of special precautions should be considered whenever the cargo to be carried has a moisture content which is likely to exceed 7 per cent in any part of the shipment. It should be borne in mind that the normal moisture content may have been seriously increased should there have been heavy rain during storage or loading, or leakage into the holds during the voyage. In these circumstances the vibration of the ship may cause the moisture to saturate and the concentrate to turn into slurry with consequent danger of the vessel taking a serious list.

GRAIN AND SEED CARGOES

The term grain includes wheat, maize, oats, rye, barley, rice, pulses and seeds.

Proper carriage and a sound and correct delivery of grain calls for constant and expert attention from all concerned, but the most onerous of the duties fall upon the Master and Officers of the vessel.

The loading and carriage of grain cargoes is governed by the International Convention on Safety of Life at Sea (SOLAS) regulations with which every ship's Master and Officers should be familiar. In May 1991 the IMO adopted a new International Code for the carriage of grain in bulk (International Grain Code 1991 edition). This replaced the original chapter VI of the SOLAS Convention 1974.

Bills of Lading

If any doubt exists as to the condition of the cargo the Bills of Lading should be suitably claused. Failure to issue Bills of Lading reflecting the true condition and description of the cargo may be construed as fraud.

Quantity Clause

As it seldom, if indeed it ever happens, that Masters or Officers are in a position to check or vouch the quantity of grain, the seeds or other cargo of like character shipped, mates' receipts and bills of lading should always be claused "weight and quantity unknown, said to be" . . . or in the case of bagged cargo taken on board "So many bags said to contain . . . weight and quantity unknown".

Settling

All grains settle in stowage, even to the extent of 5 to 6 per cent (which can be well below the deck beams, in the lower holds of a good-sized ship), so that good trimming is absolutely necessary in the interest of the safety of the vessel. The duty of seeing that grain is properly trimmed should always be attended to by the ship's officers and never delegated to others.

Ventilation

Ventilation systems on bulk carriers are not normally designed to penetrate cargoes carried in bulk. It therefore follows that efficacy of ventilation applies only to the surface of the cargo and the air space above unless transportation is from one extreme of climate to another it may well be to let well alone.

Principal Grain and Seeds of Commerce

Their properties and peculiarities are described elsewhere in this book under their respective titles. Many kinds are shipped in exceedingly large quantities, either in bulk or in bags, for use either as foodstuffs or for oil extraction; others in moderate quantities in bags only, while some, valued for their medicinal and other properties, are handled in small quantities only and usually are put in bags of superior manufacture or in well made cases. Grain in bags occupies 8 to 10 per cent more space than in bulk:

	M3/tonne			*M3/tonne*			*M3/tonne*	
Grain or Seed	*Bags*	*Bulk*	*Grain or Seed*	*Bags*	*Bulk*	*Grain or Seed*	*Bags*	*Bulk*
Ajwan	2.23		Cummin	3.62		Oats	2.31	2.12
Alsike	1.28		Dari	1.48		Onion	1.81	
Alpia	1.67		Durra	1.48		Paddy	1.81	
Alpiste	1.67		Fennel	2.65		Peas	1.39	
Aniseed	3.34		Flax	1.59	1.39	Poppy	1.98	

	M3/tonne			*M3/tonne*			*M3/tonne*	
Grain or Seed	*Bags*	*Bulk*	*Grain or Seed*	*Bags*	*Bulk*	*Grain or Seed*	*Bags*	*Bulk*
Barley	1.67	1.50	Gingelly	1.67		Rape	1.67	
Bayari	1.56		Gram	1.39		Rice	1.45	
Beans	see Beans		Grasses	1.95/2.51		Rye	1.53	1.39
Buckwheat	1.84		Guinea Maize	2.23		Sesame	1.67	
Canary	1.67		Hemp	1.90		Shursee	1.67	
Caraway	1.73		Jowaree	1.59		Soya Bean	1.39	1.23
Cardamom	2.09		Kernels	1.34		Spinach	1.95	
Carthamus	2.51		Linseed	1.62	1.42	Sugarbeet	3.76	
Castor	2.01		Locust Bean		2.45	Sunflower	2.93	
Cebadello	2.37		Maize	1.50	1.37	Surson	1.67	
Celery	2.12		Millet	1.39		Tares/Vetch	1.79 ch	
Cloves	1.34		Mirabolams	1.95		Teel	1.67	
Common seeds	2.79/3.07		Mowrah	1.70		Timothy	1.95	} Vary
Coriander	3.62		Mustard	1.67		Trefoil	1.67	greatly
Corn	see Maize		Negro Corn	1.48		Tokmari	1.64	
Cotton	2.09		Niger	1.78		Turkish Millet	1.48	
Croton	2.23		Oats Clipped	2.06	1.84	Wheat	1.45	1.31

Grain Stowage Factors — The figures given above are a close approximation only, as no figure can be relied upon correctly to express the actual space to be occupied by any grain or seed at all seasons and from all ports. The factors vary over a considerable range for, amongst others, the following reasons: the quality or the density of a grain, pea, bean or seed varies according to grade, crop, season and country of origin; whether it is shipped early or late in the season, etc.

Stowage factors of bagged grain vary also according as to whether the compartment is large or small, deep or shallow, square or pointed, the presence of obstructions, and depends, also on whether the bags are well or slack filled (see "Stowage Factor").

To attempt a fixed factor of stowage would, therefore, be but to mislead. The above factors represent fair averages and will suffice to enable the carrying capacity of a vessel or compartment to be estimated within a reasonable margin.

A fairly accurate stowage factor, for any given grade of seed, at any port or season, can be calculated by the following formula, if the weight of a bushel of the seed is accurately determined.

For Imperial bushel of 1.2837 c.f. $\frac{\text{constant } 2875.5}{\text{weight in lb}} =$ stowage factors for bulks.

For U.S.A. bushel of 1.2445 c.f. $\frac{\text{constant } 2787.6}{\text{weight in lb}} =$ stowage factors for bulks.

For bagged grain, etc., add 8 to 10 per cent.

See Appendix 9, "Grain Weights and Measures".

Preparing Holds for the Reception of Grain Cargoes

In order to pass surveys in accordance with Charter party and/or Statutory requirements at the load port it is absolutely essential when proceeding to load grain to make every effort to ensure that the holds are properly prepared for the reception of the grain cargo.

Failure to do so will, more often than not, result in the necessity to call in shore labour, if surveys are failed, as union agreements at many world grain ports prohibit the use of ships crew for cleaning work in port.

Holds must be properly cleaned and prepared and all compartments, including sides, stringers, pockets, brackets, etc., must be clean swept, well ventilated and dried. Residues of previous cargoes must be totally removed and any loose rust or scale which might contaminate the cargo must be carefully removed.

If there are signs of insect infestation these must be attended to either by spraying with appropriate insecticides or by sealing the holds and fumigating with some approved type of smoke bomb. During the cleaning process close attention should be paid to tank tops, ceilings, box beams, frames, spar ceiling, hatch beams, etc. Remember that with the movement of the vessel, old grain or other residues may fall from the box beams, etc., on to the ceiling after the hold has been cleaned.

Any timber or dunnage remaining in the holds must be stacked on billets to permit inspection and cleaning beneath the stacks.

All bilge suctions must be thoroughly clean and free from old grain.

Particular care should be taken in the re-use of old fittings particularly burlap, that there is no chance of infestation from old grain.

Approved Grain Loading Methods

The Maritime safety Committee, at its Fifty Ninth session (May 1991), adopted a new International Code for the safe carriage of grain in bulk (International Grain Code). This replaced the original Chapter VI of the 1974 SOLAS convention, which contained detailed regulations on the carriage of grain in bulk, with more general requirements and placed detailed provisions on grain in a separate mandatory code.

There are a number of approved loading methods including the following:

1. (a) A document of authorisation issued by the administration of the Country of Registration, or a Contracting Government on behalf of that Administration must be produced.
   (b) Appropriate stability data approved by the Administration of the Country of Registration, or a Contracting Government on behalf of that Administration must accompany the document of authorisation so that, if required, the Master may demonstrate the ability of the ship to comply with the stability criteria required by the regulations.
   (c) Where an equivalent arrangement is accepted by the Administration in accordance with Chapter 1 Regulation 5 of the Safety of Life at Sea Convention 1960 or 1974, as appropriate, it must be included in the document of authorisation.
2. Specially suitable ships, e.g. self trimming or partly self trimming vessels may load bulk grain provided they comply with the following conditions:
   (a) They are constructed with two or more vertical or sloping graintight longitudinal divisions suitably disposed to limit the effect of any transverse shift of grain.
   (b) As many holds and compartments as possible shall be full and trimmed full.
   (c) For any specified arrangements of stowage the ship will not list to an angle greater than 5 degrees at any stage of the voyage where:
      (i) in holds or compartments which have been trimmed full the grain settles 2 per cent by volume from the original surface and shifts to an angle of 12 degrees with that surface under all boundaries of those holds and compartments which have an inclination of less than 30 degrees to the horizontal.
      (ii) in partly filled compartments or holds free grain surfaces settle and shift to an angle of 12 degrees with that surface, or to such larger angle as may be deemed necessary by the Administration or by a Contracting Government on behalf of the Administration and grain surfaces if either overstowed by a suitable cargo, or strapped and lashed, shift to an angle of 8 degrees with the original level surfaces. If shifting boards are fitted due allowance is to be made for the divisions of the hold or compartment limiting the transverse shift.
   (d) The Master is provided with a grain loading plan and a Stability Booklet approved by the Administration or by a Contracting Government.
   (e) The Administration shall prescribe such precautions as the trimming of the ends of holds or compartments to meet the required stability criteria where "shedder plates" have not been fitted at the ends of such holds or compartments.
3. Ships without documents of authorisation by their own Country or Registry or a Contracting Government may in certain circumstances be permitted to load by the Administration at the loading port but that Administration will decided the conditions to be fulfilled.
   (See I.M.O. Code for Safe Carriage of Grain in Bulk (International Grain Code) 1991 Edition et seq.

Trimming of Bulk Grain Cargoes

Self trimming bulk carriers routinely have stability calculations for wings trimmed and untrimmed. Many countries now insist on calculations for void spaces for forward and after ends of hatches. If these are not available owners and charterers may face additional stevedoring charges for the costs of end trimming.

The IMO Resolutions governing the carriage of grain require that the void depths at the ends of compartments are taken into account in meeting the stability criteria.

The formula for calculating the mean void depth for a compartment as given in the Resolutions is as follows:

For the purpose of calculating the stability of ships carrying grain in bulk it shall be assumed that:

(i) in filled compartments of ships with hatch side girder depths between 500 and 600 mm, the average depth of the underdeck void (VD) is 460 mm.

(ii) when the depth of the hatch side girder is not between 500 and 600 mm the average void depth shall be calculated according to the formula:

$$Vd = Vd1 + 0.75\,(d - 600)\ \text{mm}$$

where Vd = Average void depth in mm
$Vd1$ = Standard void depth from the Table
d = Actual girder depth in mm.

In no case shall Vd be assumed to be less than 100 mm.

If for example the end of a compartment had a hatch end girder of 3 ft (i.e. 915 mm) and the end bulkhead was 16.4 ft (i.e. 5 m) from the hatch coaming, then using the above formula:

$$\begin{aligned} Vd &= Vd1 + 0.75\,(d - 600)\ \text{mm} \\ &= 430 + 0.75\,(915 - 600) \\ \text{Average Void Depth} &= 666\ \text{mm or } 2.18\ \text{ft.} \end{aligned}$$

TABLE 1

Distance from Hatch End or Hatch Side to Boundary Compartment metres	*Standard Void Depth Vdl mm*
0.5	570
1.0	530
1.5	500
2.0	480
2.5	450
3.0	440
3.5	430
4.0	430
4.5	430
5.0	430
5.5	450
6.0	470
6.5	490
7.0	520
7.5	550
8.0	590

If the required stability cannot then be achieved the loading Administration will order trimming of the cargo to throw the grain into ends of the compartments. In some cases it may be possible to utilise spout extensions or scoops to achieve the filling of the ends without machine trimming but this will depend on the particular design and capability of the grain elevator.

For this reason it is advisable that the stability data carried by the ship should include "trimmed moments" as well as "untrimmed moments" so that the ship's Master may be able to exercise the option whether to use "trimmed moments" and trim the ends of compartments or use "untrimmed moments" and not trim should it prove possible to meet the stability criteria in this condition.

In order to meet Charter party conditions regarding cargo quantity and trim provisions it may very well be necessary to leave one or more holds slack. As slack holds create the largest upsetting moments in order to comply with stability criteria and stay within the maximum angle of heel of 12 degrees, it may be necessary to secure the free surface of the grain.

This may be done by strapping or overloading with bagged cargo. Although both methods are generally expensive, strapping is the less time consuming.

It is achieved by setting up wires in position prior to loading at a requisite distance below the anticipated free surface of the grain, and holding them in the wings by lashings. After the required quantity of grain has been loaded and, if necessary levelled, it is covered by fore and aft and athwartship dunnage. Thereafter the wires are joined by bottle screws and tightened so that a taut strapping is formed over the grain surface.

In filled compartments where the saucering method is used to reduce heeling moments it may be done in one of two ways, either by filling a saucer shaped depression left in the grain in the square of the hatch with bagged cargo, or using the "bundling of bulk" method. The latter is generally the least expensive of the two methods. It is achieved by lining the saucer shaped depression with an acceptable material laid on top of athwartship wires or strapping and dunnage. The saucer is then filled with bulk grain and the lining closed over the top. Further dunnage is laid on top and the lashings brought together, tightened and secured.

ORES

When full or large part cargo of ore is carried, its longitudinal distribution should be carefully decided upon so as to avoid too much weight in the middle, causing sagging tendencies; and, on the other hand, in order to ensure sea-kindliness and avoid risk of hogging, not to have an excess of weight at the extreme ends. According to the nature and density of the cargo, it is frequently the practice of large carriers to distribute cargo in alternate hatches. Careful consideration must be given to get the longitudinal stresses within accepted limits. To this end masters of large carriers should be provided with loading diagrams setting out distribution of cargoes for varying conditions. The results of current investigations suggest that the alternative hold loading method should be avoided. Similarly a fast loading rate contributes to excessive stresses. Both problems exist for economic reasons at the loadport. However in order to avoid risk of loss of life both the foregoing problems should be addressed. Further investigation should be made into torsional stresses when the ship is working in a bow or quartering seaway.

If as sometimes happens the cargo is delivered from one tip, the vessel should be moved frequently so as to bring all hatches, in fairly quick alternate rotation under the tip, it being an exceedingly dangerous practice to receive all the cargo intended for one hold while other holds are empty. Many vessels have received permanent damage from this cause.

During recent years there have been a disturbing amount of losses of bulk carriers of about 120,000 tonnes deadweight carrying ores. These catastrophes have been accompanied by a disastrous loss of lives. It may be that ships scantlings have not increased commensurate with the deadweight carrying capacity. It cannot be over emphasised that strict attention must be paid to the loading rotation which at all times must be within the ships longitudinal stress tolerances. A prudent owner should carry out a careful inspection of the frames in the holds after such cargoes have been carried, particularly if they are carried on a regular basis or if the ship has been put to a contract of affreightment.

Finely Crushed Ores

When receiving finely crushed ores, it is important to bear in mind the possibility that the normal moisture content (of 8/10 per cent) may have been seriously increased by heavy rains during storage in semi-open trucks, sheds, etc. In that condition the vibration of the ship may cause the moisture saturated ore to become slurry with the consequent danger of the vessel taking a serious list. In such cases bagging has been found to be necessary to prevent the cargo from shifting.

In all ships which carry these cargoes an adequate range of stability should be preserved throughout the voyage so that, if a shift occurs, the list will be minimised (see extract from IMO Code of Safe Practice for Solid Bulk Cargoes in Appendix 13).

In two- and three-decked vessels, a proportion of a full cargo of ore, varying with the class of ship, should be received into the 'tween decks in order to reduce the strains and make the vessel easier in a seaway. The ore should be trimmed into the wings and to the bulkheads, and not left in the square of the hatch, otherwise the risk of setting down the decks will exist.

The National Cargo Bureau Inc. of New York in a circular headed "Weights in and on Deck", regarding the loading of selected commodities, advised that, in order to reduce the risk of distortion or failure of 'tween deck structures (including hatches) due to concentrated stowage of heavy cargo on any part of a 'tween deck or hatch, the cargo should be so distributed that the weight, per square foot, on any part of a 'tween deck or hatch should not exceed that for which they were normally designed.

To ascertain the approximate weight per square foot to be stowed in 'tween decks and hatches the following formula is recommended:

Height of 'tween deck × 45 = the approximate weight in Pounds per Square Foot.

With part ore cargo, the ore should be levelled off and, if overstowing with grain, seeds or other dry cargo liable to damage from moisture, it should be laid over with planks, or stout bamboo, double crossed close enough together to keep the packages from resting on the ore; these to be double matted or covered with burlap. Matting or burlap only is not sufficient. Not infrequently, especially in Indian ports during the S.W. Monsoon season, wet ore such as manganese is loaded and subsequently perishable cargo, such as oil cake, ground nuts, etc., offered for shipment on top of the water soaked ore.

No system of dunnaging and matting such as referred to in the foregoing can be relied upon to prevent damage when perishable goods are stowed on wet ores.

When Chrome Ore and Manganese Ore are carried they should be stowed in such a manner as definitely to preclude any possibility of the smallest quantity of Chrome Ore getting mixed with Manganese Ore. The admixture of even a small proportion of Chrome with Manganese renders the latter useless for important purposes in connection with the steel trade.

Particular care should be taken to prevent any loose grain mixing with the ore; numerous are the claims which, in the past, had to be met on that account.

Oils or acids should never be stowed on or over ore. When loading ore in a rainy weather, if intending to complete with dry or general cargo, hatches should be covered at night.

Some ore cargoes, when shipped in the dry seasons, are very dusty, in which case other cargo should adequately be covered up.

Hematite or Iron Ore is shipped either as "Run of mine", "Crushed and graded" or "calcined" ore (a very dusty cargo).

If Bagged Ore is shipped, the condition of the bags should be carefully noted — it happens that many are partly rotted through lying on damp ground.

Bills of Lading should be claused "weight unknown, not responsible for loss of weight". Ore shipped in the open roadsteads of India, etc., often loses 4 to 6 per cent weight between leaving mines and actual shipment — mostly lost overboard from lighters. When the swell is high the loss is sometimes even greater.

The various ores stow, approximately, as follows:

Ore	M3/tonne Bag	M3/tonne Bulk	Ore	M3/tonne Bag	M3/tonne Bulk
Aluminium Bauxite	1.11	0.84/0.98	Lead Concentrates	—	0.33/0.39
Antimony Stibnite	0.56	0.42	Magnetite	—	0.42/0.47
Asbestos	1.67/1.81	—	Manganese	0.61/0.70	0.47/0.50
Chrome	—	0.33/0.39	Manganese Perox	—	0.42/0.47
Cinnabar	0.56/0.61	—	Nickel	0.56	—
Cobalt	—	0.50/0.56	Silver	0.61	0.70
Concentrates	0.42/0.56	—	Tin	0.61	—
Copper	—	0.39/0.56	Wolfram or Tungsten	0.45/0.50	—
Corundum	1.11/1.25	—	Umber	—	1.11/1.17
Galena or Lead	0.45/0.47	0.33/0.39	Uranium	0.47/0.50	—
Garnet	0.50/0.56	—	Zinc Blende	—	0.56/0.67
Hematic or Iron	—	0.33/0.47	Zinc Concentrates	—	0.50/0.56
Kainite	1.06/1.11	0.98/1.03			

IMO Code of Practice for Bulk Cargoes is referred to in the Appendix.

Mineral Sands

Ilmentite, rutile and zircon sands are refined products and any form of contamination can be harmful to them. In consequence the following procedures should be followed prior to shipment:

1. Any residues of previous cargo must be removed. Particular attention should be given to the cleanliness of the underside of steel hatch covers, horizontal flanges of under deck beams, stringers or horizontal stiffeners, and the flanges of portable hatch beams. Particular care should be taken in removing phosphates, sulphur or ferrous residues as these are especially undesirable.
2. Any loose rust scale should be removed from the compartment prior to loading and particular attention should be given to the underside of steel hatch covers and the "hidden" flanges of frames.
3. The compartment should be washed down with a final washing with fresh water.
4. The compartment should be dry.
5. All bilges should be grain tight. All spaces on the crown of the bilge should be cemented and the limber boards covered with two thicknesses of hessian or burlap, secured by thin battens at their top and bottom edges, and an overlay of burlap be allowed to lie over the crown of the bilge. Should plastic or polyethylene material be used, it is desirable that a section of limber boards adjacent to the bilge suction be covered with burlap.
6. Where suction boxes are flush with the hold ceiling, as in most modern bulk carriers, suction plates should be covered with two thicknesses of burlap and cemented in place.
7. Should the vessel be fitted with spar ceiling it is advisable to remove the lower 6 or 7 ft as bulldozers or grabs are likely to be employed during discharge.

Ilmenite is a very heavy sand stowing about 0.36 M3/tonne (13 cu. ft to the ton), and it is almost black in colour and is found on the coast of India, Malaysia, Australia, etc.

From ilmenite titanium is also extracted from which the whitest pigment is obtained. Titanium is widely used in the manufacture of paint.

Properly trimmed ilmenite sand, which has a clinging tendency and a large angle of repose, is not likely to shift. With part shipments in large compartments, the ilmenite when in bulk should be overstowed, or if this is not possible, adequate steps taken to avoid any shifting of the stow. The cargo on account of its weight, should be well spread over the ship in order to avoid undue strains. It is dustless.

Rutile is a mineral obtained from certain sands in Australia and Brazil. It is used for hardening steel, etc. This is a very fine cargo and as such is not really suitable for grab discharge.

Zircon is another sand which is used in the process of hardening steel, etc.

PETROLEUM CARGOES

This section should be studied in conjunction with the International Convention for Safety at Life at Sea (SOLAS Convention), The IMDG Code, The International Safety Guide for Oil Tankers and Terminals (ISGOTT), also the International Convention for Prevention of Pollution from Ships, the MARPOL convention and its protocol together with the latest edition of Clean Seas Guide for Oil Tankers.

Crude Oil

A mineral oil which comprises mainly of a mixture of hydrocarbon compounds found in many parts of the world. The principal producing areas are the Middle East, Latin America, Africa, Eastern Europe, Far East/Australasia, North America and Western Europe respectively and the crude petroleum may be conveyed to the refining centres by pipeline, tankship, rail tank car or road tank car.

Crude oil is an inflammable liquid the flash points of which are very low (some are of the order of −40°C) and are therefore very rarely quoted. It has a density which may vary from 0.750 to 1.000 (see density of petroleum).

Petroleum Products

The principal products of petroleum by distillation are as follows:

Liquefied petroleum gases (L.P.G. — mainly propane and butane).
Motor gasoline.
Kerosene.
Gas oil.
Fuel oil.

In addition to the foregoing so called straight run products others which result from more complex refining processes include:

Aviation gasoline.
Premium grade motor gasoline.
Premium grade kerosene.
Lubricating oil and greases.
Wax.
Bitumen (asphalt).

Yields of each product vary over a wide range depending on many factors including the grade of crude petroleum refined and the particular processes employed.

Petroleum Products and Other Substances Giving Off Inflammable Vapours

(For substances giving off inflammable vapours see also Dangerous Goods).

The following list of "Inflammable Substances" contains only substances which hitherto, have been carried generally and is not exhaustive. Substances not included will fall into their appropriate places according to their flash points and according as to whether or not they are miscible with water.

"Benzine" or "Spirit" Class — Flash point below 73 degrees F (22.9 degrees C).

Immiscible with water:

Petroleum spirit	Lythene	Butyl acetane (iso)
Benzine	Motor Spirit	Carbon disulphide
Benzol	Naptha	Ether (sulphuric)
Benzolene	Toluol	Ethyl (acetate)
Gasoline	Xylol	Nickel carbonyl

Miscible with water:

Acetone	Diacetone alcohol	Methyl alcohol
Ethyl alcohol	Methylated spirit	Propyl alcohol
(Spirits of wine)	Pyridene	

Ordinary Class — Flash point 73 degrees F (22.9 degrees C) and upwards.

Amyl acetate	Amyl alcohol	Butyl acetate — normal
Butyl alcohol	Coal tar	Ethyl lactate
Kerosene	Mineral oil	Paraffin
Shale oil	Turpentine	White spirit (Turpentine substitute).

Variable Class — Variable flashpoint.

The following substances are of variable composition and should be grouped, therefore, according to their composition, flash points and other properties as certified by the makers:

Celluloid solution — (Collodion cotton in solution in or wet with inflammable liquids).
Paints — Bituminous paints, photogravure printing inks.
Polishes, Boot creams, etc. — Enamels.
Polishes, linoleum — India rubber and gutta percha solution.
Polishes, liquid metal — Roisin oil.
Ship's compositions — Tar oil compounds.
Varnishes, oil — Varnish, spirit.

Fuel and Lubricating Class — Those having a flashpoint of 150 degrees F (66 degrees C) and upwards.

N.B. The stowage of Petroleum Products and other Hazardous Goods in ships at U.S.A. ports is under the inspection of the National Cargo Bureau — (NCB).

BULK OIL MEASUREMENT

Density of Petroleum

The gravity or density of Crude Oil and Products are sometimes quoted in Relative Density at 60/60 F (which is the same as Specific Gravity), API gravity at 60 F or in Density (kg/Litre) at 15 C, which is defined as mass per unit volume. The relationship between API and Relative Density is based on the following:—

$$\text{API gravity at 60 F} = \frac{141.5}{\text{rel. density 60/60 F}} - 131.5$$

Tables are available for converting API to Density and Relative Density or vice versa. These are the ASTM D 1250 Tables 3, 21 or 51.

For establishing the density at 15 C (kg/Litre) of an oil having been given a relative density 60/60 F between the ranges of:

0.654–0.683	deduct	0.0001	to give	density (kg/Lt)	at 15 C
0.684–0.757	,,	0.0002	,,	,,	,,
0.758–0.786	,,	0.0003	,,	,,	,,
0.787–0.852	,,	0.0004	,,	,,	,,
0.853–0.941	,,	0.0005	,,	,,	,,
0.942–1.033	,,	0.0006	,,	,,	,,
1.034–1.075	,,	0.0007	,,	,,	,,

The Standard Temperatures within the Oil Industry are usually either 60 F or 15 C. It is emphasized that density (Kg/Litre) is at 15 C (59 F) and not at 60 F. In some Countries when Crude or Products are shipped, a standard temperature of 20 C is used when quoting the density. In a few ports other standard methods may be used.

Volume of Petroleum

Crude oil and Products are generally measured in Barrels (bbls) or Cubic Metres (M3) and when converting from one volume to the other at an observed temperature the following is used:—

Barrels (bbls) at Observed Temp. = Cu. Metres × 6.28981

If conversion at Standard Temperature is required i.e., from Cubic Metres at 15 C to barrels at 60 F then apply the following:—

(This is Table 52 of ASTM D1250. Note that in the Tables Density is usually expressed in Kilogrammes/Cubic Metre).

Density 15 C	Bbls per Cubic Metre
654.0 - 697.0	6.295
698.0 - 778.0	6.294
779.0 - 901.0	6.293
902.0 - 1074.0	6.292

Eg: Given 63000 Cubic Metres at 15 C, density at 15 C = 0.830
63000 × 6.293 = 396459 Bbls at 60 F.

When converting from barrels at 60 F, to Cubic Metres at 15 C then apply the following:—

Density 15 C	Cubic Metre per Bbl.
654.0 - 683.0	0.15886
684.0 - 722.0	0.15887
723.0 - 768.0	0.15888
769.0 - 779.0	0.15889
780.0 - 798.0	0.15890
799.0 - 859.0	0.15891
860.0 - 964.0	0.15892
965.0 - 1074.0	0.15893

Eg: Given 450000 Barrels at 60 F, density at 15 C = 0.840
450000 × 0.15891 = 71509.500 M3 at 15 C.

Ullage Survey

Bulk Liquids shipped in Tankers are measured by undergoing a manual 'Ullage Survey' or Gauging operation at each of the vessels tanks, or by remote gauge readings in the vessels Cargo Control Room. Ullage is the depth of free space above the liquid in each tank measured from the oil surface to a calibration point. This may need to be corrected for trim or list to find the corrected ullage. Ullages, temperatures and water cuts (with water-finding paste) are normally made in the presence of an Independent Inspector. Owing to the increasing demand for 'Closed Measuring Systems' in many countries both on safety and environmental grounds, portable electronic transmitting systems are often used for measuring these Ullages, temperatures and water cuts.

Sampling/Analysis

Sampling can be made by an auto in-line sampler fitted within the loading/unloading line in the Terminal installation, or by a portable sampler fitted to the Tanker manifold flange and/or by manual means. Manual sampling is performed by various means and equipment depending on the product, the individual requirement and also the type of tank openings available.

Samples of the product involved in the custody transfer are tested for density and many other criteria. If crude is being analysed a very important test is that for BSW or Base sediments and water. This is measured as a percentage and will result in a deduction to the Gross Bill of Lading or Gross Outturn making a Nett Bill of Lading or Nett Outturn. Usually the commodity is only paid for on a Nett basis and therefore this particular test can be very critical.

The density of the petroleum cargo loaded should have been determined in a shore laboratory by the Terminal staff but confirmatory measurements maybe made aboard if the correct equipment and expertise is available.

Cargo Calculations

There is a Calibration Table for each tank and these will be entered to find the observed volume of liquid, the Total Observed Volume or TOV, corresponding to the corrected ullage. The volume of water should then be deducted to obtain the Gross Observed Volume of oil or GOV. The volume then has to be reduced to a volume at the industry standard temperature of 60 F or 15 C. This is achieved by entering the ASTM-IP Tables to find the appropriate Volume Reduction Factors by using arguments of observed temperature against Density, API or Relative Density (S.G.). In some Countries when Crude or Products are shipped, a standard temperature of 20 C is used when quoting the density.

It is important that the correct information is received before the Ship's Staff complete their cargo calculations so that the cargo shortages, if any, can be recognised and an appropriate Protest Letter issued against the Bill of Lading figures.

ASTM — IP Tables for Volume Correction Factors (ASTM D 1250) (IP 200)

Generalised Crude Oils

Table 6A Volume Correction to 60 F against API gravity 60 F.
Table 24A Volume Correction to 60 F against RD 60/60 F.
Table 54A Volume Correction to 15 C against Density at 15 C.

Generalised Products

Table 6B Volume Correction to 60 F against API gravity 60 F.
Table 24B Volume Correction to 60 F against RD 60/60 F.
Table 54B Volume Correction to 15 C against Density at 15 C.

Generalized Lubricating Oils

Table 6D Volume Correction to 60 F against API gravity 60 F.
Table 54D Volume Correction to 15 C against Density at 15 C.

(These lower tables are not for generalised lubricating oils).
Table 6C Volume Correction Factors for individual and special applications. Volume Correction to 60 F against Thermal Expansion Coefficients at 60 F.
Table 24C Volume Correction Factors for individual and special applications. Volume Correction to 60 F against Thermal Expansion Coefficients at 60 F.
Table 54C Volume Correction Factors for individual and special applications. Volume Correction to 15 C against Thermal Expansion Coefficients at 15 C.

Useful ASTM — IP Conversion Tables

Table 1 — Interrelation of Units of Measurements.
Table 11 — Barrels at 60 F to Long Tons against API at 60 F.
Table 13 — Barrels at 60 F to Metric Tonnes against API at 60 F.
Table 29 — Long Tons per Barrel against SG/RD at 60/60 F.
Table 56 — Cubic Metres at 15 C to Metric Tonnes against Density at 15 C.
Table 57 — Cubic Metres at 15 C to Long Tons against Density at 15 C.

PETROLEUM — Bulk Carriage

Prior to Shipment of Crude

In the Crude Oil trade, unless the vessel has ample segregated ballast to enable her to manoeuvre effectively, it maybe necessary to arrive at the loading terminal with Clean Ballast within the ships Cargo Tanks. During the previous discharge of Crude Oil the vessel should have at least Crude Oil Washed her dirty ballast and clean ballast tanks plus 25% of the remainder for sludge control. Crude washing the Dirty and Clean Ballast Tanks will minimise the amount of bottom sludge and oil remaining in those tanks, and, with regard to the Clean Ballast Tanks, will reduce the necessity for prolonged water washing required to change from 'dirty' to 'clean ballast' condition.

It is important that Charterers Instructions regarding ballast are followed. Many Loading Terminals now have dirty ballast reception facilities which will alleviate the necessity for Clean Ballast and therefore water washing.

Specific instructions regarding any Cargo Heating requirements should be requested from the Charterer.

Prior to Shipment of Clean/Dirty Products

It is not the purpose in this section to give specific guidance on tank preparation for specific products. Oil Companies and Charterers differ in their attitudes on this subject dependant on the trade and the end use of the product.

Normally the Petroleum product grades are shipped in smaller tonnage than that allocated to Crude oil. Generally the vessels have a segregated ballast system.

It is important that the Charterers Instructions are sought regarding Tank Cleaning when a change of grade is likely. Some products are incompatible despite tank cleaning, i.e. Changing from Diesel to JP1 — JP4 or AVGAS should be avoided. Other products may require little or no tank cleaning except where sludge control is necessary.

Eg. Generally prior to loading Vacuum Gas Oil, the tanks must be free of salt and metals and therefore when tanks are washed they must either be first salt water washed followed by a fresh water rinse, or only fresh water washed. It is possible that samples of any OBQ (On board quantities), sludge and the like, will be taken and sampled in a shore laboratory and the vessel delayed if any likely contaminants are found and the vessel has to resume washing.

Tank coatings should be inspected regularly and maintained. Some products can leach out residues from previous cargoes trapped behind broken down coatings, causing cross contamination.

Cargo Heating

Steam heating coils should be pressure tested to manufacturers instructions prior to each loading of a cargo which will require heating.

Multigrade Loading

Efficient segregation is imperative when carrying multigrade crudes and products. There should be at least two valve separation within the ships pipeline system, which should not include tank valves.

Safety Guidelines

To ensure that all tanker operations are undertaken safely refer to and follow the International Safety Guide for Oil Tankers and Terminals. (ISGOTT) Strict precautions are essential when loading certain products and in particular Sour Crudes, owing to the high concentrations of hydrogen sulphide (H_2S). H_2S is extremely toxic and can be easily recognised by its offensive odour of rotten eggs.

However, the gas can have a paralyzing effect on the sense of smell almost immediately. The degree of sourness is normally indicated by the dissolved H_2S content ppm by weight. The H_2S content of various sour crudes are listed below; this list is by no means exhaustive:—

Crude Name	Loading Port	H_2S ppm by weight
Isthmus	Mexico	Up to 60 ppm
Iranian Light	Kharg Island, Iran	20 ,,
Kirkuk	Turkey	40-70 ,,
Iranian Heavy	Kharg Island, Iran	70 ,,
Brega	El Brega, Libya	260 ,,
Es Sider	Es Sider, Libya	260 ,,
Qatar Land	Umm Said, Qatar	200-300 ,,
West Texas	Baytown/Houston, Texas	Up to 1000 ,,

Stress Criteria

It is imperative that hull stresses are checked at intermediate stages of any loading or discharging operation and also for the ballast passage

Stability Criteria

Common tankers which have cargo tanks 98% full would not normally have any stability difficulties. The metacentric height would be large in comparison to General Cargo vessels, but when many tanks are only to be partly filled then the GM should be precalculated as a precaution. With OBO's (Oil, Bulk, Ore Carriers) extreme care is necessary during loading and discharging of Oil cargoes or during ballasting and deballasting owing to the large unobstructed surface area within the holds. The free surface effect in slack holds (i.e. filled to below the hatch coaming) can allow unrestricted movement of the liquid, resulting in both loss in stability and "sloshing". It is important that Loading and Discharging Instructions, together with the Stability Data supplied to all Combination Carriers, are rigorously followed and before arrival in port a plan of operation is prepared bearing in mind these critical issues. (See ISGOTT).

Carriage of Petroleum Products in General Cargo Vessels — Cases/Drums Etc.

If quantities of this class of cargo are carried in an ordinary cargo vessel, the type and marking of cases or containers, handling, stowage and carriage of the same will be governed by the IMDG (see "Dangerous Goods" "Case Oil" etc.).

The risk of fire and explosion is ever present when this cargo is carried, more especially as regards the Benzine or Spirit Class, so that the greatest care is necessary to avoid such. Smoking and the use of naked lights, forges, etc., should be strictly be prohibited.

When compartments are filled with the Spirit Class of these goods, it is recommended that they be kept, as nearly as possible, hermetically sealed, but when it forms a proportion only of the cargo in any compartment, efficient ventilation is of first importance, to ensure which the "downtake" ventilators should be carried to the bottom of compartments. To guard against sparks, the mouths of ventilators should be covered with fine gauze wire, especially when situated near the ends of deck houses accommodation, etc.

It should be borne in mind that the risk of explosion, etc., persists after the cargo has been discharged and until all traces of vapour have been expelled from the holds and bilges.

After discharge is completed, special attention should be given to the ventilation of all compartments, for which purpose hatches should be kept uncovered, all ventilators including wind sails, when available, kept on the wind, limber boards lifted, all oil and/or oily water in bilges removed, bilges cleaned — the use of naked lights, or defective electric fittings or cable being avoided.

Delicate or edible goods liable to damage by tainting should not be stowed with or near petroleum products, neither should the latter be stowed with or near to commodities which are liable to spontaneous combustion from whatever cause, such as coal, cotton, etc. Stow away from sources of heat.

Carriage in Deep Tanks

Apart from the large volume of these oils carried in bulk by tankers., consignments of certain classes of Mineral Oils are carried in the deep tanks of general cargo ships, some of which are:

Lubricating Oil, Water White Kerosene and Batching Oil — All three are readily contaminated by dirt and foreign matter so that, not only must the tanks be scrupulously clean and dry, the pipe lines must also be thoroughly steamed out, cleaned and all traces of oil, water, or other residue removed.

This requirement is best met by steaming and cleaning tanks, etc., immediately after the discharge of palm or other oil content, the plentiful use of caustic and thorough washing down with hot water.

According to shipper's requirements, the temperature should be raised slowly prior to pumping, usually 38 to 49 degrees C, however check shippers carrying instructions.

Batching Oil is of straw colour and, as it is sold on its colour, merits special pipelines, usually provided by consignees for discharging same as the slightest trace of fuel, lubricating or other oil contaminates and lowers its value.

Another safeguard against contamination resorted to when discharging these petroleum products is to provide a small tank on the deck of receiving lighter into which the first pumping is received, the oil thus received being later clarified at the consignee's shore plant.

LIQUEFIED PETROLEUM GAS

The term Liquefied Petroleum Gas (LPG) is generally applied to the commercial grades of Propane, predominately C_3s, and Butane, predominately C_4s. Chemically pure liquefied petroleum gases are carried by sea but are usually referred to by their particular names, Ethane, Propylene, Butadiene, etc.

Commercial grades of LPG are not required to be chemically pure; the end use of the LPG's produced is mainly for heating purposes or as a motor vehicle fuel, an expanding market.

LPG's are maintained in the liquid phase either by pressurisation or by refrigeration. Commercial Propane may be kept in its liquid form by maintenance of a pressure of approximately 8 bar gauge or temperatures reduced to below approximately −40° C, commercial Butane by maintenance of a pressure of approximately 2.5 bar gauge or temperatures reduced to below 0° C.

Vessel Types

Vessels which carry LPG's are of three types:—

(a) Pressurised vessels, which carry liquefied gases under pressures up to 10-17 bar; these vessels are normally capable of carrying liquids with temperatures up to 45° C.

(b) Semi-refrigerated vessels, which carry liquefied gases up to pressures of about 3 to 5 bar and, dependent on the type of metal and construction of the containers, down to − 50° C.

(c) Fully refrigerated vessels, which carry liquefied gases at, or just below, their boiling points. The cargo is kept liquid in the tanks by an efficient insulation system rather than a refrigeration plant. The tanks are constructed to contain pressures of about 0.25 bar, i.e. slightly "over pressure" to prevent any ingress of air (oxygen).
Such vessels are normally intended for the carriage of liquefied chemical gases and are constructed of metals capable of containing liquids down to − 50° C.

Note: All three types are usually equipped with a compressor/heat exchanger plant which reliquefies any "boil off" gas and returns it to the tank.

LPG's are normally carried in either pressurised or semi-refrigerated vessels, the type used on any particular trade being dependent on the type of facilities at the load and discharge ports which must match the ship type.

The vessels vary between the very small coastal types of 5-600 cubic metres to those of 70-80,000 cubic metre capacity. The smaller vessels are usually carry LPG' under pressure, the larger vessels are virtually all semi-refrigerated.

Cargo Measurement

Measurement of LPG's can either be carried out by static means, gauge systems, or by dynamic methods, shore based meters, when loading. In the past LPG's were also measured manually but this is no longer the practice.

Measurement in ship's tank is normally by automatic gauges of which there are many different types in use, from the float gauge through to ultra sonic and laser systems. Temperature and pressure in each tank are equally as important as the liquid level and are measured by "in tank" instruments. All automatic "in tank" systems should be calibrated and reset to their datums at every opportunity, to ensure accuracy.

Cargo Handling

LPG's are handled under a closed system and should not be either seen or physically handled. All measurement should be by automatic gauging and analysis carried out under controlled laboratory conditions.

Care must be exercised when sampling that the samples are as representative as possible. The connections on the sampling equipment and the sampling points in the ship must match.

Cargo Quality

LPG's are used as fuels and as such are vaporised before ignition. If the composition of the LPG meets the specification laid down then no serious problems should be encountered in use, other than the attempted use of the wrong product e.g. the use of butane when ambient temperatures are at or below 0°C will be practically impossible as the butane will remain in its liquid phase rather than vaporising.

Contaminants in LPG are:—

(a) Sulphur, other than dimethyl sulphide or ethyl mercaptan which are used to stench the product with an objectionable odour to make any leaking gas readily noticeable, will cause corrosion to the metals used in pipework carrying the gas from storage to burner. The standard test for corrosion is the Copper Corrosion test, in which a clean, bright strip of copper is suspended in a sample of the gas for one hour at 40°C. The copper strip is inspected on completion of the test and its visual appearance compared with standard discoloration strips. Discoloration indicates a constituent in the gas which is corrosive to copper and would eventually make any installation using copper piping unsafe due to leakages.

There are a number of empirical tests (tests which are qualitative but not quantitative) for sulphur compounds e.g. Hydrogen sulphide may be detected by the use of moist lead acetate paper.

(b) Oily mattter and gums will cause sooting when burnt and are undesirable. They also reduce the calorific value of the LPG.

The empirical test for the presence of oily matter and gums known as the weathering test. The test involves the "weathering" or opening a sample of the LPG to atmosphere under controlled temperature conditions and allowing it to evaporate for a set period of time. On completion of the time, the quantity remaining is measured giving some measure of the oils and gums present in the LPG.

(c) Water is another undesirable contaminant. Its presence may lead to the formation of either ice or hydrates which may block the pumping system.

Water, in small quantities, may be dealt with by the addition of small quantities of methanol which is hygroscopic. Care must be taken in the quantities added to ensure that the characteristics of the cargo are not changed to the extent that it becomes unacceptable.

An empirical test for water relies on the effects of water vapour on cobalt bromide.

(d) Ammonia; LPG's may be carried in vessels which have previously carried ammonia. Ammonia contamination may result in a product which is highly corrosive to copper or copper alloy fittings.

Extreme care must be exercised when cleaning tanks and associated pipelines, pumps and compressors from ammonia to LPG. At one stage of the operation it is preferable that the tanks are placed under breathable air and then opened up for internal inspection. In some instances it may be preferable to wash the tanks out with fresh water to ensure that all traces of the ammonia and any ammonia salts, which appear as white crystals on the inner tank walls, are removed.

Specifications and Standards

There are various specifications for LPG's. Typical is that issued by the British Standards Institution, another commonly used specification is that issued by the NGPA (National Gas Producers Association of the U.S.A.). The International Standards Organisation (ISO) also issued standards applicable to the handling, measurement, carriage and analysis of LPG's.

LIQUEFIED NATURAL GAS

Liquefied Natural Gas (LNG) is a clear, water white liquid carried at or just below its boiling point of approximately − 163°C. Its major constituent is Methane (CH_4).

Natural gas is found either in association with crude oil deposits or, for the commercially exploited deposits, in reservoirs which only contain gas and possibly gas liquids.

Marine carriage of LNG requires specially constructed vessels with containment systems capable of withstanding the extremely low temperature at which the cargo is transported. The liquid is maintained in the liquid phase by a very efficient insulation system preventing heat leakage into the cargo. Whilst maintenance of LNG in liquid form by pressure is technically possible it is considered non viable, commercially, due to the high financial commitment required to build both ships and shore plant capable of containing the high pressures involved.

Containment Systems

An LNG carrier has to meet IMO Class I standards of construction; all such ships are double hulled and the containment system is either inserted or built into the inner hull. There are four main types of containment system:—

(1) Conch Self Standing System:—Tanks are built of a low temperature aluminium alloy and either built in the hull during construction or built ashore and lowered into the holds of the ship. Insulation is fitted externally between the tank and the inner hull.

(2) Gaz Transport System:—A system built up on the inner hull consisting of two layers of insulant, perlite in special plywood cases, each layer of insulation having a covering of a high nickel steel metal sheeting, some 1.5 mm thick, called INVAR. The coefficient of expansion if INVAR is approximately 13×10^{-7} m/m/°C; for all practical purposes, zero.

(3) Technigas System:—A system made up of waffled stainless steel plates laid on an insulating system which has a secondary barrier consisting of a thin metal layer some 15 cms below the primary barrier. The waffling is very carefully designed to allow for the shrinkage of the inner barrier metal under LNG temperatures so that the barrier waffles provided sufficient metal to allow the contraction stress to be taken up during the shrinkage process.

(4) The Kvaerner-Moss System:—A system of large spherical tanks which sit in "egg cups" in the ship's holds. The tanks are either built of aluminium or stainless steel. Insulation is attached to the external surfaces of the spheres.

Stowage

Stowage is governed by the rules concerning loading levels; these are set down to ensure that, whilst at sea, the cargo cannot generate large surface waves and cause structural damage to the containment system, especially in tanks of types (1), (2) and (3). These tanks are built with a chined section in the top and bottom which effectively reduces the surface area of any liquid, thus reducing the risk of "sloshing".

Cargoes must be carried with their surfaces in the chined sections. Each vessel is provided with specific instructions concerning the maximum and minimum levels at which she may carry cargo at sea.

LNG vessels do not have to contend with the problems of expansion or contraction of the cargo as do normal oil tankers. Any heating of the cargo will result in the liquid "boiling off" (vaporising) and having either to be processed through the compressors, in an attempt to liquefy the cargo, or, more likely, to be vented to atmosphere.

Cargo Measurement

Measurement of LNG is possible in shore tanks but the loading or discharge process requires that the product is handled in a "closed circuit" to avoid introduction of air (oxygen) into the system; any over pressurisation of the system is corrected by returning liquid/vapour to the cooling process, or burning it at the shore flare. Because of the difficulties in accurately measuring the quantities returned to shore it is normal practice to rely on the ship's calculated cargo for the Bill of Lading quantity.

LNG carrier's cargo tanks are accurately calibrated as are their cargo measurement systems which, by the nature of the product, are all automatic. The methods of calculation are normally contained within the Contract of Sale/ Purchase/Carriage but there are measurement and calculation standards published by the International Standards Organisation (ISO), The Institute of Petroleum (IP) and the American Petroleum Institute (API).

Most LNG vessels will use any "boil off" gas as fuel in the main boilers whilst at sea and, in fact, most sea passages are fuelled by the "boil off". This use of the cargo for ship's purposes is acknowledged in the carriage contracts, the amount used being the difference between the cargo quantity loaded and the cargo on arrival at the discharge port.

The vessel will retain a small "heel" of cargo for use in cooling down the cargo tanks on the ballast passage. This saves time, effort and LNG prior to commencement of loading. Again, account is made of the quantity retained on board and an adjustment made when calculating the cargo delivered.

Handling

Great care and attention to detail must be given to handling this product. In its liquid form its temperature is so low that any contact with the normal metals from which ships are built will result in instant metal failure, with potentially catastrophic results for the integrity of the hull. Special care must be taken to ensure that any loading or discharge connections between ship and shore are fully drained and contain no liquid whatsoever before they are opened. Drain/vent openings are provided and both ship and shore should make certain that the lines are completely free of any liquid.

Monitoring

All LNG ship's are provided with an automatic monitoring system which samples all spaces surrounding the cargo containment system for the presence of LNG vapours. Some are provided with "cold spot" temperature detection, but it may be necessary to institute a system of visual checks of the outer surfaces of the inner hull for "cold spots". These easily can be detected by ice formations which occur in the vicinity of any containment system breakdown. Great care must be exercised when entering the spaces between inner and outer hulls and the Regulations for Entry into Closed Spaces must be fully observed.

LIQUID CHEMICAL CARGOES

Introduction

The wide range of liquid chemical cargoes transported in bulk by tank-ship can be divided into the broad categories viz: chemicals which are liquid at normal carriage conditions and those which are gases under ambient conditions but are liquefied for bulk shipment purposes by the application of higher pressure or lower temperature (eg. Chlorine & Ammonia respectively). Liquefied Petroleum Gases (LPG), most notably Propane and Butane are usually classified separately.

As the name 'bulk chemical' implies, the subject commodities are generally intended for further downstream processing in the chemical industry, often involving expensive and sensitive catalysts which are susceptible to irreversible damage by contaminants in the feedstock. Consequently, most chemical commodities are manufactured to a high level of purity and custody transfer schedules applied during bulk shipment encompass stringent measures for cargo quality control.

GUIDELINES ON CARE OF CHEMICAL CARGOES DURING SEA-BORNE CARRIAGE

Key Points to Consider Prior to Loading

Prior to loading, careful planning is essential. Onboard consideration must be given to a number of technical and operational factors. A summary checklist of key points is given below.

Tank Capacity: Check that there is sufficient volumetric capacity, making due allowance for possible expansion of cargo due to likely increase in temperature.

IMO Tank Type: Select correct tank type (IMO I, II, or III) based upon cargo requirements.

Tank Coating: Refer to tank coating resistance list issued by manufacturer. Refer individual enquiries to the manufacturer if necessary. Ensure that the integrity of the tank coating is satisfactory for the safe care of the goods.

Adjacent Cargoes: Refer to the US Coast Guard guidelines regarding compatibility of adjacent cargoes. Check whether heated adjacent cargoes are present and, if so, if the situation is acceptable.

Temperature Requirements: If the cargo is to be heated or cooled, stow in a tank capable of performing the task. Test the system before loading the cargo. Establish that the stowage is acceptable with respect to adjacent cargoes.

Tank Cleanliness: Ensure that cargo tank cleanliness is adequate. Refer to in-house guidelines and published tank cleaning references.

Vessel's Trim: At all times, the vessel must operate within acceptable operating limits for trim and stability; factors which must be considered for proposed cargo loading and discharge rotation of cargo parcels.

FURTHER EXPLANATORY NOTES

Tank Cleaning and Inspection

Cargo tank cleaning recommendations for most commonly shipped bulk liquid chemical products are set out by Dr. A. Verwey Chemical Laboratories and Superintendence Company reference manual 'Tank Cleaning Guide' (latest edition 1992). The guide has been developed along the rules and recommendations set out in the International Maritime Organisation (IMO) Code for the Construction and Equipment of ships Carrying Dangerous Chemicals in Bulk (BCH-Code) and the IMO International Code for the Construction and Equipment of Ships Carrying Dangerous Chemicals in Bulk (IBC) and lists chemical products contained in those Codes and in Annex II. Appendices II and III of MARPOL 73/78.

Tank Coating

The suitability of a vessel's particular tank coating for coating for loading a specific chemical cargo can be ascertained by reference to the coating manufacturer's "resistance list" which may also advise on special tank cleaning procedures.

The vessel's officers are also expected to have an up to date knowledge of the condition and integrity of the cargo tank coatings, since degraded and/or perforated/missing coating may cause the vessel's tanks to be rejected for loading a particular chemical grade when inspected by an independent cargo surveyor.

Cargo Lines and Fittings

Inspection of the vessel's cargo pumping, piping and tank heating facilities is an essential part of the preloading survey for determining a vessel's suitability to load a sensitive chemical product. Although many chemical parcel tankers have dedicated cargo piping and individual submersible cargo pumps and heating coils for each tank, there still exist possibilities that prior cargo residues, tank washing solutions, water and/or heat exchange fluid may be retained in undrained piping, cargo pump cofferdam spaces and leaking heating coils. To this end it is essential that tank cleaning operations also include thorough line cleaning and draining of cargo piping and heating coils, blowing of cargo pump cofferdam spaces and pressure testing of heating coils and submersible cargo pump seals.

Care During Loaded Passage

Special attention to instruction details concerning chemical cargo stowage conditions during the loaded voyage is essential in order to maintain product quality integrity which may be affected by variations in conditions of temperature, inhibitor levels, moisture and/or oxygen content of the containment tank headspace vapours.

Certain chemical products are shipped with specific stipulations with regard to the minimum and/or maximum temperature levels for onboard stowage of the grade in question and those parcels stowed in adjacent cargo tanks.

Chemicals which are sensitive to degradation by oxygen are usually required to be loaded into tanks which have previously been purged with Nitrogen to a specified reduced oxygen level in the cargo space atmosphere. Further, the inert atmosphere has to be restored on completion of loading and, by routine monitoring, must be maintained inert throughout the voyage.

Chemically unstable products, notably those which polymerise during passage of time are normally shipped with an inhibitor added to prevent the reaction. The inhibitor system may require specific tank atmosphere conditions and sometimes there is a need to add extra inhibitor when, during a long voyage, the routine cargo tests performed by vessel's officers indicate decline in inhibitor activity. It is essential to monitor the temperature of such cargoes.

Safety Advice

There is an extensive body of information and data sheets on safety and emergency procedures in the handling of hazardous and noxious chemicals. IMO has published an 'Index of Dangerous Chemicals Carried in Bulk'. The index, prepared by the United Kingdom Laboratory of the Government Chemist, lists chemicals referred to in the BCH and IBC Codes (Chapters VI, VII and XVII and XVIII respectively) and provides alternative names and abbreviations by which certain chemicals are known. In addition to providing the chemical United Nations serial number (UN No.) (where assigned), where possible the number assigned by IMO 'Medical First Aid Guide for Use in Accidents Involving Dangerous Goods' (MFAG No.) is also provided for easy reference. Another source of health and safety data on hazardous chemical cargoes is the International Chamber of Shipping (ICS), "Tanker Safety Guide (Chemicals)" which contains individual data sheets on approximately 300 bulk chemical products. The data sheets also provide details of fire and explosion data, physical and chemical data and handling/storage recommendations on the individual chemical grades.

Pollution Prevention

This is an area of intense public concern and legal regulations worldwide are becoming increasingly stringent. The International Convention for the Prevention of Pollution from Ships (MARPOL 73/78) has published regulations with respect to shipboard handling and discharge at sea of harmful substances. Annex II contains details regarding bulk liquid noxious substances and includes liquid chemical cargoes.

All vessels engaged in the chemical trade are required to provide a Procedures and Arrangements Manual (P & A Manual) which is customised to the individual vessel's cargo handling and stowage systems. Amongst several other sections concerning the vessel's equipment, arrangements, procedures and performance characteristics, the P & A manual also lists those chemicals permitted for onboard stowage, relevant restrictions with respect to the pollution category, cargo arrangements and discharge of residues and tank washing solutions to sea and shore reception facilities.

MAJOR CHEMICAL COMMODITIES SHIPPED IN BULK

Caustic Soda (Sodium Hydroxide Solution)

There are several grades of Caustic Soda traded internationally, some of which require special shipboard handling/stowage conditions in order to eliminate product contamination to prevent crystallisation of sodium hydroxide from the liquid phase ("freezing") and to improve pumpability. 'Rayon' and 'Mercury' grades in particular require stowage in perfectly coated tanks (not zinc based) or stainless steel tanks. Some shippers may require protection against iron pick-up. Stowage temperature/heating requirements may vary from no heating in warm climates to heating to 43° C in severely cold conditions. The cargo specific gravity may be as high as 1.5.

Sulphuric Acid

At all concentrations, bulk shipment of Sulphuric Acid requires containment in tanks of stainless steel construction materials. However, the more dilute solutions of Sulphuric Acid (common ones are 65%-75% H_2SO_4) are highly corrosive to certain stainless steel alloy formulations. In all cases, the stainless steel manufacturer should be consulted prior to shipping Sulphuric Acid. Mild steel and coated tanks are not suitable for Sulphuric Acid shipments. Commonly encountered, high purity grades (98-99.5% H_2SO_4) have an SG of about 1.84 and an approximate freezing point of + 5° C.

Methanol (Methyl Alcohol)

Methanol is regularly shipped as a high purity commodity requiring strict preloading assessment of cargo tank suitability with respect to coating compatibility (consult coating resistance list/manufacturers) and coating condition/cleanliness (wall-wash tests for Chlorides and Hydrocarbon residues) in order to comply with the 'water-white' tank cleanliness requirement. Specific Gravity 0.79.

EDC (Ethylene Dichloride)

EDC shipment normally requires that vessel's cargo tank condition is strictly assessed to comply with the 'water-white' cleanliness standard by means of wall-wash testing. Tank coating compatibility should also be investigated by reference to the coating manufacturer's resistance list. EDC is a water-critical cargo requiring thorough drying of vessel's cargo handling and containment systems prior to loading. Specific Gravity 1.25.

Styrene (Styrene Monomer)

Styrene monomer is shipped as an inhibited cargo and requires stowage away from heated cargoes. The product is traded at 99.5%+ purity levels and requires a high level of cargo tank cleanliness ascertainable by visible inspection. Certain types of coatings and bare, mild steel tank surfaces may exclude those tanks from carriage of Styrene (check with shipper and coating resistance list). Monitoring and maintenance of inhibitor levels may be required during long voyages. Post-discharge tank cleaning requires special attention to prevent Styrene residues from polymerising.

Benzene, Toluene, Xylene (BTXC)

Benzene requires a closed loading arrangement, restricted adjacent cargo temperatures (maximum 37°C) and a minimum stowage temperature of 8°C.

Toluene should not be stowed with adjacent cargo temperatures in excess of 38°C.

Xylene has three isomers — ortho, meta and para (o, m, p) for which the maximum adjacent cargo temperatures are respectively 37, 37 and 48°C. P-Xylene has a recommended carriage temperature of 23°C, thus frequently requiring heat.

MTBE (Methyl Tertiary-Butyl Ether)

MTBE, a gasoline blendstock (water critical), cargo tanks generally require visual inspection and checking for water in vessel's cargo handling and containment systems. Certain tank coating materials are not compatible with MTBE and resistance lists should be referred to.

MEG (Mono-Ethylene Glycol)

Polyester Fibre Grades (FG) and sometimes Antifreeze Grade, require wall-wash inspection of vessel's tanks to comply with stringent requirements of cleanliness ('water-white standard'). Carriage under Nitrogen blanket is recommended for FG product during ocean voyage. An adjacent cargo temperature limit of 38°C maximum is to be observed and a high degree of tank coating integrity is essential to prevent deterioration of MEG UV light transmittance characteristics by way of pick up of traces of contaminants.

Phosphoric Acid (Ortho-Phosphoric Acid)

Phosphoric Acid is shipped in several grades (eg Food, Fertiliser, Commercial) which range in purity from 40-98%. Certain grades are corrosive to stainless steel alloy formulations. The commodity should always be shipped in compatible stainless steel stowage. Manufacturers of the stainless steel construction materials should, if necessary, be consulted before accepting cargo. The high purity (food) grades will require heating to prevent freezing whereas the fertiliser grades contain suspended particulate matter and are often delivered to the vessel direct from the manufacturers at high temperature but do not require on-board heating. A maximum carriage temperature of 50°C is specified otherwise corrosion of stainless steel becomes excessive and certain grades are recommended to be circulated during stowage in order to prevent sludge settlement.

Ethanol (Ethyl Alcohol)

Ethanol is shipped at different proof levels (200 and 190) and in denatured and unadulterated purity forms. The Potable grades, frequently require especially stringent quality control and schedules, including "water-white" tank cleanliness and wall-wash inspection together with specialised organoleptic (taste and odour) inspections. Ethanol is aggressive to some types of cargo tank coatings — check resistance list.

IPA (iso-Propanol, iso-Propyl Alcohol)

IPA normally requires "water-white" tank cleanliness and wall-wash inspection and cargoes frequently have an odour-critical specification.

Adjacent stowage heat must be limited to 38°C maximum and coating resistance list must be consulted when considering this sometimes aggressive cargo.

OILS AND FATS

1. *The Nature of Oils and Fats*

The oils traded internationally can conveniently be divided into several categories. From the point of view of end-use, it is useful to divide edible (fatty) oils into vegetable and marine oils, with the vegetable oils comprising those obtained by the processing of seeds, eg. sunflower oil, or of fruit, e.g. palm oil, the oil recovered from the fruit of the oil palm. Marine oils are basically all fish oils, but the oils obtained from different species of fish differ considerably in composition. The category of edible oils also include oils of animal origin, including butter oil, lard and tallow.

A second form of categorisation involves classifying the oils on the basis of the extent of processing which they have undergone prior to shipment. Most oils are shipped in the crude state, but the shipment of refined oils, often described as 'Refined, bleached and deodorised' (RBD) has increased considerably as Malaysian palm oil production has grown. Refined oils are in some respects more sensitive to poor storage and handling conditions than crude oils.

The division of oils into those that are liquid at normal temperature and those that are either semi-solid or fully solid in this temperature range is important from the point of view of the shipper in view of the effect on handling the oil from the time of loading to that of discharge. Many of the seed oils are liquid at ambient temperature and are often referred to as 'soft' oils, whereas oils such as palm oil and coconut oil are sufficiently solid at ambient temperatures to require some heating before being pumped into or out of ship's storage tanks. Castor oil is an unusual seed oil, which is traded internationally by the two most important producing nations, Brazil and India. The chemical structure of its principal component makes if far more viscous than all other vegetable oils and as a result, despite the fact that it remains liquid throughout normally encountered temperature ranges, it requires some heating prior to pumping.

Palm oil and its products, mainly palm oleine and palm stearine, account for a large proportion of the oil traded internationally, mainly using parcel tankers. These oils are shipped for the most part from various ports in Malaysia and Indonesia, but also from Singapore. Large quantities of these oils are shipped to various Asian destinations. The seed oils which are encountered most frequently in international trading are soya and sunflower oils, which are produced in and shipped from South American countries. In addition groundnut (peanut) oil is regularly exported from West Africa. A by-product of palm oil refining, Palm Fatty Acid Distillate (PFAD), is another important cargo.

Although it is possible to transport refined oils over long distances without significant loss in quality, provided proper precautions are taken, oil users in most cases prefer to purchase crude oils, as they can then refine the oil using their own preferred procedure. In the case of palm oil the Malaysian government has encouraged producers to export processed oils, and for this reason large quantities of RBD oils are exported from Malaysia.

2. *Edible Oils and Fats Contracts*

FOSFA International (the Federation of Oils, Seeds and Fats Associations Limited) concerns itself with all aspects of the international trade in oils and oilseeds and provides guidance on many questions relating to the sea-borne transport of oils.

FOSFA International provides contracts for use in the buying and selling of edible oils and fats *in bulk*. The contracts available cover all vegetable and marine oils traded, in most cases on a CIF basis but in some cases also in FOB form. The major role of palm oil products in the international oil trade has led to the development of a contract form specifically for these oils (Contract No. 81) in collaboration with the Palm Oil Refiners Association of Malaysia (PORAM).

The contracts, which may be modified by agreement between the parties to the contract, call for information to be included regarding oil quality, specifications agreed and quantities to be shipped. In the case of Contract No. 80, which covers crude palm oil, analytical characteristics of the oil both at the time of loading and on arrival at the port of discharge form part of the contract documentation. Other clauses in the FOSFA contracts cover tolerance with regard to quantity of oil delivered and the use of superintendents for surveying and sampling. Standards for the vessel, the ship's tanks used for the contracted cargo and conditions for transshipment also form part of the contract. Insurance and shipping documents are covered by other clauses. The clauses on Sampling and Analysis should be read in conjunction with that on Superintendents.

3. *Acceptable and Banned Cargoes*

The International trade in vegetable and marine oils generally entails the carriage of oil from the producer country to the main user countries. Since the user countries are in most cases not shippers of edible oils this situation leads to the dual use of shipping space, vegetable or marine oils being carried in one direction and other liquid cargoes, mainly chemical cargoes, being carried in the other direction. As vegetable and marine oils carried are predominately used for edible purposes purchasers are increasingly concerned about any indication of contamination, even at extremely low levels.

This concern has led FOSFA to create categories of non-oil cargoes for the protection of the end users of edible oils. Two categories of cargoes carried by vessels before loading a cargo of vegetable or marine oil are now recognised. The main category comprises chemicals or other liquid cargoes which are considered unacceptable from the point of view of possible contamination. FOSFA has, therefore, in collaboration with the (American) National Institute for Oilseed Products (NIOP), created a list of Banned Prior Cargoes, and has at the same time produced a list of Acceptable Prior Cargoes. The chemicals on the Banned List are known to have adverse toxicological or carcinogenic properties, and in some cases the banning extends to the two cargoes carried immediately before the carriage of an edible oil cargo.

4. *Ship Suitability, Tank Coatings and Tank Cleaning*

Ship's Suitability

- The ships shall have top class rating e.g. 100A1 from an internationally recognised Classification Society.
- Heating coils should be of stainless steel construction only; copper and its alloys such as brass, bronze or gun metal should not be used for any part of the cargo handling system that has contact with the oils or fats, such as piping, pipe connections (pumps), valves, heating coils etc.
- Tank access/cleaning hatches must be staunch and tight with suitable packing and gaskets compatible with the cargo and its carriage temperature.
- Cargo lines should preferably be of stainless steel constructions with sufficient drain valves to ensure complete clearing and draining of the system. The internal structural members in cargo tanks should be self draining.

Tank Coatings

Tanks, other than those of stainless steel, are normally coated, although mild steel tanks are employed for short sea passages. Only coatings fit for food grade products and suitable for the carriage of the oils or fats to be loaded shall be used.

- No closed blisters or loose splits should be allowed in a coated tank. If after initial cleaning any blisters or loose splits are present they should be opened by the ship's staff to determine whether any cargo residue remains behind. Areas of mild steel exposure in coated tanks shall be minimal and in all cases shall contain no loose scale.

- Zinc silicate coated tanks are not favoured and should not be used for crude oil unless the acid value is 1 or under. specialist advice should be sought if clarification is required.

Tank Cleaning

If there are tell-tale signs of previous cargo residues, the tank should be re-cleaned. With regard to tank cleaning, comprehensive guidance is published by Verwey "Tank Cleaning Guide" which is well known in the tanker trade. Additionally, manufacturers of proprietary tank cleaning chemicals should be consulted for advice in conjunction with tank coating manufacturers for permission to utilise the various tank cleaning products. If clarification is required specialist advice should be sought.

5. *Cargo Maintenance*

The range of physical properties of vegetable and marine oils, particularly with respect to their tendency to solidify at ambient temperature, means that the conditions for handling and storing oils cannot be standardised for all oils but must be considered and specified for each type of oil. The International Association of Seed Crushers (IASC) has adopted a temperature regime for the handling and storage of a large number of oils, and details of the conditions adopted as suitable are given in Table 1. The sensitivity of all oils to high temperature makes it essential that where oils are to be heated prior to pumping, this should be done with the lowest realistic temperature differential between heating medium and oil. In consequence, extended heating times prior to discharge are necessary (where the consistency of the oil is such as to require heating).

IASC RECOMMENDED SCHEDULE OF TEMPERATURES FOR CARRIAGE OF FATTY FOODS

OIL	During Voyage		On Discharge	
	Min Deg C	Max Deg C	Min Deg C	Max Deg C
Sunflower		Ambient	Ambient	20
Soyabean		Ambient	20	25
Safflower		Ambient	Ambient	25
Groundnut		Ambient	20	25
Rape Seed		Ambient	Ambient	20
Maize (Corn)		Ambient	Ambient	20
Palm	32	40	50	55
Palm stearine	40	45	60	65
Palm oleine RBD	25	30	50	55
Coconut	27	32	40	45
Fish	20	25	30	35
Palm Fatty Acid Distillate	42	50	67	72

6. *Type of Damage*

Damage to a fatty oil cargo, leading to it being in an unsatisfactory or even unacceptable condition on discharge, may be due to one of several causes, these being.

(i) poor temperature control during voyage.
(ii) sea-water ingress.
(iii) contamination with chemicals or other substances miscible with oils.
(iv) adulteration.
(v) admixture.

It is possible to encounter a change in the characteristics of an oil cargo which may appear as damage but is no more than a natural change in the oil. After a period of some weeks a tank full of, for example, sunflower oil will have a deposit of foots on the tank bottom, and a cargo of such an oil may appear cloudy on discharge despite having been clear at the time of loading. In this case the oil has not been damaged, but claims concerning oil quality are sometimes experienced where the *appearance* of the oil has changed between loading and discharge.

6.1 *Damage Due to Inadequate Care of the Cargo*

Even the presence of natural antioxidants in a crude fatty oil will not protect the oil from gross abuse and for this reason guidelines have been set out for the temperatures to be maintained during carriage and discharge of a fatty oil cargo. Exposure of the oil to high temperature can lead to oxidative damage, which has several ramifications. However, some deterioration in quality can be expected during a voyage lasting three to four weeks in conditions where considerable temperature cycling can occur, and for this reason loading of a cargo the characteristics of which are close to the specified upper limit will sometimes lead to the oil at the port of discharge being outside the specification. Oil colour and Peroxide Value are particularly prone to this problem. Normally only a modest increase in free fatty acid content is found over the average length of journey.

6.2 *Seawater Ingress*

Ingress of water into the ship's tanks constitutes another form of cargo damage, though in this case the damage caused is for the most part more readily reversible than in the case of overheating. A combination of high temperature and excessive moisture content may lead to the formation of free fatty acids, which can have an adverse effect on cargo quality.

6.3 *Contamination*

Contamination with chemicals has in the past constituted the most serious problem for cargo owners as well as the shipping industry, and this was particularly the case where the contaminant is documented as being toxic or carcinogenic. This form of contamination almost always arises due to incomplete removal of the residues from a previous cargo, but may also be caused by the use of incompletely cleaned road tankers transporting oil to the loading terminal. Yet another source of contamination is caused by inadequate processing at the oilseed extraction stage — if the oil recovered is not properly desolventised hexane residues will be found in the oil. The advances made in analytical chemistry have made it possible to detect minute traces of most residues, including those from the second-last cargo carried by the ship. The techniques available can now detect some contaminants at a level of 1 ppm and below.

This type of contamination can again be divided into two categories — contamination with a single compounds, e.g. ethylene dichloride, styrene, and contamination with a material which itself consists of a number of compounds. In this latter category diesel oil or other hydrocarbon mixtures derived from petroleum constitute the best and most frequently encountered examples. Contamination with a hydrocarbon mixture derived from petroleum refining presents a complex set of problems, as the principal components comprise an extensive range of molecular sizes, but the mixture normally also contains polycyclic aromatic hydrocarbons which, though present at a relatively low level present considerable difficulty as far as removal is concerned.

Contamination can also take the form of admixture with a fatty oil of another description, sometimes referred to as comingling. This has occurred in the past where a ship has carried two different oils, for example palm oil and palm kernel oil, and insufficient care has been taken to segregate the two cargoes, particularly during loading and/or discharge. This type of contamination obviously does not give rise to any risks as far as oil edibility is concerned, but it may make it difficult to use the oil for the specific purpose originally intended.

6.4 *Adulteration*

Adulteration — mixing of two lower-valued oils to simulate the composition of a more expensive oil — has been known in the past but has been largely eliminated by the development of more sophisticated techniques for the characterisation of oils. By its nature adulteration does not involve the ship, as any adulteration will have occurred before loading.

6.5 *Admixture*

Ships carrying parcels of different vegetable oils must exercise caution to prevent cross-contamination due to faulty valves or a failure to clean lines between discharges of different parcels. Although oils are compatible with one another cross-contamination can have serious consequences for certain uses of vegetable oils and can therefore lead to substantial claims.

PART 3

ALPHABETICAL SECTION

The commodities in this section have been arranged in alphabetical order. Where alternative names exist cross referencing to those names has been made.

The illustrations given against each commodity indicate the major problems that may be encountered when stowing or storing that commodity. Greater detail is given in the paragraph below each commodity.

Those using this book should independently check the commodity with regard to local legislation, international regulations and requirements, and any requirements there may be in the country of destination and those countries which may be transitted. This is particularly important in the case of Dangerous goods.

 FRAGILE

 USE NO HOOKS

 KEEP DRY

 KEEP AWAY FROM HEAT

 MAY REQUIRE TEMPERATURE CONTROL

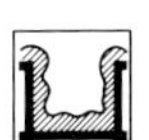 LINE CONTAINER OR ADJACENT DECK

 POSSIBLE TAINT PROBLEM

 POSSIBLE FIRE RISK

 POSSIBLE POISON

 POSSIBLE CORROSIVE

 POSSIBLE INFESTATION PROBLEM

 MAY SIFT OR GIVE OFF DUST

 MAY GIVE OFF MOISTURE

MAY BE DANGEROUS GOODS

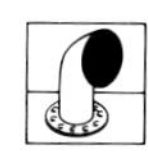 MAY REQUIRE VENTILATING

Commodity | *Characteristics* | *M3 per Tonne* | *Packaging*

ABACA | See Hemp

ABALONE | 2.23/2.37 | Bags
2.51 | Cases

Shellfish. The shell is used for inlay work and the body used for foodstuffs. Dry stowage. May also be carried hard frozen. See Fish, shellfish.

ACETATE OF LIME  | 2.23/2.51 | Drums. Bags

Wet Stowage.

ACETATE TOW | Bales

Synthetic fibre used for manufacture of tipped cigarettes. Must be kept clean and taint free.

ACETONE  | 2.23/2.5 | Drums. Casks.

Colourless liquid valuable as a solvent of organic compounds. Wet storage. Hazardous. See IMDG Code.

ACHIOTE | See Annatto

ACID ARABIC   | 1.11/1.25 | Kegs. Drums

A crude gum of an acid nature; the natural exudation of a species of Acacia tree. Cool dry stowage.

ACID OIL | See Coconut oil products

ACIDS  | See Part 2, Dangerous Goods

AEROPLANES
Light Aircraft  | Cases

Packed in cases. They are shipped in the partially dismantled condition and owing to the light construction and size of the cases require very careful handling to avoid damage.

Sling by both ends, placing the slings in positions which, ordinarily, are marked on the cases and adjusting their lengths so that the case is level and not tilted when slung, using spreaders to avoid crushing the upper edges of the case.

Stow on a solid and perfectly level flat making provisions where necessary for reeving slings underneath the case for discharging without lifting or tilting the case. Secure the chock off well with suitable packages.

Aircraft may also be carried in knockdown form within containers. Particular care should be taken in securing the parts without stressing the fuselage and wings.

See also Motor Cars.

Commodity	*Characteristics*	*M3 per Tonne*	*Packaging*
AGAR-AGAR		1.95/2.09	Bales

A gelatinous substance obtained from certain tropical seaweed. Dry stowage.

AGRICULTURAL MACHINERY		1.39/2.23	Crates/Cases

Crates may be insufficiently robust — avoid heavy overstow. Beware leaking oil.

AJWAN SEED **Ajowan** 	 	2.15/2.23	Bags

An oil bearing aromatic seed from which stearoptine or thymol is obtained also an essential oil. Dry cool stowage remote from tea and edible goods.

ALABASTER	 	1.10/1.28 1.06	Bags Bulk

Exported from Spain, Italy, etc. See Gypsum.

ALBUMEN (MOIST)	 	1.28	Tins in Cases

Wet cargo. Usually shipped in small quantities from Far East. Packed in tins. Sometimes shipped as frozen cargo requiring a temperature of −15 degrees C or lower. See Egg Albumen, also Part 2, Refrigerated Cargoes.

ALBUMEN (DRIED)		1.53/1.62	Tins in Cases

Packed in tin lined cases sometimes carried in refrigerating chambers when a temperature of −15 degrees C is considered to give best results. See Part 2, Refrigerated Cargoes.

ALBURNAM		1.53/1.67	Bags

The hardened heart of wood of the ebony, mahogany, etc., trees; used in the tanning processes.

ALCOHOL	 	2.09/2.12 1.95/2.23	Barrels Drums

Spirit of wine is obtained by the fermentation of substances containing saccharine or sugar. Pure or absolute alcohol has a specific gravity of 0.80 and boils at 78.8 degrees C — flash point 4.4/18.7 degrees C. Wood alcohol flash point 8.7 degrees C. See Dangerous Goods. See IMDG Code.

ALE	  	See Beer	

ALFALFA

A fodder grass. Gives off green dust. Affected by moisture and strong odours. See Hay.

Commodity	*Characteristics*	*M3 per Tonne*	*Packaging*
ALGAROBA		See Carob	

ALKANET		1.34/1.39	Cases

A valuable root from which is obtained a deep red dye. Cool stowage.

ALKYL BENZINE		1.39	Drums
		1.11/1.17	Bulk
			Bulk Bags

Non-corrosive, non-hazardous liquid used in the making of detergents.

Absolute cleanliness of the tanks is essential prior to loading. After discharge, tanks can easily be cleaned by washing down with fresh water.

ALSPICE		3.34/3.62	Bags

See Pimento, also known as Jamaica pepper.

ALMACIGA			

Incense made from the latex of the Almacigo tree. May taint.

ALMONDS		1.95/2.09	Cases
		2.03/2.09	Bags
		3.34	Hogsheads

Mostly shipped from Spain, for unshelled add 20 per cent more space. Dry fine cargo, clean stowage. Bags shipped from India have a stowage factor as high as 2.5/2.65 M3 per tonne.

ALOES		1.11	Cases
		1.31/1.39	Mats

A vegetable product. Shipped chiefly from Socotra, Aden, West Coast of India, Barbados, etc. Are exceedingly bitter to the taste — to be stowed apart from dry delicate goods.

ALPISTE		As for Canary Seed. See also Seeds.	

ALUM	 	1.48	Cases
		1.62	Casks
		1.95	Baskets

A mineral salt. It is colourless and has an acid taste. Dry stowage, well dunnaged, clear of metallic and textile goods which it attacks, also of goods which are susceptible to damage by moisture as this commodity produces a condition of dampness. B/L should protect ship from claims for rotted bags.

ALUMINA, calcined	 	0.61	Bulk

Angle of repose 38-40 degrees. Light to dark grey, small particles, lumps and dust.

Commodity	*Characteristics*	*M3 per Tonne*	*Packaging*
ALUMINA SILICA	 	0.70 0.78/0.84	Bulk Crystals Bulk Pellets

Angle of repose 35 degrees. White to off-white. Very low moisture content.

ALUMINA Trihydrate	 	0.70	Bulk

An aluminium bearing earth or clay. Keep dry. See Ores.

ALUMINIUM		0.80	Slabs/Ingots Wire Bars

Usually in strapped bundles. Liable to corrosion if in contact with concrete for any time.

Separation cloths or Polythene sheeting should be used to protect the aluminium from bagged or bulk commodities.

ALUMINIUM DROSS	   		Bulk

Contact with water may produce flammable and toxic gases. Load in dry conditions.
(See IMO Code of Safe Practice for Solid Bulk Cargoes).

ALUMINIUM NITRATE	  		

Non combustible but if mixed with combustible materials may burn fiercely and give off toxic fumes.
(See IMO Code of Safe Practice for Solid Bulk Cargoes).

ALUMINIUM FERRO SILICON	   		Bulk

Contact with water may produce highly flammable and toxic gases. Condition Certificate must be produced prior to loading.
(See IMO Code of Safe Practice for Solid Bulk Cargoes).

ALUMINIUM SILICON	   		Bulk

Contact with water may produce flammable and toxic gases. Shipment certificates to be supplied by shipper.
(See IMO Code of Safe Practice for Solid Bulk Cargoes).

ALUMINIUM ORE		1.11 0.84/0.92	Bags Bulk

Be careful to note condition of bags and, if necessary, clause mate's receipts and Bs/Lading. See Bauxite.

ALUMINIUM SCRAP, FOIL		1.39	Bags, Bales

Foil, in cases on skids. For cigarette packing. Highly susceptible to taking taint.

Ammonium Nitrate in stow.

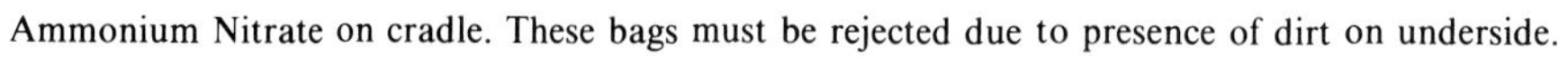

Ammonium Nitrate on cradle. These bags must be rejected due to presence of dirt on underside.

Commodity	*Characteristics*	*M3 per Tonne*	*Packaging*

ALUMINIUM TURNINGS  Drums, etc.

Hazardous cargo when not carried in hermetically sealed tins. See IMDG Code.

ALUNITE 0.70/0.75 Bulk

See Ores.

AMBER 1.00/1.06 Cases
GREY AMBER

A yellow semi-transparent fossil resin, mostly obtained in the Baltic area. Used in the jewellery and pipe making trades. Stow in special cargo locker and watch for signs of pilferage when receiving cargo.

AMBERGRIS 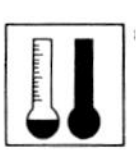  Lined Cases

A solid, inflammable, fragrant substance found on the coasts of, or floating on the sea around India, Africa, Brazil, etc. sometimes referred to as "Grey Amber". Used in perfume manufacture and is a very valuable article. Stow in strong room; shipped only in small quantities.

AMBLYGONITE  0.71/0.84 Bulk
0.78/0.84 Bags

Ore, dusty, but not odorous (similar to Lepidolite).

AMBOYNA WOOD 1.45 Bulk

A valuable wood. Unless B/L claused otherwise, it can be used for dunnaging fine dry cargo or filling broken stowage amongst same.

AMMONIUM NITRATE   1.00 Bulk

Fertiliser. IMO Class 5-1. Check chemical composition. Avoid any possible contact with oil. Risk of explosion. If heated may give off toxic gases. Carefully check specification/shippers description as may also be non hazardous.

(See IMO Code of Safe Practice for Solid Bulk Cargoes).

AMMONIUM SULPHATE   1.00 Bulk

Liable to cake. Corrosive to steel if sweating develops.

AMMUNITION 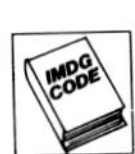 See Part 2, Dangerous goods.
See IMDG Code.

ANATTA 2.42/2.65 Barrels

A dye product shipped in liquid or powder form.

Commodity	*Characteristics*	*M3 per Tonne*	*Packaging*
ANGOSTA BARK		See Barks	
ANILINE DYES	  	1.81/1.95	Casks
Oil		1.50	Drums

A by-product of coal-tar which emits fumes of a very dangerous character. Generally shipped either as oil in drums, or crystals in casks. The oil fumes are very penetrating and taint goods stowed with or near this cargo; furthermore, the oil leaves damaging stains on whatever it contacts with. Ships have been put to very great expense in freeing holds of the taint of Aniline Oil, entire compartments having had to be scrubbed with soap and water. Stow in poop, or forecastle, away from foodstuffs and crew's quarters, and well removed from Bleaching Powder (which see) as the mixture of their gases is dangerous. See IMDG Code.

ANILINE SALT	  	1.62/1.73	Casks, Drums

A poisonous liquid which must be packed in strong and sound containers — it gives off vapour. Any spillage stains. Stow remote from foodstuffs and delicate goods. See IMDG Code.

ANIMAL MEAL	  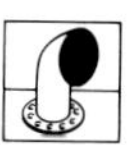	1.45/1.53	Bags

A dirty odorous cargo, stow well away from fine goods. Adequate ventilation. See IMDG Code. (See Seed Cakes).

ANIMAL OILS		See Part 2, Oils and Fats. See Vegetable and Animal Oil Constants. Appendix 11.	
ANIMALS		See Part 2, Livestock.	
ANINE **African Copal**		A yellow transparent Rosin. See Copal.	
ANISEED	 	3.34	Bags

A very light star-shaped seed which gives off a pungent odour. Dunnage used should not be used for subsequent taint-sensitive cargoes. Compartment or ISO container may need special cleaning or ozonating after use.

ANISEED OIL		1.39	Tins in Cases

Obtained by distillation from the fruit of the "Pimipinella Ansium" — shipped from India, Malta, Spain in small lots. Treat as Essential Oils, which see.

ANNATO	  	1.67/1.78	Cases

A reddish yellow pulpy vegetable substance obtained from the seeds of a tree; largely used for colouring chocolate, butter and in the dyeing process. Strong odour; keep away from foodstuffs and goods liable to taint. Cool dry stowage.

Commodity	*Characteristics*	*M3 per Tonne*	*Packaging*
ANTHRACITE		A non-bituminous coal — see Coal.	

ANTIMONY COMPOUNDS  

All are poisonous. Stow clar of foodstuffs. Particular care must be paid to ensuring satisfactory separation from other cargoes. All cargo spaces must be thoroughly cleaned after carriage of antimony. See IMDG Code.

ANTIMONY ORE		0.78/0.89	Drums
		0.56/0.70	Bags
		0.61/0.75	Cases
		0.50	Bulk
		0.84	Bulk (Africa)

May be liable to sifting. Be careful to note condition of bags and, if necessary, clause mate's receipts and Bs/Lading. See also Ores.

APPLES	 	2.37/2.65	Cases, Cartons.

Thorough ventilation is an absolute necessity. Not suitable for stowage in deep lower holds on account of the difficulty of getting adequate ventilation and of the "topweight" on lower tiers. See also Fruit, Green.

Apples are carried under temperature control conditions in—

Cartons, South Africa 2.51/2.79 M3/tonne
Cartons, U.S.A. 2.94 M3/tonne
Cartons, British Columbia 3.21 M3/tonne
Cartons, Australia and N.Z. 2.10/3.07 M3/tonne
Boxes, Australia and N.Z. 2.65/2.78 M3/tonne.

Apples carry best at a temperature of 0.5 to 1 degree C, but up to 4.5 degrees C according to variety, and withstand fluctuations of temperature much better than soft fruits and pears.

Pre-cooled apples and apples loaded at orchard temperature should whenever possible be given separate stowage.

Because of risk of taint, it is recommended that wherever possible apples should not be stowed within the same bulkheads as meat, butter or cheese.

A large proportion of the damage sustained by apples during transit is due to "brown heart" which is a functional disease or disturbance caused by their absorption of or inhaling the carbon dioxide previously given off by the fruit itself. Notwithstanding that to be the case, it has been demonstrated that the presence in the chamber air of about 2 per cent of carbon dioxide exercised a decidedly beneficial effect. These two, apparently, contradictory conditions call for constant attention to ensure that the CO_2 content of chamber air does not fall below or rise above that quantity.

Apples, like all refrigerated goods, require to be stowed on carefully laid dunnage or gratings so laid or designed as to ensure about 50 mm of air space to assist circulation of air; vertical surfaces should also be laid over with 50 mm battens when the cargo is in boxes or cartons. Stow fore and aft and tight against airscreens to prevent short circuiting of air. Requires at least 75 mm from deckhead or overhead fittings.

To prevent bruising of contents, cases should be stowed with the bulge on the side. Careful attention to handling and provision of suitable protective walking boards, etc., is essential.

Dunnage should be laid in line with the air flow and in normal circumstances about 50 mm should be sufficient. In vessels equipped with a blower refrigeration system it should not be necessary for any other additional dunnage. In vessels equipped only with the brine grid system, laths should be laid at least between every second tier of cases.

Commodity	*Characteristics*	*M3 per Tonne*	*Packaging*

Where cartoned apples are concerned, advice should be sought as to the number in height that can be stowed. If deep stowages have to be obtained over the prescribed number in order to use maximum space, false decks, break dunnage or tables should be built to obtain maximum lift.

From Australia, some fruit is carried in bins. These bins are approximately 2500 kg weight. They have the advantage of largely eliminating the considerable bruising which is incurred during the normal handling of apple cargoes. The most advantageous stowage is for these bins, if sufficiently robust, to be block stowed in a deep compartment in order that they can be placed with the aid of a fork lift and discharged in the same manner. Care should be taken to ensure that they are stowed in such a manner that there is no impediment to air flow. A mix of boxes and cartons with bins is an unwise practice, since it severely slows down the rate of loading or discharge.

ISO containers — depending on the packaging, vertical battens every second row may or may not be required. Bins may also be suitable for ISO containers, as may bulk loading under certain controlled conditions. If container loads of apples have to remain for any time without refrigeration (e.g. Port-hole insulated containers without clip-on Units), port-holes should be left open to assist ventilation.

Apples and pears may be stowed together under certain conditions, e.g. when carriage temperatures are compatible. See Pears.

APPLE JUICE — Specific Gravity 1.35 — Bulk Bags
Drums Plastic.

Suitable for carriage in Bulk Bags in ISO general purpose containers, and in liquid containers.

APPLE PEELINGS (DRIED)  — Bags. Cases.

May contain insects. Container may require fumigation after emptying. See Also fruit, dried.

APRICOT KERNELS   — 1.67 — Bags or Cases

Dry stowage remote from moist and odorous goods. May itself take taint.

APRICOTS 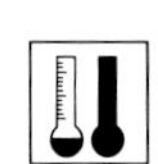 — 2.56/2.63 — Cartons

May be carried as dried fruit — see Fruit, Dried. May be shipped fresh for carriage under temperature controlled conditions. See Part 2, Techniques, Refrigeration.

ARABIC — See Acid Arabic.

ARACHIDES (Groundnuts)

	M3 per Tonne	Packaging
	1.90	Shelled in cases.
	1.81/1.95	In Bags
	2.79/3.34	Unshelled

A ground nut which yields about 45 per cent of oil. See Nuts.

Commodity	*Characteristics*	*M3 per Tonne*	*Packaging*
ARACHIS OIL (Groundnut Oil)		0.89	Bulk
		0.89	Drums

This oil is obtained from certain varieties of ground nuts, such as Monkey Nuts, Peanuts, Arachides (which see) from which it takes its name. It has a very nutty taste and smell, and is of a very pale yellow colour. See Vegetable and Animal Oil Constants, Appendix 11. See Part 2 Oils and Fats.

ARAGOES		1.06/1.11	Cases

A valuable stone use in the jewellery trade, varying in colour from white to deep red. Stow in special cargo locker.

ARCHIL		1.11/1.17	Casks
		2.37/2.51	Bales

Extract of the Archil (orchil) weed, used for dyeing purposes. A weed which must be stowed well clear of moist or wet goods.

ARECA NUTS	  	1.62/1.67	Bags

This, like almost all other nut cargoes is very much given to heating, a case having been known where the temperature of the hold was raised to the extent of 40 degrees C through this cause. If at all damp or wet, they should be rejected as they are totally unfit for shipment in that condition. They may also be infested.

Stow away from heat sources and apart from chillies and from all cargo liable to be affected by humid heat. Good ventilation is a first essential to their proper carriage.

The nuts lose weight up to about 10 per cent, on long passages. See Nuts.

ARNOTTO			See Annatto

ARRACK		2.01	Drums
		1.53	Cases

A spirit obtained by distillation from rice, sugar and the sap of the palm. Shipped principally from Indonesia. Leakage seldom less than 5 per cent; any loss above that is generally debited to ship. Should be carefully watched to avoid pilferage. Special stowage. See IMDG Code.

ARROWROOT		1.95	Cases
		1.48	Bags

Is a starchy food extracted from the tuberous roots of various plants. Shipped in considerable quantities from Ceylon, West Indies, S. America. A very delicate cargo which absorbs moisture very readily. Must be stowed in a dry compartment containing no cargo likely to throw off moisture or odours.

ARSENIC	 	0.67	Cases
		0.84/0.98	Kegs

A tin-white metallic powder. Generally packed in paper lined canvas covered cases. Highly poisonous. See Dangerous Goods and IMDG Code.

Commodity	*Characteristics*	*M3 per Tonne*	*Packaging*

ARSENIC COMPOUNDS 

Stow away from foodstuffs of all kinds and should any packages be damaged and leakage of contents result, the greatest care is to be exercised in removing all traces of arsenic before receiving grain or similar cargo. See Dangerous Goods and IMDG Code.

ASBESTOS    

		M3 per Tonne	Packaging
		1.53/1.67	Cases
		2.51/3.62	Bags
		3.34/3.80	On Pallets

A fibrous white or grey mineral product which is incombustible and stows well with the finer cargoes. When shipped in sheets packed in crates, stow on edge and avoid any weight above. See IMDG Code.

Asbestos is the common name given to a number of naturally occurring inorganic silicates with complex compositions and of fibrous crystaline structure. There are four main types:—

1. Chrysotile (known as white asbestos) is the commonest type. It is a fine silky flexible white to grey/green fibre.
2. Amosite (known as brown asbestos) is a straight brittle fibre light to grey to pale brown.
3. Crocidolite (known as blue asbestos) is generally accepted as being the most harmful type of asbestos. It is a straight flexible blue fibre.
4. Anthophyllite. It is a brittle, white to brown fibre.

Unprotected exposure to asbestos dust may produce a risk of severe injury to the respiratory system if inhaled.

Packaging. Asbestos fibre of all types should be securely packed in impermeable bags. If deliveries are made up in to unitised loads these should be securely fastened and may be further enclosed in a cover of impermeable material, properly secured. If ISO containers are used the shipper should ensure that all projections inside the container are removed or adequately shielded to avoid damage to the bags in transit. Ideally should be loaded on last in, first out basis. Stow well away from foodstuffs.

Hooks or other sharp equipment should not be used.

Respirators of an approved type should be worn by those engaged in the collection and re-bagging of any loose materials. Suitable protective clothing should be worn and arrangements made for its cleaning and storage after use. Damaged bags should be over slipped as soon as possible into suitable impermeable bags, a supply of which should be available where asbestos consignments are being handled. The bags should be securely tied, clearly identified and forwarded with the consignment. Any remaining spilled material should be finally cleared either by vacuum equipment or by thoroughly wetting and sweeping. All spillage should be collected and disposed of in accordance with local regulations.

Other transport including ISO containers used for the carriage of asbestos may become contaminated by asbestos. It is recommended that these containers be thoroughly cleaned by means of an approved vacuum cleaner after discharging each load.

Asbestos, and asbestos products, any be considered Dangerous Goods, if they are likely to release significant quantities of asbestos. Such products would require control under the asbestos entry in class 9 of the IMDG Code. These would include raw asbestos, treated asbestos (still in the form of free fibres) and furniture, cushions, etc., stuffed with asbestos.

Non-hazardous products would be cement/asbestos sheet or pipe, brake shoes and furniture where the asbestos is unlikely to escape as free fibres, e.g. when used as a toughening agent in structural plastics rather like glass fibres.

If in any doubt as to the classification the product should be submitted to the appropriate Authority.

ASBESTOS CEMENT 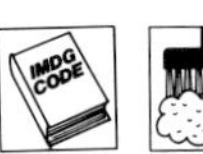 1.78/1.84 Double Ply Bags

A crude fibre mixed with Portland Cement. Very Dusty. See IMDG Code.

Commodity	*Characteristics*	*M3 per Tonne*	*Packaging*
ASBESTOS POWDER		2.23/2.37	Bags

See Asbestos

ASBESTOS SHEETS, TILES, PIPES		1.59/1.73 0.84/1.11	In Cases or Loose Crates

Very brittle and easily damaged if stowed or handled incorrectly. No heavy overstow. See also Asbestos.

ASBESTOS ORE		1.95	Bags

See Ores.

ASPARAGUS (Canned)		1.23/1.28	Cases

Tinned vegetable. Cool stowage indispensable to avoid fermentation. See Part 2, Canned Goods. Carried under temperature control in crates and bundles. Carriage temperature 1.5 degrees C.

ASPARAGUS (Fresh)			Trays. Cartons.

May be carried chilled at 0 - 2 degrees C.

ASPHALT Asphaltum		0.94/0.98 1.25/1.39 1.28/1.39	Bulk Barrels Drums

Native asphalt is a mineral resin formed by the natural drying up of rock oil or petroleum in its bed, deposits of which are found in Trinidad, also Cuba, Venezuela, Peru, etc.

Asphalt or Asphaltum (Bitumen) is also obtained from petroleum by distillation and is exported in large quantities from the Caribbean and certain other oil producing and refining countries. It is principally used as paving material and shipped in bulk, in barrels, open ended or otherwise and in drums.

In some form Asphalt is inflammable, see IMDG Code. Owing to its objectionable character the carriage of Asphalt, except in barrels or drums, is frequently excluded from Time Charter party.

Trinidad crude asphalt averages 10 per cent water content.

Bitumen is graded R.C. 5 to R.C. 0. All to a greater or lesser extent being diluted with naptha.

R.C. 5 to R.C. 4 can only be carried in especially constructed tankers. Carrying temperature being about 132 degrees C.

R.C. 3 to R.C. 0 can be carried in bulk in normal tank ships. Carrying temperature being about 140 degrees C.

R.C. 3 and R.C. 0 is also carried in Drums or Barrels. Many of the drums are open headed and must of course, be stowed on end. Cool storage is desirable and remote from all edible and other goods liable to taint. Holds should be limewashed before loading. Care should be taken during handling to avoid "leakers". This is a most difficult cargo to clean up after, if the contents leak, the naptha evaporates in the atmosphere and the residue sets solid. All bilges must be thoroughly protected. Cases have been known where the asphalt or bitumen has even penetrated and filled the bilge suction pipes which had to be renewed. Drums should be well dunnaged every tier and packed hard with dunnage or cordwood to avoid any movement of the stow. 200 litre drums should be restricted to eight high.

Commodity	*Characteristics*	*M3 per Tonne*	*Packaging*

Emulsified Asphalt — is asphalt rendered into liquid form which involved the addition of water, and is not inflammable. Is shipped from U.S.A. ports in steel drums of 250 litre capacity measuring about 0.3 cubic metres and weighing about 230 kg. It is also carried in bulk in deep tanks.

Rock Asphalt — is exported from Cyprus etc., and consists of limestone rock with an average bitumen content of 12 per cent.

ASSAFOETIDA		1.17/1.23	Cases

The sap of a tree grown in India, Iran, etc., which has a very pungent offensive smell and a very nasty taste thus rendering this useful commodity wholly unsuitable for stowage anywhere near or with fine goods of whatever kind. Stow for Essential Oils, which see.

ATTAR OF ROSES **OTTO OF ROSES**			Protected Drums or Tins in Cases

A very valuable extract obtained from various species of roses. It is mainly a product of the East and is generally adulterated with grass oil before packing for shipment.

A very valuable cargo, stow in strong room. See Essential Oils.

AUTOMOBILES		See Motor Cars	

AVOCADO OIL		See Part 2, Oils and Fats.	

AVOCADO PEARS	 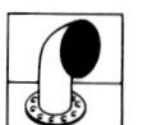	2.50/2.38	Cartons, Trays.

Suggested carriage temperature + 7 degrees C (45 degrees F). However master should seek and obtain specific instructions from shipper regarding carriage. This fruit is time temperature sensitive and all effects are cumulative in the final product presented to the consumer.

There are two compelling reasons why avocados should be cooled promptly and thoroughly after harvest to maintain satisfactory quality—

(a) reduce the rate of respiration.

(b) retard the development of decay caused by fungi.

Avocados are very susceptible to chilling injury, and mechanical damage from rough handling. Fresh air ventilation is required to maintain a low CO_2 atmosphere.

ISO containers — control after discharge. Port-hole closures of insulated containers loaded with avocados should be closed after discharge from the vessel and should remain closed thereafter (while without refrigeration). This is the exact opposite of what is required for citrus fruit.

AWABI		See Abalone	

BACON	 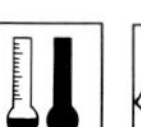 	1.53/1.78	Cases/Cartons

Salted or smoked. When packed in salt it is apt to give off drainings and is to be treated as wet goods. Stow on ground or floor and not on or over dry goods.

When dry, stow in a cool place and avoid overstowing with heavy goods.

May be carried under temperature control, in cases, 1.73/1.84 M3 per tonne. Dunnage as for frozen goods, in cases, battens spaced to suit dimensions and weight of packages.

Liable to give off moisture which if permitted to percolate into the floor insulation will damage same.

Commodity	*Characteristics*	*M3 per Tonne*	*Packaging*

BAGGED CARGO See Part 2, Techniques

BAGGED MEAT — See Meat, Sundries.

BAGONG 

Local name (Philippines) for small shrimps. Treat as shrimps/prawns for carriage and stowage.

BAIRA See Barjari

BALATA  1.39/1.67 Bags

A gum not unlike rubber obtained from the "bully tree" of S. America. Used very largely as an insulating material. Varies very considerably in its stowage factor. Stow in a dry and cool place, well removed from moist articles. See Gum, also Rubber.

BALES See Part 2, Techniques

BALLAST  

Double bottoms, deep tanks and special ballast tanks are the most common means by which ballast is introduced into a vessel. The ballast may take the form of fuel oil, fresh water, salt water. Recent legislation in some countries may prohibit the pumping out of liquid ballast which may be considered to be detrimental or harmful to the maritime environment. The practice of taking on solid ballast has largely disappeared with the exception of permanent ballast. The following may be of historical interest.

In the event of solid ballast being required, reference should be made to appropriate government regulations and advice, particularly where there is risk of ballast shifting (i.e. necessitating the use of shifting boards, etc.).

Solid ballast may be easily installed in the form of concrete blocks which must be properly secured and fixed to prevent any lateral or fore and aft movement. Other forms of solid ballast (which have been used successfully in the past) Includes:—

Blue Billy — A species of quicklime refuse from gas works which is liable to take fire if overstowed with damp material.

Red Marl and Stones — Shipped as dry rubble ballast; is to a large extent soluble in water.

Coal Ballast — Especially anthracite peas, and small nuts. Shifting boards should be fitted.

Bog Ore — a loam containing oxide of iron — very liable to shift.

Sludge Iron Ore — shipped at Spanish ports, etc., very liable to shift. See Sludge Ore.

Flint Stones and single ballast very liable to shift. See Slurry, Mud, Coal or Duff.

If ballasting under dry cargo, the ballasting material should be dry and devoid of objectionable properties.

	M3 per Tonne
Dry rubble makes very suitable ballast	0.70/0.84
Shingle	0.64
Dry Fine Sand (bilges should be carefully protected)	0.53
Coarse Wet	0.56/0.59
Common Earth (see note below)	0.64/0.67
Clay — very messy	0.50/0.67
Gravel	0.56/0.61
Pit Slag	0.50/0.56

Attention is drawn to the danger of introducing foreign insects and plant disease by the use of top soil, such as that obtained from gardens, greenhouses and fields and even sand, as ballasting material. Advice should be sought as to what materials are unacceptable at the country of destination. To ignore such a requirement may involve a ship in heavy expense.

Commodity	*Characteristics*	*M3 per Tonne*	*Packaging*
BALL CLAY **See China Clay**		1.11	Bulk

This is liable to deteriorate into a slimy mass and cannot be trimmed. It cannot be overstowed and the proper cleaning of holds after this clay involves a great deal of labour and of time. A very high degree of hold or compartment cleanliness is required.

May also be shipped in bulk bags of 2-6 m.t.

BALSA WOOD		9.36	Bundles

Grown in Central and S. America, it is lighter even than cork. Used as an insulator, etc. Stow in a dry and cool space remote from moist or oily goods. Do not overstow with heavy goods.

BALSAM COPIVA	 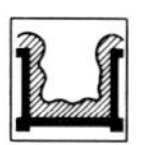	1.67/1.81 1.53	Bottles In Cases

A very bad tasting oil, having a most offensive odour. Wet cargo. Should be stowed remote from all foodstuffs and crew quarters.

BAMBARRA	 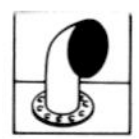	1.81/1.95	Bags

A ground nut, native of Zanzibar and E. Africa. See Ground Nuts.

BAMBOO POLE	 	1.95	Bundles

Liable to split, do not overstow with heavy cargo nor should they be used for broken stowage. May be infested.

BAMBOO REEDS		1.81/1.90	Bundles

Very useful for broken stowage and interlaying bag cargo provided they are not liable to split.

BAMBOO BLINDS		3.34/3.62	Bales
BAMBOO SPLITS		2.79 5.57	Bales Bundles

See Rattans

BANANAS	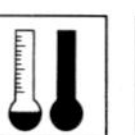 	3.63/3.90 3.62/4.18	Cartons Bunches

Most bananas are now being carried in cartons. The introduction of this method largely being dictated by cheaper handling methods. Cartons are perforated and fitted with handholds to facilitate handling. Cartons usually contain 7/8 cut hands, each being wrapped separately in cellophane. Prior to shipment, sample cartons should be opened to ascertain conditions. Stow on grating dunnage up to 8/10 cartons height.

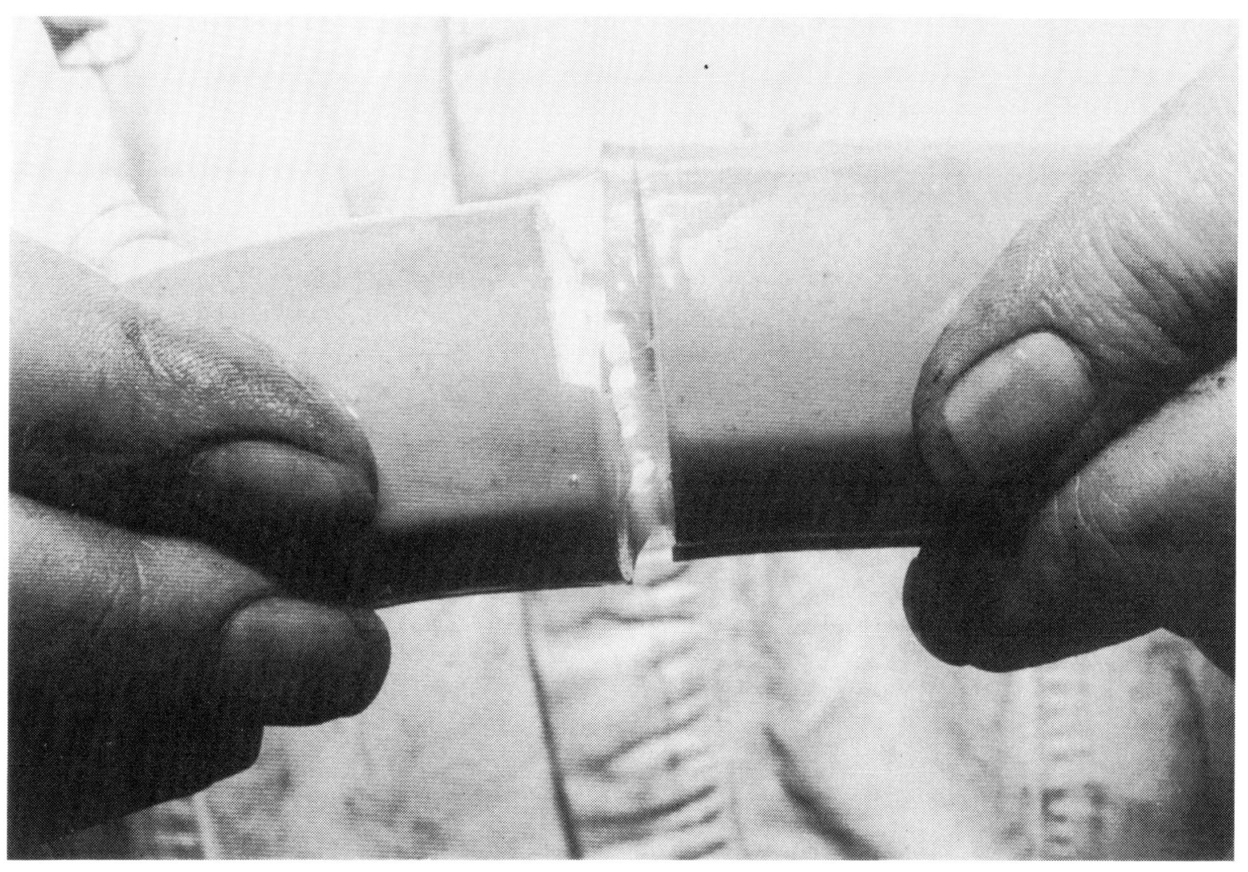

Note latex like strands linking together the two halves. (above)

Testing for ripening

Skin of same banana peeled back. (below)

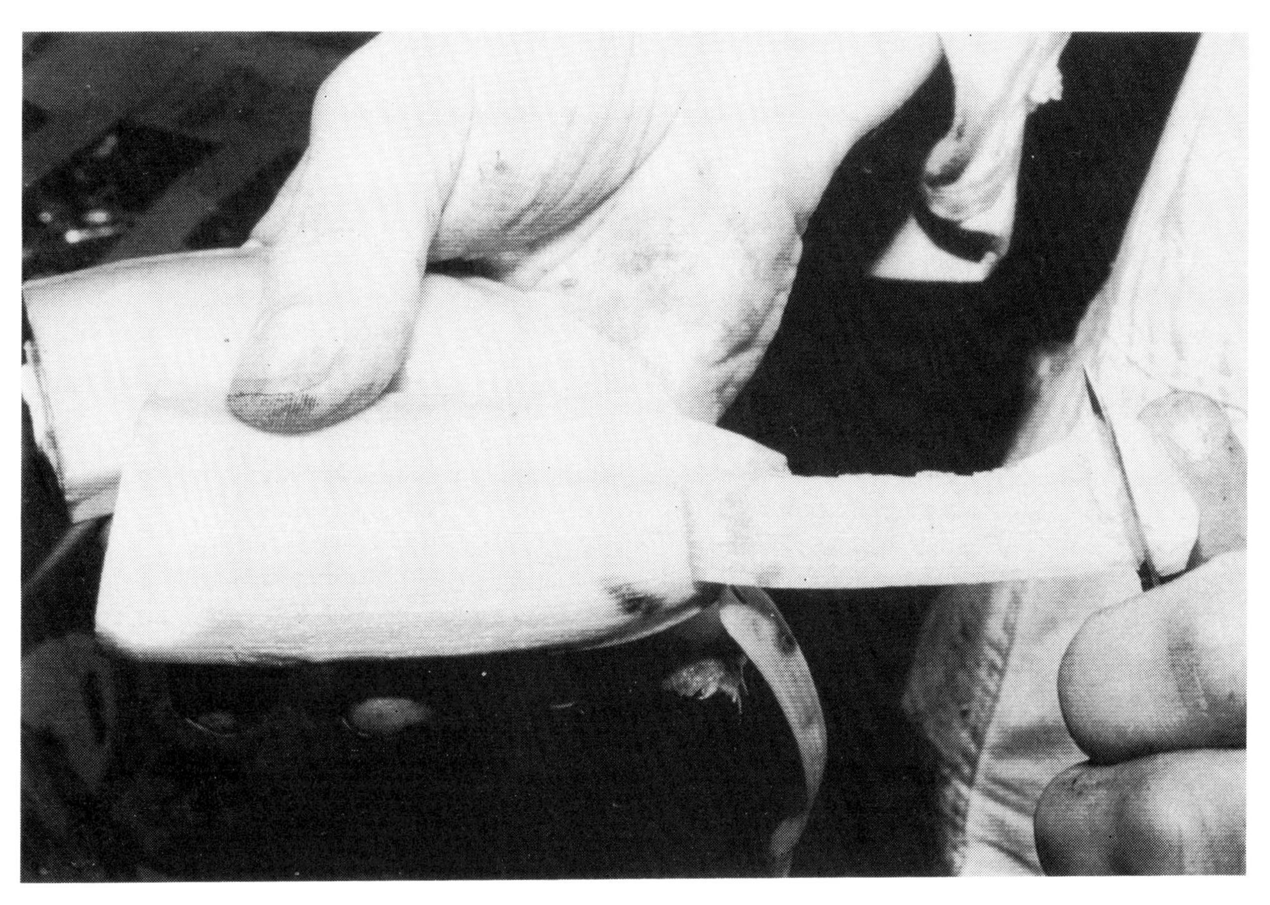

Bananas may be shipped in their natural state attached to the stalks or in straw or paper wrappings, or in perforated polythene bags. Some trades require the hands to be packed in cases. Care also should be taken to ensure that the fruit is not suffering from stem and rot. Pre-shipment condition is vital to successful transport of this fruit, i.e. fruit should be all green (it is not acceptable to just remove single ripened fingers from a bunch. The whole bunch should be rejected even if one banana has turned yellow). Bananas should not be accepted where it is known that they have not been held in a controlled atmosphere and, have been cut 18-24 hours previously. To determine the suitability for transport, sample fingers should be cut in half with a sharp knife. On separating the two faces gently, filaments of a latex-like substance should link the cut surfaces until some distance apart. If this does not happen the bananas are ripening and should be rejected. A further test may be to shave the toe of the banana the flesh should be dry and have a slightly grainy feel. The presence of moisture may indicate the ripening process has commenced. Careful handling is necessary to prevent bruising when picking, grading and packing.

The successful carriage of bananas requires much more fan power than is necessary for meat or other refrigerated carriers, so that while the banana carrier is quite suitable to carry citrus fruits, apples, etc., a vessel designed and equipped solely for the carriage of the latter would not necessarily carry bananas satisfactorily over a long voyage.

Strict attention should be paid to shippers carriage instructions. Normally delivery air should be at + 12.5 degrees C. Once return air reaches + 14 degrees C the delivery air temperature should be raised to 13 degrees C.

Banana carriers which deliver their cargoes in cold northern ports require to be fitted with means for warming the cargo spaces when cold weather prevails, especially when discharging during cold weather.

Some bananas are still shipped in the green condition a few hours after cutting, the cut ends being smeared with vaseline, and are stowed in bulk, bins of portable horizontal sparring being provided to avoid crushing due to the motion of the vessel.

Prior to loading, compartments should be thoroughly cleaned out and the compartment well aired. The compartment should be ozonated and if necessary fumigated in order to remove any trace of odours, more especially from citrus fruits. Because of the large amount of condensation experienced from the coils in carriage, special attention must be paid to seeing that all scuppers are clear.

The bunches are stowed on end, generally two heights with one tier on the flat on top.

While deep stowage is to be avoided, especially for voyages exceeding 10 or 12 days in duration, bananas carry well, under favourable conditions when stowed to a depth of 3 m.

Gratings of a depth which will ensure not less than 75 mm air space should underlie the cargo, and all vertical plain surfaces overlaid with 50 mm battens, to facilitate air circulation.

Every care should be taken to avoid bruising.

The respiration rate increases dramatically after cutting. This produces ethylene and is followed several hours later by carbon-dioxide. Unless these gases are removed the ripening process will continue inexorably.

Spaces should normally be lowered to a temperature of a 4.5/7 degrees C prior to loading; this can be higher in some circumstances. During carriage the temperature should not be allowed to drop below 11 degrees C otherwise bananas are subject to a form of peel injury known as chilling. It should be emphasised that it is essential again to obtain and observe shippers instructions.

It is essential that a detailed and accurate log of delivery and return temperatures is maintained during the voyage. Comprehensive preloading checks should be carried out and recorded in the reefer log. These should include the condition of the cargo holds, clean and free from odours, air ducts free from obstruction, bilge wells clean and suctions freed, cooling system run and proven.

Single cartons weigh about 60 kg.

For voyages of a few days duration, bananas, if harvested at the right time, are carried on deck, loosely stowed on end with good results. In such cases they are entirely at shippers risk. Protect from sun's rays by awnings which, however, should not damp the circulation of air through the cargo.

Cartons of bananas in stow.

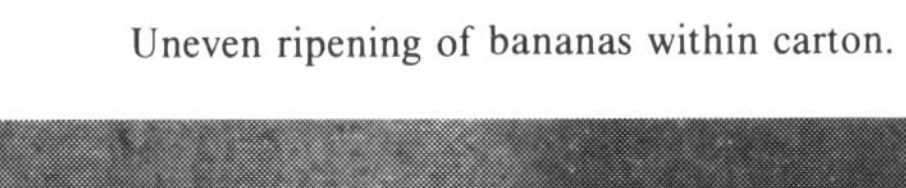
Uneven ripening of bananas within carton.

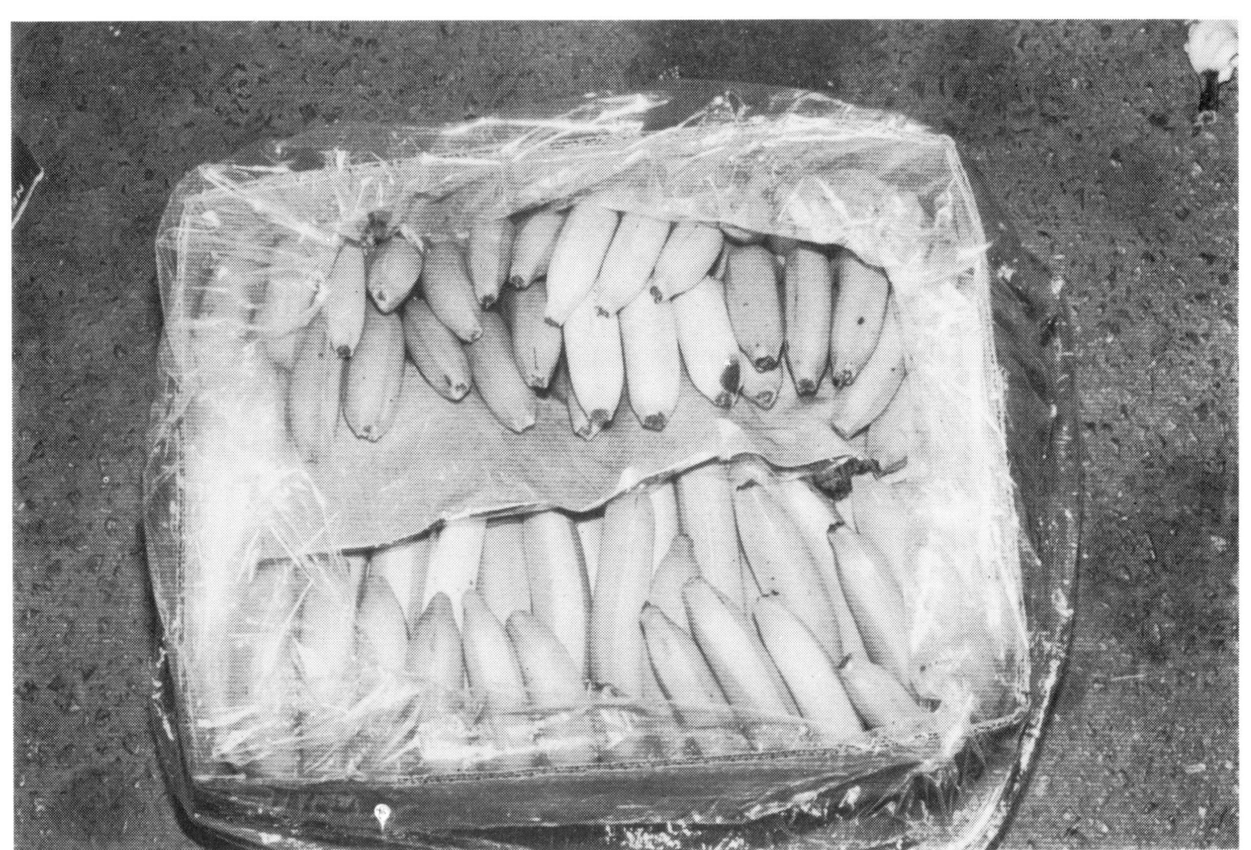

Commodity	*Characteristics*	*M3 per Tonne*	*Packaging*

ISO containers.

Bananas may be shipped in closed box containers without refrigeration for very short journeys. Container doors should remain open and fruit green at time of stuffing. Up to 25% wastage must be expected over 2 or 3 days carriage.

In insulated containers under refrigeration, payload of cartoned bananas in 20 ft unit can be 7.5 tonnes. Suggested carriage temperature for Pacific Island fruit 12 to 13 degrees C air delivery, with air freshening.

Stuffing containers with fruit at field heat may be possible for limited transport distances but is not recommended because:—

(a) It puts a great strain on container refrigeration equipment.
(b) Temperature control of all parts of the stow becomes less likely.
(c) Exact timing between picking and stuffing is too open to chance — e.g. ship delays, harvest dates coming forward etc.

BANQUE See Hasheesh

BARBED WIRE See Wire, Barbed

BARILLA  0.92/0.98 Casks

An alkali obtained from the ashes of a plant largely cultivated in Spain. Dry stowage remote from edible goods.

BARIUM NITRATE     Bulk

Will intensify burning and give off toxic fumes. Toxic. Avoid inhalation of dust.
(See IMO Code of Safe Practice for Solid Bulk Cargoes).

BARJARI 1.48/1.53 Bags

An Indian pea. See also Seeds.

BARKS   2.51/5.37 Bales

A great variety of barks are shipped from various places and for various purposes. Most of these are for medicinal purposes, others for tanning, notably oak bark.

If free from smell may be stowed with fine goods, but all should be stowed clear of odorous, oily, or moist cargoes.

A large number of varieties used for medicinal and other purposes are valuable commodities and being, for the most part, put up in bales, require gentle handling. Do not overstow with heavy cargo.

Australian, ground bark 2.09/2.15 Bags

Australian, chopped bark 2.65/2.79 Bags

Various barks are dealt with herein under their specific names. They are all very light cargo — varying from 2.51 to 5.57 cubic metres per tonne, and should not be overstowed with weighty cargo.

Commodity	*Characteristics*	*M3 per Tonne*	*Packaging*
BARLEY		1.34/1.50 1.45/1.67	Bulk Bags

Is principally used for malting; valuable also as a foodstuff for man and beast, and is extensively grown in most countries.

See IMO Grain Rules.

Bagged barley should be separately from other grain with burlap or other suitable separation.

BAROLYTE			See Witherite
BARRELS			See Part 2, Techniques
BARWOOD			See Camwood
BARYTE		0.34	Bags, Bulk

Industrial Sand, crystalline ore mineral; a sulphate of Barium. Used in paints, textiles and as filler for paper. Angle of repose 37 degrees. See Sand.

BASIC SLAG		See Slag, Basic	
BASIL	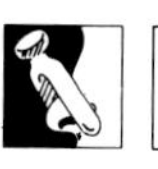 	2.79/3.90	Bales

Aromatic herb used for culinary purposes. Keep dry.

BASILS	  	2.79/3.90	Bales

Soft tanned sheep and lambskins. Liable to damage by wetting. See also Leather.

BATHS		3.07	Crates

Top with light cargo to avoid damage from crushing.

BATCHING OIL		See Part 2, Petroleum	
BATTENS		See Timber Measurements, Appendix 9.	
BATTERIES	   	1.48	Cartons, Boxes

Lead Acid. Nickel Alkaline. Dry Cells. Vulnerable to damage by excessive heat or damp. If filled with acid may be subject to D.G. regulations. See IMDG Code.

BAUXITE **ALUMINIUM ORE**		0.98/1.11 0.70/0.85	Bags Bulk

The principal ore of aluminium. An earthy ore, both lumps and powder. Moisture content may be 0 to 10 per cent.

Commodity	*Characteristics*		*M3 per Tonne*	*Packaging*
BDELLIUM			20.9/2.23	Cases

A pungent aromatic gum resin known as Olibanum or Frankincense. See Gums.

BEANS		Alexandria	1.34	Bulk
		Black	1.45	Bags
		Castor	1.67	Bags
		Cocoa	21.7	Bags
		Dried	1.39	Bags
		Horse	1.73/1.90	Bags
		Jarry	2.79	Bags
		Lima	1.87	Bags
		Locust	2.34	Bulk
		Soya	1.39/1.53	Bags
		Soya	1.28/1.39	Bulk

Almost all varieties of beans are apt to heat. They very quickly deteriorate if shipped in a damp condition which will cause them to heat, sweat and ferment. If in doubt as to their condition, the B/L should be qualified accordingly. Generally clean, but inspect for bean beetle which will attack bran, etc.

I any case, B/L should contain the written endorsement "Shipped in apparent good condition", if nothing stronger is endorsed. The above applies in a special degree to Soya beans.

Be careful that the bags are not bled and empty bags removed on shore during loading. Many cases of short deliveries are to be attributed to this cause.

Dunnage well and pay careful attention to ventilation. See Part 2, Ventilation, also Bag Cargoes.

BEAN CAKE (Soya Bean)		1.11	Cake Bulk

See Seed Cake.

BEAN OIL		1.08	Bulk
		1.39/1.53	Cases
		1.76	See Part 2, Oils and Fats

Shipped in bulk from Far East, U.S.A., etc. Solidifies at about 12 degrees C.

It may also be shipped in barrels.

The cases, sometimes, are tied with straw cord which renders good firm stowage impossible; the cross hitch at top and bottom pierces the cases above and below under pressure. If the roping is cut off, the cases will stow much firmer, with consequent less leakage.

Leakage of this oil, whether in barrels or cases, is very considerable and great care should be taken, by suitable boarding over, to keep other cargo from getting into contact with the oil containers.

BEAVER BOARD		2.23	Skeletal Cases

Beaverboard and plaster board may be easily damaged if overstowed with heavy cargo. Normally 'tween deck stowage only. Shape and size give poor utilisation in a closed box ISO container.

BECHE-DE-MER Trepang		3.79/4.18	Bale

A large sea slug, caught off New Guinea and East Australian coasts, cured and dried before shipment — generally put up in bales or in bags. Dry stowage, away from goods likely to be tainted by odours.

Commodity	*Characteristics*	*M3 per Tonne*	*Packaging*
BEEF		1.67/1.73	Barrels or Tierces
		1.53/3.76	Cartons
		2.79/3.76	Chilled
	Hard Frozen	2.37/2.79	Frozen
		3.90	Quarters
		1.39/1.67	Frozen/Cartons
		2.65/2.93	Bagged Boneless

Pickled, Salt. Wet Cargo. Stow well apart from dry and odorous goods like turpentine, etc., and in a cool place, not more than 7 heights. See also Part 2, "drums, barrels, etc.".

Carried under temperature control, beef may be chilled or hard frozen; boneless or on the bone; bagged or cartoned. Tempered beef is a combination of frozen and chilled, i.e. the beef is shipped in the hard frozen condition, and stowed in a chilled locker (often in the dead space beneath hanging chilled beef), and allowed to reach the chilled condition during the voyage. The main reason for this technique is to take advantage of the freight rate which does not take into account the dead space mentioned above.

Chilled Beef is probably the most difficult of all meat cargoes to carry successfully. Unlike frozen meat, chilled meat is carried as near to its freezing temperature as possible, without actually freezing it. The reason is that if ice is allowed to form in the meat cells they will rupture, and on thawing allow the blood to escape on which the flavour of a good juicy steak so richly depends. However it is necessary to keep the temperature as low as possible without actually freezing in order to limit the growth and multiplication of micro-organisms — all of which calls for very careful temperature control. Normally shippers require 0.0 - 0.5 degrees C. Pay careful attention to shippers instructions. Since the meat is not frozen it is not hard, and must either be hung or packed in cartons to prevent bruising, crushing and inhibiting the air flow. When hung, special rails — insulated from the locker deck-heads — must be used, and never the overhead grids (if any). If ISO containers are to be used for hung beef, portable rails have to be installed to carry the weight of the meat, and this of course reduces possible payload by adding weight and reducing cube. The meat must not be packed too tight for fear of bruising, nor too slack to allow movement and therefore chafe. Often the fore quarters are stowed around the periphery as they are of less value and more suited to take the brunt of any pressure or movement. If height of compartment permits, quarters are hung on hooks, with fores beneath them on chains.

Because of the nature of the cargo and the possibility of continued mould growth in the relatively high temperature, great cleanliness must be observed at time of preparation and loading. Before shipment, locker or container must be tested for air-tightness, thoroughly cleaned and washed out with an approved disinfectant. After drying the space should be fumigated to remove all bacteria and moulds. During this time the appropriate fans should be run to ensure air trunks and fan spaces are similarly treated. The spaces should be precooled to remove all heat from the insulation — this may be 24 hours before loading in the case of lockers; a few hours in the case of containers. Each space should be loaded as quickly as possible in one continuous movement, then the doors closed and sealed to ensure air-tightness. The space should then be brought down as swiftly as possible to the carrying temperature using full fans to ensure an even temperature throughout. With carrying temperature achieved the fans should be slowed and frequently reversed to ensure even temperature throughout. Too great an air circulation tends to dry out the meat to its detriment. Very fine temperature control is essential and warm brine may be needed on occasion to ensure the cargo never reaches its freezing temperature. The stowage life of chilled beef under these conditions is strictly limited, but may be extended by holding the cargo in an atmosphere of about 10% CO_2 — hence the necessity for air-tightness of the compartment. It is important that at any time of discharge the gas concentration is removed by ventilation before entering.

Chilled beef in cartons is usually boned, and each cut, hermetically sealed in plastic film. This keeps a minute concentration of CO_2 within each pack next to the meat itself, thereby inhibiting growth of mould and bacteria. It is important that this film is never broken. Cartons should be stowed to ensure that the refrigerating air circulates and reaches at least one side of every carton — with the insertion of vertical and horizontal battens where necessary. Because the cartons are not packed out with meat (as is the case with frozen meat in cartons), the cartons are susceptible to crushing and collapsing damage. Tiers should not, therefore, be more than 10 high, and care has to be taken when stowing in ISO containers to avoid those cartons at the sides collapsing into the air channels in the floor.

Carton Beef from South America stows at about 2.51 M3 per tonne.

Hard frozen beef may be shipped in bulk, the quarters being wrapped in hessian cloth. Clean canvas or duraluminium slings should always be used when loading frozen meat. Mouldy meat should be rejected as should meat with a bone temperature warmer than −7 degrees C. Beef being considerably heavier than lamb or mutton should, from consideration of stability be stowed in the lower holds and the latter in the 'tween decks. If stowed in the same compartment the beef should be lowermost.

Quarters of beef should be stowed on the flat, as if stowed on edge, air circulation will be impeded. As fore and hind quarters do not stow well together, better results will follow if they are stowed at different ends.

In stowing at ends or sides, care should be observed to ensure that the shanks do not protrude into and interlay the brine pipes, which if permitted will bring strain to bear on the pipes through settling of cargo as weight is superimposed and the natural settling of cargo occurs, neither should the meat be allowed to get into contact with pipes or insulation.

Stow to reaching height from each end towards the square of hatchway, after which the meat in way of hatchway should be covered with a clean canvas screen and a platform of wood laid over same to land the incoming cargo, repeating as necessary. Walking boards should also be provided and used to avoid damage and disfigurement of meat by walking on it.

The breaking of stowage and the further use of dunnage battens is not necessary with frozen quarters of beef or mutton and lamb carcases, as their regular forms of such, when hard frozen, does not permit of stowage sufficiently compact as to impede air circulation.

If over stowing quarters of beef with other cargo, a separation of timber battens should be made.

Boneless beef in Cartons — laid on dunnage, 50 m clear of floor, to allow cold air circulation. 25 mm board dunnage laid over (and possibly nailed to) to 50 mm. Cartons if bulged are best stowed fore and aft on edge. If overstowed with butter, bagged beef, carcase meat, etc., at least 50 mm dunnage to be laid in line with the air flow over the cartons. Also see Offal, etc.

ISO refrigerated containers — frozen beef must be at or below carriage temperature when being stuffed. No intermediate dunnage normally required if container has permanent floor and wall battening. See Part 2 Refrigeration.

BEER

Commodity	Characteristics	M3 per Tonne	Packaging
BEER		1.50/1.56	Casks
		1.56/1.62	Hogs'd
		1.95/2.09	Bottles
		1.39/1.84	In Cases
			Canned in Cartons
	Specific Gravity 1.005		Bulk Bags

Beer in Casks should always be stowed in a cool well ventilated space, well removed from engine and boiler spaces and from cargo (or bunkers) that are apt to heat, as beer deteriorates if heated. May also be damaged by excessive low temperature, e.g. sub-zero C.

Beer is also readily damaged by strong scent or fumes, which should be safeguarded against when arranging the stowage.

All stained, leaky or damaged casks (whether repaired or not), should be rejected and a sharp watch kept for part-empty casks.

Bottled and canned beer is shipped in cases and cartons. Cartons of bottles are liable to collapse if wet from condensation or breakage. Stow in special locker if possible, in any case, cool stowage is essential. Watch carefully to prevent pilferage.

See also spirits.

Commodity	*Characteristics*	*M3 per Tonne*	*Packaging*

Beer and wine are sometimes carried under refrigeration at shipper's request. Generally wines are carried at 10 degrees C, and beer at about 7 degrees C. Beer and wine must not be frozen. Usually packed in cases or cartons. Watch for broaching.

BEESWAX	  	1.95/2.09 1.81/1.95	Barrels Cases

This commodity melts at about 62.8 degrees C (145 degrees F). It is clean cargo, stow in a dry, cool place.

BELLADONNA	 	2.51/2.65	Bags

Or Deadly Nightshade, a poisonous herb valued for its medicinal properties. Dry stowage apart from moist and edible goods. See IMDG Code.

BENJAMIN		1.39/1.53	Bundles

A gum resin used in the manufacture of incense, benzoic acid, etc. See Gums.

BENNIN SEED		1.45/1.59	Bags

A fine oil bearing seed, shipped from West Africa. See Bag cargo and Grain.

BENTONITE		2.20/1.59	Bags

A collodial clay used as a lubricant mud for oil well drilling and various industrial purposes. Stow away from cargoes that emit moisture.

BENZINE PETROLEUM		See Part 2, Chemical Cargoes.	

BENZOIN		See Benjamin	

BERYLLIUM ORE		0.56 See Part 2 Ores.	Bags

A hard metallic element used in the making of alloys. See IMDG Code.

BETEL NUT		1.67/1.81	Bags

See Areca Nut.

BIDI LEAVES		6.97 and over	Bags

Clean dry stowage, away from tea.

Commodity	*Characteristics*	*M3 per Tonne*	*Packaging*
BILLETS		See Steel	
BINDER TWINE		See Twine, Binder	
BIRIE LEAVES		6.97 and over	Bags

Dried leaves of an Indian tree used as a wrapper for cheap cigarettes. Dry stowage, away from tea.

Commodity	*Characteristics*	*M3 per Tonne*	*Packaging*
BISCUITS— **"Cookies"**	 	Variable	Tins or Cartons

May be damaged by taint, high temperatures (chocolate coated and filled) or crushing. Stow upright to prevent "crumbling".

Commodity	*Characteristics*	*M3 per Tonne*	*Packaging*
BISMUTH ORE		0.84	Bags
BISMUTH METAL		0.28/0.33	Boxes
		0.20/0.22	Ingots

A brittle yellowish or dish white metal used medicinally and as an alloy in the manufacture of solder. Stow clear of fine goods and of copper and zinc. See Ores. See IMDG Code.

Commodity	*Characteristics*	*M3 per Tonne*	*Packaging*
BISULPHIDE OF CARBON		See Part 2, Petroleum.	
BITUMEN	 	1.25/1.35	Barrels
		1.28/1.39	Drums
		1.53	Casks

A Bitumnen tar, an inflammable commodity, shipped in the solid and in the liquid form. Stow away from taintable cargo. See Asphalt.

In some forms it is classified as Dangerous Goods. See IMDG Code.

Commodity	*Characteristics*	*M3 per Tonne*	*Packaging*
BLACK LEAD		1.34/1.39	Paper Lined Kegs

See Plumbago

Commodity	*Characteristics*	*M3 per Tonne*	*Packaging*
BLACK SAND		See Ilmenite Sand	
BLACK SEEDS			Bags

Very fine seeds, will perculate, minutely. Although dry should not be stowed with dried fruit, grain or other bagged cargoes. See Seed.

Commodity	*Characteristics*	*M3 per Tonne*	*Packaging*
BLACK SHEETS		0.56	Envelopes

In bundles and envelopes. Very heavy, care may be needed to spread load in 'tween decks and on ISO container floors.

BLACKWADD — The ore of Manganese, which see

BLACKWOOD		1.35/1.53	Logs

A dark hardwood used for furniture making. Stow clear of grease and oils.

BLEACHING POWDER	  	2.12/2.23	Drums

Chloride of Lime, Sodium and Potash, Calcium Hypochlorite

Packed in plastic or steel drums. A strong disinfectant and deoderant white powder; it releases corrosive pungent fumes (chlorine), and oxygen.

Cases have occurred where packages, carefully stowed, were discovered to be smouldering, eventually breaking into flames. Calcium Hypochlorite is shipped in various degrees of purity. The strongest product (70%) is an oxydising agent and will decompose explosively if heated. It reacts strongly with some types of organic matter. Leaking containers must be rejected and all spilled material swept up.

The most suitable stowage is "on deck", or in poop, or forecastle well clear of crew's quarters.

See Aniline Dyes and Oils. See IMDG Code.

BLENDE		0.56/0.67	Bulk

The principal ore of Zinc. See Part 2 Ores. See Appendix 13.

BLOOD, DRIED    See Dried Blood

BLUE ASBESTOS — See Asbestos

BLUE COPPERAS — See Copper Sulphate

BLUE VITRIOL — See Copper Sulphate

BOLTS AND NUTS		0.47	Casks. Bags.

Very heavy. Distribute weight as necessary — particularly in ISO containers.

BONES	  	1.67	Bags

Generally shipped in plastic or hessian bags. B/L should be claused "Not responsible for loss of weight".

Commodity	*Characteristics*	*M3 per Tonne*	*Packaging*

Bones accepted for shipment should be clean, dry and free from extraneous matter and may have to be accompanied by a Sanitary Certificate visaed by the Consul of the country of destination. Stow in well ventilated spaces away from foodstuffs. Thoroughly clean and ventilate upon completion of discharge. See IMDG Code.

BONE ASH	 	See Fertilisers.	Bags

Very Dusty. Heat when moist. See IMDG Code.

BONE GRIST	 	1.81/1.95	Bags
BONE MEAL		1.11/1.25	Bags

Dry stowage, clear of edible goods. Shipped in bags, condition of which should be carefully noted. Is very clear in handling, and odorous.

B/L should be claused "Not accountable for cost of rebagging, consequent upon deterioration of bags through the action of their contents".

Stow clear of dry goods and below other cargo to avoid mixing with same owing to the rotting of bags. See Manures.

BONE MANURES		1.67/1.35 See Fertilisers	Bags

BORACIC ACID		1.91	Barrels
		1.59	Cases

Quite harmless cargo. Ordinary dry stowage. Shipped in considerable quantities from West Coast of South America, Canada, etc.

BORAX ANHYDROUS	 	0.78	Bulk

Highly refined white crystalline appearance. Can be dusty. Dust is irritating but not toxic if inhaled. Hygroscopic and will cake if wet. Very abrasive.

BORAX **(Pentahydrite** **Crude, "Rasorite 46")**	 	0.92	Bulk

Fine powder and granules. Dusty, irritating but not toxic if inhaled. Hygroscopic and will cake if wet. Angle of repose 37 degrees. Used as the major source of borax and boron products.

BORAX		1.50	Casks
(Tincal)		1.17/1.25	Bags

Powdered. A chemical compund of Boracic Acid (which see) and soda, used as a flux for metal working and many domestic purposes, including the preservation of foodstuffs. Obtained in California, where it is known as Tincal — F. East, S. America, etc. Keep dry.

Commodity	*Characteristics*	*M3 per Tonne*	*Packaging*
BOUSSIR		1.78/1.84	Bags

A rice offal used as camel food. See Rice.

BOTTLES (Empty)		2.00/4.04	Bags

May be suitable for stowing or stuffing with temperature controlled cargo as "space filler" provided any cross-taint cannot adversely affect packaging. Caution: condensation may form on packaging and contents when unloaded or unstuffed to ambient conditions.

BOXWOOD		1.53/1.81	

This heavy, close grained hardwood, which does not float in water, is exported from Central and S. America, West Africa, etc., in logs approximating Pitprops in length, etc. Also in short billet form varying from 610 mm to 1200 mm in length and up to 152 mm to 178 mm in diameter.

Dry stowage — avoid oil and acid stains and mixing of billets with dunnage wood.

BRAID		See Hemp, also Strawboard	

BRAN		2.79/3.34	Bags
		2.23/2.37	Bales

A very light cargo — usually put up in bags, but sometimes compressed into bales. Dry stowage. Requires to be well ventilated to prevent heating and caking internally as distinct from external caking due to sweat and moisture. Prone to red beetle infection which will attack all agricultural produce. See Rice Bran.

BRANDY		See Wines	

BRAZIL NUTS		2.29/2.37	Bulk
		2.51/2.65	Bags

An edible high value nut used largely for sweet making, etc., it is also oil bearing. Shipped in large quantities from Amazon and other South American ports. Must be carefully protected from rain, and should be rejected if at all in a damp condition. Shipped in bulk and on occasion in bags. When the former is the case and a number of consignments are shipped by the same vessel, it is necessary to keep them from mixing by erecting bins or dividing compartments into sections of a suitable size. These nuts are very liable to heat, and require maximum air circulation to avoid deterioration. See Nuts.

BRICKS		0.56/0.61	Bulk
		0.70	Crates
	Firebricks	0.70/0.84	Bulk

Weigh approximately 3 tonnes per 1000. Stow on ceiling, deck or firm platform; if spar ceiling fitted, board same over to keep the block firm in seaway.

Bricks should not under any circumstances be used for chocking off cargo.

Commodity	*Characteristics*	*M3 per Tonne*	*Packaging*
BRIMESTONE		See Sulphur	
BRISTLES		1.91/2.09	Cases

Pig bristles, etc., shipped in considerable quantities from Far East, Russia, Denmark, etc.

A Sanitary certificate issued by the Health Authorities, endorsed by the appropriate Consular Official is usually necessary for this commodity.

Sometimes packed in cases containing naptha, therefore treat as odorous cargo. Bristles are very valuable and should be carefully guarded to prevent pilferage.

Examine the packages and seals carefully when delivered alongside for any indications of tampering.

Tally on board by marks and numbers and stow in strong room, special locker or amongst fine cargo, provided they can be over stowed with other cargo immediately.

They require the same careful watching on discharge.

BROWNTOP SEED		2.84/3.09 1.39	Sacks Bulk

See Grass Seeds.

BRUNAK **(Poonac)**	  	2.17/2.23 2.09/2.15	Bags Bales

An oil cake, the residuum of the coconut (Copra, which see) after as much of the oil is extracted as possible. Liable to spontaneous combustion. Stow in a cool well ventilated space away from fine goods and do not overstow with weighty cargo. See Oil Cake. See IMDG Code.

BULBS	 		Cases Casks and Cartons

A great variety of bulbs are shipped during the winter season, especially from Holland. Packed in well made cases or cartons, sometimes containing husks, dried mould, sand, etc. Being a perishable cargo it is carried at Shipper's risk and B/L should be claused accordingly. For stowage for long voyages, see Lily Bulbs. There may be claims from receiver that the bulbs have matured on the voyage. Care should be taken to separate real damage from a commercial claim which may be based on the anticipated "shelf life".

BULLION

Gold and silver bullion, i.e. uncoined gold and silver respectively. Shipped in ingot or bar form, and sometimes packed in strong well made boxes which usually are fitted with strong rope beckets for handling. Unless vessel is fitted with a proper strong room or safe of suitable capacity, bullion should not be received on board.

See Part 2, Specials.

BULK, DRY AND LIQUID CARGOES

See Part 2.

Commodity	*Characteristics*	*M3 per Tonne*	*Packaging*
BUNDLES		See Part 2, Timber.	
BURLAP	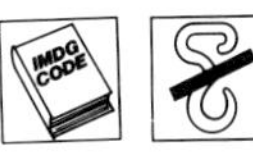	1.76/1.81	Bales

This is the name by which Hessian is commonly referred to by seamen.

Burlap should be carefully dried before stowing away in peaks, etc., after use, otherwise it will heat and quickly rot. Cases are known where men have lost their lives through fumes emitted by this jute material having rotted in badly ventilated compartments. Burlap is very liable to spontaneous combustion if stained with vegetable oil, especially linseed oil. Cloths or bags so stained should be destroyed and never stowed away in peaks or holds.

As cargo, the use of hooks in handling should be strictly prohibited. See Hessians. Also Gunnies. See IMDG Code.

Commodity	*Characteristics*	*M3 per Tonne*	*Packaging*
BUTTER	  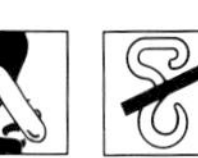 	1.45/1.50	Cases/Tins
		1.67/1.78	Kegs
		1.95/2.09	Tins
	Hard Frozen	1.34/1.39	Cartons

Butter is packed in cases, boxes or cartons, and occasionally in kegs, and is very suitable for lower hold stowage on account of its weight, added to which is the fact that it stows safely with frozen meat.

Australian cartons. Stow at 0.44/0.47 L.H.'s, 0.52/0.55 T.D.'s.

New Zealand cartons. Stow at 0.45/0.42 L.H.'s, 0.52/0.55 T.D.'s.

Butter which shows signs of having melted should be rejected if claims are to be avoided.

Loading temperature should not exceed −1 degree C.

Stow on 50 mm × 50 mm battens, suitably spaced to ensure adequate support to the cases or alternatively stow on 22 mm battens at about 450 mm centres, overlaid with 150 mm × 25 mm boards spaced to ensure adequate support to the cases. Where there is a turn of bilge, it is necessary to build timber bridges or steps to ensure sound stowage.

Plain boxes or cartons of butter stowed in brine cooled compartments should be interlaid with wooden lathes at least every four tiers high.

It is often required that parcels of butter be separated and the most effective way of doing this is to mark off with well diluted colour wash marking paint. Ordinary paints must never be used.

Butter is extremely susceptible to taint; butter therefore should not be stowed in the same compartment as fish, calf vells, rabbits, hares or poultry. It is recommended that whenever possible butter should not be stowed between the same bulkheads as fresh or citrus fruit.

The most prevalent damage to butter — if carried loose — is caused by the practice of dock workers' hooks holing the boxes or cartons during loading or discharging. It is essential to avoid considerable claims, to have adequate supervision in order to check labour from the use of hooks.

Canned butter may also require refrigeration (chilled probably 5 degrees C, 41 degrees F) to maintain texture. Cargo sweat may form on the cans immediately on discharge from vessel or refrigerated container. Adequate ventilation during subsequent storage should be sufficient to remove this without ill effect.

ISO containers — Butter is suitable for refrigerated containers. Must be at or below carriage temperature at time of stuffing. No requirement for intermediate dunnage provided container has adequate fixed battens. Typical payload for 20ft Insulated Container 16.8 tonnes (New Zealand).

Cases of tinned butter, also kegs of butter are sometimes shipped on certain short routes as ordinary cargo. Stow in a cool place and well removed from all odorous cargo, such as fish, manures, turpentine, etc. Very liable to taint. See Part 2, Refrigerated Cargoes.

Commodity	*Characteristics*	*M3 per Tonne*	*Packaging*
BUTTERFAT DRIED	 	2.09/2.23	Cases

Processed butterfat used in the manufacture of foodstuffs, etc. Cool, dry stowage.

BUTTERNUTS 2.79/2.93 Bags

Shipped from America. See Nuts.

CABLE (Chain) See Chain Cable

Stow in holds by flaking athwart.

CABLE (Insulated) 1.11 Drums

Stow fore and aft, on firm platform, as solid as possible, and chock off with timber to preclude any possibility of movement and chafage.

Insulated cable should not be stowed over dry or fine goods, as drainage of the tar from insulating material is to be anticipated in warm weather. The liberal use of sawdust to absorb the tar is advisable to protect wood ceiling.

Sling reels by means of stout bar passed through hole in wood core provided for that purpose. Protect wire if exposed from chafing by overstow. In ISO containers, if possible stow fore and aft. May constitute "overheight" cargo — especially if carried in half height containers. Good lashings and chocking nailed to container floor.

CABOGE 

See Gamboge.

CACAO BUTTER  1.81 Cartons

Obtained from the Cocoa bean. See Vegetable Fats.

CAJAPUT  See Essentail Oils

CAKE LAC  See Lac

CALABA BEANS   2.23 Bags

Poisonous. Stow away from cargoes intended for human or animal consumption. See IMDG Code.

CALAENA ORE   0.45/0.50 Bags

A lead concentrate, shipped from E. Africa. Absorbs moisture and is very liable to heat. Deck stowage. See IMDG Code.

Commodity	*Characteristics*	*M3 per Tonne*	*Packaging*
CALCINED PYRITES (Pyritic Ash, Fly Ash)	  	0.43	Bulk

Must be kept and loaded dry. Harmful if dust is inhaled. Stow away from foodstuffs. (See IMO Code of Safe Practice for Bulk Carriers).

CALCIUM CITRATE			Bags

Non hazardous. Subject to heavy condensation.

CALCIUM CYANAMIDE		See Cyanide	
CALCIUM NITRATE	 	0.91/1.12	Bulk

Fertiliser. Granules consisting mainly of double salt with ammonium nitrate and water. Non-combustible but if mixed with combustible material will burn fiercely and give off toxic fumes. Check specification as commercial grade. May be non-hazardous. (See IMO Code of Safe Practice for Solid Bulk Cargoes).

CALNITRO		See Fertilisers	
CALVES		2.51/2.65	Sides
		2.51/2.79	Qrs.
		2.23/2.51	Bags

Calf carcases are shipped in the hard frozen condition covered in cloth. Inspection and stowage as for mutton, which see. Bags of boneless veal should be given stowage similar to boneless beef, which see.

CAMATA			See Valonia
CAMATINA (CAMATA)			See Valonia
CAMELS		See Part 2 Livestock.	

On deck, at shipper's risk, B/L to be claused "Ship not responsible for mortality".

For short passages in fine weather latitudes they are just tethered to a line spread fore and aft along the decks.

Some of these animals are very vicious and given to biting any stranger within reach. Wounds so inflicted are very apt to become septic, and however slight, should be given early and careful treatment. To prevent accidents of this kind the animals should, where possible, be confined to one side of the deck. Camels should be well watered before embarkation. For other than short passages, allow 30 litres of water, about 2kg of grain, in addition to green foodstuffs per day. See Part 2, Livestock.

CAMOTE			

Local name (Philippines) for Sweet Potatoes. See Potatoes.

Commodity	*Characteristics*	*M3 per Tonne*	*Packaging*
CAMPHINE	  	1.67/1.81	Drums

A very dangerous liquid. Owing to its highly inflammable properties, the utmost care should be taken to keep lights away from its vicinity, smoking etc. It has also the additional property of being one of the strongest smelling articles likely to be offered for shipment — the smell persists long after the stuff itself has been removed. For this reason it is not at all suitable for under deck stowage in a vessel likely to carry fine goods shortly after, and owing to its extreme inflammability, its carriage on deck is attended with considerable risk. See IMDG Code.

CAMPHOR	 	1.67/1.81	Tin Lined Cases

Has a strong persistent odour which can contaminate timber, insulation and ship's paintwork.

CAMPHOR OIL	   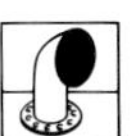	1.95	Drums

A white, solid, semi-transparent oily substance with a bitter taste, obtained from the camphor laurel trees, native to Far East, India, etc. May also be synthetic. Very pungent smell which will spoil any foodstuff stowed in the same compartment; the oil is highly volatile. Should not be stowed in holds. Stow in a dry, well ventilated peak, and if essential oils are stowed in the same compartments, stow camphor oil below and not over essential oils. Residual odour may be discernable for some considerable time after discharge. See Part 2, Obnoxious Cargoes. See IMDG Code.

CAMWOOD (BARWOOD)			

A red dye wood. Dry stowage. Salt water causes discolouration.

CANADA BALSAM		1.48	Casks

A thick gum resin — odorous and bitter. Stow away from foodstuffs and delicate cargoes.

CANARY SEED		1.39/1.67	Bags
		1.36/1.37	Bulk

A small seed.

CANDLES	 		Boxes, Cartons.

Fragile. Avoid crushing. See also Paraffin Wax.

CANELLA ALBA		3.62	Bales

The inner bark of the white cinnamon tree — apt to taint fine cargoes. See Barks.

CANES		2.79/3.34	Bundles or Bales

Tsinglee, Malacca, etc. Dry stowage. See Rattans.

Commodity	*Characteristics*	*M3 per Tonne*	*Packaging*
CANNED GOODS		Variable	

Usually in cartons. May be palletised or on skids. Cans provide inherent strength to carton and the stow should be tight and even to prevent crushing of chafe. See Part 2, Cartons. Subject to damage by sweat. Stow in separate compartment to moisture inherent cargoes. See Part 4, Damage and Claims.

CANVAS	 	1.39	Bales and Rolls

Stow as for carpets. Tarred canvas may be classed as Dangerous Goods.

CAOUTCHOUC		See Rubber	
CAPSICUMS		4.18/4.32	

Dried pods used in cooking. Shipped from East and West Indian Ports in bags, baskets, etc. See Pepper.

CARAWAY SEED		1.67	Bags

The seed of a small plant — shipped from Germany, Holland, England. See Seeds.

CARBON BLACK	  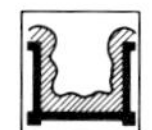	1.67	Bags

See Lampblack. See IMDG Code.

CARBORUNDUM		0.56	Bulk

A hard black crystalline compound of carbon and silicon. Odourless. Slightly toxic by inhalation. Used as an abrasive and for refractory purposes. Angle of repose 40 degrees.

CARDAMOMS		2.65 2.09/2.93	Cases Bags

A plant seed used for medicinal and culinary purposes. Shipped from Sri Lanka, Malabar Coast, Singapore, etc. They are odorous and should be stowed away from tea, tapioca, sago and other delicate foodstuffs.

CARNARINA	 	1.25/1.39	Bags

A manure composed of powdered bones — strong smelling. See Manures.

CARNAUBA		See Vegetable Fats.	

CARNE SECA (DRIED MEAT)
May give off grease.

Commodity	*Characteristics*	*M3 per Tonne*	*Packaging*
CAROB		2.34/21.51	Bulk

A bean shipped in fair quantities from Cyprus, etc. See also Locust Beans.

CARPETS		2.79/3.34	Bales

These constitute very valuable cargo at times. The more expensive carpets are packed in cases or made up in rolls, while the common kind are made up into tightly packed bales. Rubber/foam backed carpets are susceptible to damage by crushing. Should not be stowed more than 3 rolls high. The use of hooks in handling this cargo should be strictly forbidden. Ordinary dry stowage. *ISO container* stowage — bales at front of container and rolls in way of doors allows more speedy container unpacking.

CARTHAMUS SEED		See Safflower	
CARTONS		See Part 2, Techniques	
CASE OIL		See Part 2, Petroleum Cargoes	

This term is applied to cargo, consisting of various kinds of oil — mostly Petroleum Products, classed as "Dangerous" or "Hazardous" Goods — contained in 5 gallon cans packed, usually two in a case the following being the most important of such—

	Size of U.S. Cases	*Approx. Wt. Kg*	*M3/Tonne*
Kerosene	1.6 c.f.	37.19	1.39/1.45
Refined Petroleum	1.6	34.02	1.50/1.53
Petroleum	1.6	36.74	1.39/1.45
Gasolene	1.6	36.29	1.39/1.45
Mineral Lubricating	1.6	40.02	1.25/1.31
Mineral Turpentine	2.0	36.29	1.53/1.56
Benzine	2.0	34.02	1.56/1.59

Holds must be thoroughly clean and provided with adequate means of ventilation before they will be accepted as fit to receive case oil. Clean dunnage only will be accepted.

Electric power mains should be cut off from all compartments in which inflammable liquids are carried. All leaky or stained cases should be rejected.

In all instances, prior to loading low flash point liquids either on or under deck, the IMDG Code or other appropriate reference consulted. Similarly it will be necessary to establish local port authority regulations for ports of call as well as port of discharge. It will also be necessary for vessels transitting the Suez or Panama Canals, etc., to conform to their requirements on stowage and carriage.

The Authorities in most countries require to satisfy themselves as to the flash point of case oil before granting permission for discharge to commence. In some cases they demand samples of each brand to be sent on shore. Time will be saved by arranging for a sample case of each brand to be on hand for landing immediately on arrival.

Petroleum products give off vapour at ordinary temperatures, which when combined with air, forms an explosive and inflammable mixture. The use of naked lights, smoking, etc., should never be allowed in or near compartments containing case oil. Ventilation of holds should receive constant attention and conform to the requirements as set out in the IMDG Code for the particular category of cargo concerned.

Before receiving any delicate or edible goods into a compartment which has recently contained case oil, all oil stains should be removed by the use of limewash and bilges thoroughly cleaned and washed out.

Commodity	*Characteristics*	*M3 per Tonne*	*Packaging*
CASEIN	 	1.84/2.26 1.76/1.84	Bags Cases

The coagulated cheesy substance of milk in powdered form, used for a variety of purposes including as food stuff. Dry stowage — treat as Flour. Obnoxious smell when damp. See Milk Powder.

CASES — See Part 2, Techniques

CASHEW KERNELS		1.95	Tin Lined Cases

Valuable cargo, shipped from East Africa and other tropical ports. The Malabar cases weigh 30 kg, and measure 0.05 M3. Inflammable.

CASHEW NUTS	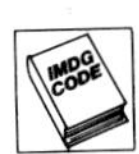		Bags

The kidney shaped, greyish fruit of the Cashew tree. A tropical product containing a highly inflammable black oil. See Nuts. See IMDG Code.

CASHEW NUT OIL		1.48 0.92 S.G.	Cases Drums Bags Bulk

This nut oil is corrosive to the flesh and clothing and requires special chemicals to clean the tanks after discharge. See Vegetable Oils. See Part 2 Oils and Fats.

CASHMERE 

See Mohair.

 **CASINGS**	  	1.39/1.67	Cask Kegs

Animal intestines shipped in the pickled condition packed in barrels, which should be stowed as Wet Cargo liable to leakage; also as Refrigerated Cargo, which see. Cool stowage, liable to fermentation if heated.

Treat as Special Cargo and valuable. *ISO containers.* Line container floor with plastic and sawdust to absorb leakage. Protect container from direct sunlight on board and in the terminal stack.

CASKS — See Part 2, Techniques. Also Appendix 6.

CASSAVA  — See Manioc

Commodity	*Characteristics*	*M3 per Tonne*	*Packaging*
CASSIA **(Cassian Lignea)** **(Cassia Fistula)**		2.79/3.62	Cases, Bundles Bales

The dried bark of the "false cinnamon" used as a cheap substitute for cinnamon. Stow clear of tea, and, as it is put up in frail packages, do not overstow with heavier packages. It is sometimes put up in mats when it comes in very useful for beam fillings with clean cargo. Can be dusty and is sometimes weevil infested. See Barks.

CASSIA BUDS		3.62/3.76	Cases

Are somewhat heavier than the bark. Stowage as for Cassia.

CASSIA LEAVES		See Senna Leaves	

CASSITERITE		See Tin Ore	

CASTOR OIL		1.67/1.73 0.945/0.965 S.G.	Drums Bulk

The composition of castor oil makes it unique amongst the vegetable oils. Major producers of the oil are India and Brazil, with both countries very active in the export market. The oil has a viscosity much higher than all other vegetable oils, and this makes it more difficult to pump than other vegetable oils. It has therefore been recommended that the temperature for loading and discharge should be 30-35 degrees C. The unusual composition of the oil is also the cause of it having a higher density than other vegetable oils, this being approximately 960 kg/m^3 at 25 degrees C. The oil is normally traded with a maximum Acid Value of 2 mg KOH/gm, equivalent to 1% free fatty acid.

Castor oil is used as a lubricant, in the production of cosmetics and a range of chemicals and also in the production of various surface coatings. These applications make it important to avoid discolouration of the oil during handling. See Part 2, Oils and Fats.

CASTOR SEEDS **(Beans)**		1.95/2.37	Bags

The seeds from which castor oil is obtained, which being of a poisonous nature must not be stowed with or near grain and foodstuffs. Heavy claims have had to be met where such stowage was permitted. Separate stowage compartment if possible. Avoid stowing in a 'tween deck when there are foodstuffs in bulk in the lower hold. Clean stowage space or container thoroughly after discharge. Contact with the dust or crushed beans can cause severe irritation to the eyes or skin. Respirators should be used when the dust is agitated. See Seeds.

CATECHU		See Cutch	

CATTLE		See Part 2, Livestock	

Commodity	*Characteristics*	*M3 per Tonne*	*Packaging*
CATTLE MEAL CAKE	  	1.95/2.09	Bags, Bulk.

Can be dusty and odorous. Dry, well ventilated stowage. See also Seed Cake.

CAUSTIC SODA LIQUOR (Sodium Hydrate)	 	0.95	Drums Bulk

Is classified as Dangerous Goods. It is strongly caustic to the skin and attacks organic matter, e.g. wood fibres, clothing, food, etc., also aluminium and zinc or their alloys.

In small quantities it is packed in glass and earthenware containers, suitably protected; in tin cans, also in steel drums of 100 gallons capacity.

Stow on deck or under deck (See Part 2, Techniques, Deck Cargo). Not in contact with ammonia compounds, aluminium, zinc or their alloys.

In Bulk — this commodity can be carried in bulk in Deep Tanks which must be free of aluminium and zinc or their alloys.

On discharge the tanks should be filled with water and pumped out. Any dilute solution left on the sides, etc., of the tank will lose its causticity by chemical reaction with CO_2 in the air, leaving behind carbonate of soda which is harmless. See IMDG Code. See Part 2, Chemical Cargoes.

CAVIARE		1.39/1.50	Cases

The prepared roe of the sturgeon and certain other large freshwater fish. Shipped in considerable quantities from Russia and Iran. A valuable food delicacy which should be protected from pilferage. Dry cool stowage.

CELERY SEED		2.09/2.12	Bags

A valuable seed. Bags should be well made of good material. Double bagged at times. Dry stowage.

CELLOPHANE (Regenerated Cellulose Film)			Rolls. Cylinders.

Stow upright. Ensure reels are not snagged or compressed.

CELLULOID GOODS	 		

Inflammable. See IMDG Code.

CEMENT	  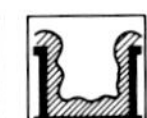 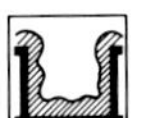	0.72/0.79 0.65/0.70 0.98/1.11 0.60/1.00	Unitised Bags Drums Bulk

Major exporters include Spain, Greece, Japan, Korea, France, Cyprus, Poland and Romania. Large shipments tend to be in bulk rather than bags. When in bulk cargo, should be trimmed to near level stow and cargo allowed to "settle" before embarking on voyage. Many kinds of cement are in use and are obtained from various carbonates, the different kinds varying considerably in their specific gravities and therefore in their stowage factor. Most bag shipments are pre-slung. Clean stowage. See cement, Portland.

Commodity	*Characteristics*	*M3 per Tonne*	*Packaging*
CEMENT BOILER		1.76/1.81	Double Bags

A mixture of vegetable fibre — sometimes asbestos fibre is used — with special heat insulating earths. See IMDG Code.

CEMENT CLINKERS		0.61/0.84	Bulk

Unground cement. Angle of repose 24 to 45 degrees.

CEMENT COLOUR		1.67/1.81	Drums

Ordinary drum stowage.

CEMENT, PORTLAND	 	1.03/1.11	Casks
		0.61/0.64	Bulk
		0.98/1.11	Steel Drums
		0.65/0.70	Bags

When pure, cement sets only in dry air, but when mixed with clay it sets when immersed in water, from which it follows that the latter variety is useful for stopping underwater leakages in ships, etc.

It is a very dusty cargo and fine cargoes should be carefully covered when cement is being worked into or from the same compartments, etc.

Cement is, on occasions, packed in barrels (which should always be paper lined to prevent siftage) in Kraft paper bags or gunny bags (called "pockets") and in steel drums, which, in some cases, weigh as much as 181.44 kg.

The loss of contents when carried in casks is, usually, very considerable, and estimated to be slightly less than 2 per cent, with Kraft pockets.

For the most part large parcels of cement are carried in bulk with suitable mechanical loading and discharging facilities.

When pumped into vessel through hoses, cement flows like a liquid whilst in motion, and it is important that the ship remains upright at all times. On completion cargo should be reasonably level. A period of time should be allowed for the cement to settle and any air to escape. Once settled cement is not liable to shift and shifting boards, or bagging, is not considered necessary. Discharge normally by elevator or grab.

Should sugar have been recently carried in a compartment into which cement is to be loaded, the cleaning of floors, brackets, stringers, etc., should be thorough, and an officer should inspect every part of the compartment before loading commences.

Sugar — to the extent of only .001 per cent — mixing with cement renders the cement worthless as a binding mixture. Sugar should never be stowed on top or above, cement, neither is cement to be stowed with or near Ammonia or its Sulphate as their fumes or gas alters the character of the cement to a quick setting cement.

Pockets and Kraft paper packages are made up to a net weight of 1 cwt or 50 kg according to country of destination. To avoid excessive losses must be handled carefully and must not be dragged over packages in stow.

When overstowing other cargo with bag cement a firm and level floor or platform composed of boards suitably disposed to withstand the weight is indispensable, while care must be exercised not to overstow cargo which will be adversely affected by cement dust with any class of cement package.

To prevent the collapse of ground tier, barrels of cement should not be tiered more than nine high, and the dragging of slings of barrels by the winch from ends and wings of hold is to be avoided.

Commodity	*Characteristics*	*M3 per Tonne*	*Packaging*

Parcels of cement should be separated by mats or old separation cloths, but in most cases when cement is shipped in paper bags, a separation is not necessary because the bags are differently coloured for the various deliveries. Should not be stored in the same compartment/container as sulphate of ammonia, because of resultant fumes.

If not infrequently happens, when cement is loaded direct into ship alongside factory's premises that, the cement, having but recently passed through the kilns, is shipped at a high temperature, cases being known where the temperature varied between 160 and 165 degrees C.

Exception to such a condition need not be taken when loading a full cargo of cement, but, if only parcels of cement are received, it becomes necessary to avoid stowing any other cargo in the same compartment or in one connecting with same which will be affected by the high hold temperature thus artificially produced.

Before receiving cement the holds should be well swept (and the limber boards made dust tight) to enable cement siftage, of which usually there is a great deal, to be recovered as clean as possible. *ISO containers.* Possible to achieve a payload of 18 tonnes in 20 ft closed box containers.

CEMENT, PALLETISED 0.72/0.79

Shipped on pallets with poly-wrapping, may either be strapped or shrunk wrapped.

CERASIN  Cases

A gum which exudes from the cherry and plum tree. See Gums for stowage.

CERASINE 

 1.81/1.95 Barrels

A flammable wax like substance — a petroleum product. See Petroleum Products.

CHAIN CABLE (See Cable)

Stud Link Cable

Approximate weight and space required for stowage:—

DIAM	per 100 metres		DIAM	per 100 metres	
mm	KG	M3	mm	KG	M3
13	389	0.14	43	4000	1.52
18	666	0.26	45	4305	1.62
19	805	0.31	50	5278	2.25
21	944	0.35	55	6667	3.25
24	1250	0.48	60	7445	3.87
25	1417	0.54	65	9111	5.19
27	1610	0.60	70	10472	6.12
30	2000	0.76	75	11973	7.12
33	2445	0.93	80	14084	8.36
38	3166	1.24	85	15195	8.98
40	3444	1.33	90	17000	10.07

CHALK  1.17/1.25 Drums
1.00/1.11 Bulk

Familiar form of limestone, shipped in barrels and in bulk.

Do not stow over fine goods and when carried in bulk protect other cargo from the dust which arise during handling.

Commodity	*Characteristics*	*M3 per Tonne*	*Packaging*
CHALCO PYRITE		0.40/0.60	Bulk

May liquify if shipped wet. Relevant information including moisture content should be obtained prior to shipment.

CHAMOTTE		1.50	Bulk

Burned clay. Shipped in the form of fine crushed stone. Used by zinc smelters and in manufacture of firebrick (road metal). Angle of repose 32 degrees.

CHAR	 	1.60/1.70	Bulk (IBC) Briquettes

Processed and crushed brown coal briquettes. May give off a fine dust.

CHARCOAL **(Wood)**	  	2.79/5.02	Bags. Bulk.

Wood burnt in a high temperature with as little exposure to air as possible. Sometimes packed in bags, baskets, or made up into trusses. Usually shipped in bulk. Very dusty and light cargo: protect fine cargo from dust when handling. Oak charcoal is fairly heavy. Beware coconut husks which may be shipped as charcoal. Not permitted for transport in bulk if class 4.2. If carried in bulk must be accompanied by a certificate from shipper stating cargo is not class 4.2 based on definitive test.

Stow clear of all chlorates, and not over any cargo liable to be damaged by charcoal dust.

This cargo absorbs moisture equal to about 18–20 per cent, of its weight. See IMDG Code. (Also see IMO Code of Safe Practice for Bulk Cargoes).

CHASAM		See Chussums	

CHEESE	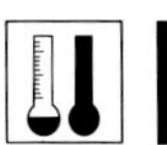 	1.48/1.62 1.00/1.34 1.20/1.25	Crates Cartons Cases

Is either packed in crates having solid heads, either round or elongated in shape, usually fitted with a centre division — two or three cheeses in a crate. Or more usually in cases or cartons. NZ cartons 21.23 kg, stowing at 1.00 M3 per tonne. Carrying temperature 4–7.5 degrees C.

When stowing crates of cheese on an insulated or sheathed deck no dunnage is ordinarily required. However, because of the possibility of heating from the oil in double bottom tanks some shipping lines recommend that in lower holds 150 × 25 mm boards should be laid over the ceiling to support ends and centre of crates. In unsheathed decks 150 × 25 mm boards should be laid to support ends and centres of crates.

When cheese is stowed over a hard frozen compartment beneath, at least 125 mm of dunnage and about 50 mm of sawdust on the deck, should be used to avoid damage from condensation or the effects of the low temperature in the compartment below striking through to the cheese stowage.

Case and carton cheese should be stowed on a floor of dunnage of 50 × 50 mm spaced at approximately 450 mm centres, covered with 150 × 25 mm boards, spaced to give adequate support to the bottom tier of packages.

Commodity	*Characteristics*	*M3 per Tonne*	*Packaging*

When cheese is carried in a compartment immediately below low temperature goods, special precautions are needed to reinforce the insulation of the intervening deck, in order to avoid the cold striking through from the low to high temperature compartment, and to guard against condensation, and its adverse effects, especially when the cheese is in the lowermost position. At least 125 mm of dunnage and some 50 mm of sawdust should be laid on the deck immediately above the cheese stowage.

Cheese in close proximity to brine pipes must be protected from injury by the low temperature there prevailing, by vertical battens over which are nailed boards or hard mats in such a manner as not to impede circulation.

The internal temperature and condition of cheese at loading should be determined and recorded. Cheese is occasionally loaded at atmospheric temperature without any pre-cooling, and it is preferred by many refrigeration engineers to complete loading the compartment and seal hatches before commencing to cool down, this in order to reduce, as much as possible, the formation of snow on the brine pipes.

The roof grids (if fitted) should not be used with cheese if the proper temperature can be maintained without them on. Owing to the relatively high temperature at which cheese is carried the use of roof grids produces "dripping" which renders necessary the fitting of elaborate arrangements of zinc lined wooden trays with pipes leading from same to convey the water to the bilges — a very costly item.

Felt lined wooden troughs are usually fitted to collect the water which is formed on side and end grids with pipes leading to the bilges.

See Preparing Holds for Refrigerated Cargo. (Part 2, Techniques, Refrigeration).

Fluctuations of temperature cause cheese to become friable which reduces its marketable value, and the injurious effect on cheese of freezing or low temperature should be borne well in mind.

Under certain conditions cheese gives off poisonous fumes. Men should not be permitted to enter chambers in which cheese has been for some time before they are cleared of fumes.

Handling — where conveyors are not used in handling cheese, it should always be slung on trays. Avoid rough handling to avoid bruising contents. Damaged packages should be rejected. Every possible supervision must be exercised to avoid hook hole damage to cartons. If the wrapping within the cartons is pierced mould will form and the cheese will have no marketable value. In order to avoid hooks being used to "break out" the stow, it is often the practice to insert clean hessian strips for the first few tiers in the square of the hatch.

Stowage — crates of cheese should be laid in line with the airflow of the compartment to enable the air to circulate through the mass of cheese. Cases and cartons stowed fore and aft.

ISO containers — suitable for carriage in ISO refrigerated containers. Typical payload 20 ft insulated containers 17.5 tonnes (Australia).

Cheese must be at correct carriage temperature at time of stuffing.

CHEESE **Processed**		1.53/1.67	Cases Cartons

Processed cheese shipped in small quantities from Australasia. Cool, dry stowage with adequate ventilation. Avoid stowage in compartments with other cargoes which are liable to give taint.

CHEMICALS		See IMDG Code	

CHESTNUTS		3.34/3.48 5.02/5.57	Cases Bags

From the Far East. Stow in a cool place. Preferably under refrigeration. Carrying temperature 0–1 degree C.

Commodity	*Characteristics*	*M3 per Tonne*	*Packaging*
CHICK PEA		See Gram	
CHICKLE		1.53/1.67	Bags

A gum largely used in the manufacture of "chewing gum" usually shipped in cake form packed in bags. See Gums.

Commodity	*Characteristics*	*M3 per Tonne*	*Packaging*
CHICORY (CRUDE)		1.62/1.67	Bags
(KILN DRIED)		1.78/1.84	Bags
		2.09	Tins in Cases

A plant with a carrot-like root which is shipped in the crude form, kiln dried and in roasted form ground up and ready to mix with coffee. The kiln dried variety absorbs moisture very readily, gains weight and is apt to develop mould if stowed in a damp place. Chicory (fresh) carried as refrigerated cargo at 0–1 degree C.

Commodity	*Characteristics*	*M3 per Tonne*	*Packaging*
CHILLERS (DRY)	  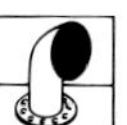	2.65/3.07	Bags
		5.57/6.97	Bundles

Pods of the capsicum annual, shipped from the East and West Indies, etc., in the dry and green condition for use as condiments. Stow in a well ventilated place away from delicate goods, onions, areca nuts and like goods.

Commodity	*Characteristics*	*M3 per Tonne*	*Packaging*
CHINA BARK		See Cinchona	
CHINA CLAY	 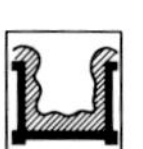	1.06/1.11	Bulk
(Kaolin)			

Is decomposed felspar of granite, sometimes called Kaolin. Shipped from England, Belgium, France, Germany, USA, etc., in bulk or I.B.C.

Charterers stipulate that holds must be perfectly clean to receive clay in bulk. The clay is shipped in the form of China clay, China stone, or Ball clay — which see. Container stowage — particular care must be taken to line the container floor with impervious (e.g. plastic) material, taking the lining up the container walls and taping it in place. This will permit China clay to be carried in Reefer containers — when used on a "general" leg of the voyage.

China clay slurry is usually shipped in specialist carriers on account of its liability to shift.

Commodity	*Characteristics*	*M3 per Tonne*	*Packaging*
CHINA GRASS		1.34	Presses
(Rhea fibre)		1.95	Bales

A nettle fibre used for fine cloth making. See Fibres.

Commodity	*Characteristics*	*M3 per Tonne*	*Packaging*
CHINA ROOT		2.56/2.68	Cases
		2.79/2.93	Bags

Ordinary dry stowage.

Commodity	*Characteristics*	*M3 per Tonne*	*Packaging*
CHINA STONE		See China Clay	
CHINAWARE		3.34/5.57	

Porcelain. Put up in various kinds of packages — crates, cases, casks, baskets, tubs, etc. Very fragile cargo, requires careful handling. Avoid dragging out of wings and ends of compartments by winches — the packages usually will not stand that form of handling — or overstowing with weighty goods.

CHINA WOOD OIL	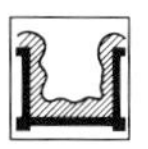	1.78/1.81	Drums
Also see Tung Oil		1.59/1.62	Tins in Cases
		1.07	Bulk

This oil, which readily dries on exposure to the air is extensively used in the paint and kindred industries and in the manufacture of artificial silk, is obtained from the nuts of the tung tree, extensively grown in North China, Japan, etc. It is also produced in Florida, New Zealand and East Africa. The leakage of this oil is considerable and sometimes very heavy, resulting in vexatious claims. A very sharp watch should be kept, when loading this cargo, to detect leaky containers and those which have been but temporarily stopped or patched up. It not infrequently happens that the proportion of recepticles brought alongside which are leaky is very high. All leaky packages should be rejected.

Special care should be exercised in stowing and plenty of suitable dunnage wood should be to hand.

Avoid stowing too many heights — never exceed six with second-hand recepticles, five tiers is enough in the case of some drums.

Over stow with the lightest cargo after boarding the drums, etc., over to keep cargo above from oil damage.

It is best to arrange for a cargo surveyor of good standing to superintend the stowage and to certify same as having been properly carried out.

The condition of drums should be carefully endorsed on mate's receipts and on Bs/L which should contain a clause exempting ship from all claims for leakage and the cost of re-coopering.

Bulk shipments — Wood oil is also shipped in bulk; it has a specific gravity of about .938, solidifies at about 3.0 degrees C and occupies about 1.07 cu. m per tonne. See table of Vegetable Oil Constants.

This oil is capable of holding water in suspension for a very considerable time.

A cloudy condition of the oil, which is sometimes experienced, may be due either to contamination or it may result from the oil having been exposed to very low temperatures, such as is experienced in country of origin, etc. In the latter case normal colour may be restored by raising the temperature of the oil to a certain high point.

Steam can seriously damage the oil and if heating coils are used every care must be taken to ensure that all joints are tight.

If the discharge is to be made under low temperature conditions steam heating will be necessary, otherwise not. For other conditions cleaning, testing and heating of tanks, etc. See Vegetable Oils in Bulk. Owing to the quick drying nature of this oil the tank should be cleaned with paraffin as soon as possible after discharge and before the film of oil left adhering to steel is allowed to harden.

ISO containers — Suitable for carriage in bulk bags inside ISO closed box containers. Compatibility of the oil with the bag material must first be established.

CHINESE GROCERIES 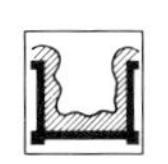 

Wet cargo — generally not too well packed. Be careful to stow marks up and guard against damage to other cargo from leakage, also against broaching and pilfering during loading. Must be stowed in a cool place.

Commodity	*Characteristics*	*M3 per Tonne*	*Packaging*
CHINESE VEGETABLE TALLOW	 	See Vegetable Fat See Part 2, Oils and Fats.	
CHIRETTA		9.75/11.15	Bundles

An aromatic herb valued for its medicinal properties — dry stowage remote from odorous goods.

CHLORIDE OF LIME		See Bleaching Powder	
CHOCOLATE	 	See Confectionery	
CHOW CHOW		See Chinese Groceries	
CHROME ORE **(Chromic Iron)** **(Lead Chromate)**	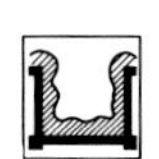 	0.33/0.42 0.42/0.45	Bulk Cases

A heavy ore from which a very brittle almost infusible metal is obtained. When Chrome Ore and Manganese Ore are carried they should be stowed in such a manner that precludes any possibility of the smallest quantity of Chrome getting mixed with the Manganese. Admixture of even a small proportion of Chrome with Manganese renders the latter useless for important purposes in connection with the steel trade. History of stowage prior to shipment should be ascertained as this cargo if stored for long periods in open may contain excessive moisture content which may sift out on passage.

The ore is used in its natural state and is not smelted or otherwise treated before use. It is therefore most important that no foreign matter becomes mixed with it. If loading over especial care is needed in effecting adequate separation. See Part 2, Ores.

CHROME PELLETS		0.60	Bulk

Angle of repose 23 degrees.

CHUNAM		See Lime, Hydrated	
CHURRAH			

A gum extracted from the hemp plant. See Gums.

CHUSSUMS		2.09/2.23	Bales

Silk waste used in the textile trade. Dry stowage remote from moist goods.

CHUTNEY		1.06/1.11	Casks

A condiment much used in the East. Wet stowage remote from odorous goods.

Commodity	*Characteristics*	*M3 per Tonne*	*Packaging*
CIGARS		4.74/5.02	Cases

Very carefully packed in strong cases — require careful watching for broaching. Stow in a dry cool place, strong room for preference, clear of goods such as tea, tapioca, sago, etc., which are likely to suffer from the smell thrown off by these goods. Examine cases carefully before receiving. Cases with damaged seals or with signs of tampering should be rejected.

CIGARETTES		3.34	Cardboard Cartons Tin Lined Cases

See Cigars.

CHINCHONA		3.90	Bales
(Peruvian Bark)		2.79/3.34	Bales
(China Bark)		2.09	Hydraulic pressed

A bark valuable for its medicinal properties. Packed in bales usually and shipped from Bolivia, Peru, also from India, Indonesia, Japan, Mauritius, etc. This bark is free from objectionable properties. Stow in a dry place and clear of odorous goods. See Barks.

CINNABAR		0.50/0.56	Bags

A quick silver bearing ore, shipped from California, Peru, Austria, Germany, etc. See Part 2, Ores.

CINNAMON		3.62/3.90	Bundles
		2.79	Cases

The inner bark of an evergreen tree native of Sri Lanka, shipped also from Indonesia, W.C. of India, China and from S. America. Highly scented. Stow in a dry place clear of all foodstuffs and other cargo liable to suffer from the scent. See Barks.

CINNAMON SEEDS		2.79/3.07	Double Bags

See Cinnamon for stowage.

CITRON		1.95	Bags, Cases.

A fruit which is shipped in considerable quantities from Madeira, either pickled in salt or candied ready for consumption, also as fresh fruit.

Citron oil, obtained from the rind, is used for the manufacture of perfumery.

CITRONELLA		2.09/2.23	Drums

Fragrant etheral oil, obtained from tropical grass.

Commodity	*Characteristics*	*M3 per Tonne*	*Packaging*
CITRUS FRUITS		2.50/2.68	
ORANGES		2.50/2.65	Cartons
LEMONS/GRAPEFRUIT etc.		2.39/2.65	Cartons

Oranges, lemons, grapefruit, etc., are packed in open boxes, or more often in cartons and require the same care and precaution in stowing as are set out under Fruit, which see.

Suggested carriage temperatures, if they have not been provided by shipper or government authority.

Oranges 4–4.5 degrees C.
Lemons 10–11.5 degrees C.
Grapefruit 10–11.5 degrees C. though this may be as low as 6.5 degrees C for fruit from S. America or Australia. See also below.

Carrying temperatures vary with source and variety. Basically it is essential that shippers carrying instructions are obtained and temperatures maintained.

It is often the practice for oranges, lemons and grapefruit to be stowed within the same space. In such conditions carrying temperature should be 6.5/8.0 degrees C. Shippers permission should be obtained for this type of mixed stowage.

Particular care should be taken to ensure that compartments having carried oranges are thoroughly de-odourized after discharge, see Part 2, Techniques. It is a penetrating odour and difficult to dissipate, especially from insulated compartments, and has been responsible for heavy claims for taint damage to such commodities as meat, eggs, flour, butter, etc., subsequently carried in those compartments. Loose dunnage used in citrus cargo stowage must not subsequently be used under other refrigerated cargoes.

Citrus being generally a hardy fruit, may carry for long distances without refrigeration. However ventilation of the fruit is still essential, and where *insulated containers* of citrus have to stand without refrigeration (e.g. port-hole containers without Clip-on Units, port-holes should be left open to assist ventilation.

In certain trades the practice of loading citrus at ambient temperature and reducing to carriage temperature on voyage has been successful. However instructions to this effect must be received from the shipper or fruit authority before such a course of action is embarked upon, and full understanding (in writing) that some fruit may not reach carriage temperature prior to discharge.

Mandarines (suggested carriage temperature + 6 to + 8 degrees C) may be damaged by crushing. Height of stow should, therefore, be limited to about 10 cartons high. When stuffing *ISO containers* care should be taken that cartons do not collapse into the side channels due to the excessive top weight and moisture. Great care at time of picking and packing of this fruit is necessary to ensure good out-turn.

CITRUS PELLETS		1.61/1.64	Bulk

Shipped in bulk in considerable quantities from Florida to Europe for cattle feed. Dry stowage, no ventilation required. May heat to combustion if wet.

CIVET LEAVES	 	6.41/6.97	Bags

Very strong smelling leaves from which is obtained the essential oil Citronella. Dry stowage remote from edible goods and other liable to taint damage.

C.K.D.		2.79/3.07	

See Motor Cars.

Commodity	*Characteristics*	*M3 per Tonne*	*Packaging*
CLAY		0.50/0.61	Bulk

Ballast Clay. See China Clay and Ball Clay.

CLINKER		1.5	Bulk

A partially fused product of a kiln which is ground, usually with gypsum, to make cement. In appearance cement clinker is greyish black, nodules about the size of golf balls. Frequently handled by conveyors.

CLOTHING — See Garments, hanging.

CLOVER SEED	 	1.39/1.53 1.39	Bags Barrels

See Grass Seed. Should be packed in well made or double bags of good material to avoid loss. Must be stowed in a thoroughly dry place otherwise it will quickly deteriorate.

CLOVES	 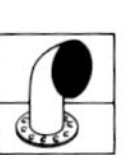  	3.07/3.21 3.07/3.34 3.38/3.42	Chests Bales Bags

The unexpanded flower buds of a small tropical tree. Shipped from Mauritius, Zanzibar, East and West Indies. Are highly aromatic and are particularly liable to damage by moisture.

Stow in a dry well ventilated space, away from tea and all goods liable to be damaged by the odours of cloves — also well removed from moist or wet goods.

ISO containers; may be carried in open-sided containers, with floor dunnage or pallets laid to permit cross-ventilation. Tilts rolled up when weather permits, and stowed on board below decks in spaces where ventilation may be effected.

CLOVE OIL		1.57/1.81	Drums

Liable to taint other cargoes, give separate stowage. Valuable, examine all drums before accepting.

COAL	   	1.08/1.39	Bulk

See IMDG Code.

The carriage of coal is attended by the risk of both fire and explosion. Whilst many millions of tonnes of this commodity are safely carried annually, the attendant risks should not be forgotten.

Coal is a mineral of organic origin, formed from the remains of vegetation which has, over the course of millions of years, been transformed by the effects of heat and pressure from overlying rock or water. As might be expected, therefore, the term "coal" covers a wide range of products with a correspondingly wide range of properties. In spite of this, coals of all types may be carried safely if they are handled, loaded and stowed properly with due regard being given to their particular properties.

The principal hazards associated with coal which are of importance when shipping the commodity are:

1) The potential for self heating, in extreme cases to ignition,
2) The potential to emit methane (a flammable gas),
3) The potential to cause corrosion to the ship's structure.

Discharging coal at Rotterdam.

Commodity *Characteristics* *M3 per Tonne* *Packaging*

Whilst all classes of coal are susceptible to self heating under appropriate conditions, some coals, particularly the lower rank (or geologically immature) coals, have a particular propensity in this regard: the blending of different types of coal may also enhance this property. Very little can be discerned about the potential properties of a cargo simply by visual inspection at the time of loading. Thus, given the wide variability in the properties of coal, it is essential that full details of the specific characteristics of the cargo to be carried, and recommended procedures for its safe carriage, are obtained from the shipper prior to loading.

For cargoes which are described as "liable to self-heat" ventilation should be restricted to an absolute minimum, and it is particularly important that cargoes of this type are trimmed as level as is reasonably possible. An increase in temperature is clearly one manifestation of self-heating and temperature probes in the cargo can be of assistance in monitoring the process. However, the nature of the cargo is such that a hot spot in the stow is unlikely to be detected in the early stages unless there happens to be a temperature probe in precisely the right location. Evidence of self-heating, therefore, is generally most readily detected in the early stages by monitoring carbon monoxide (CO) levels in the hold atmosphere (CO is a by-product of the self-heating process), and equipment suitable for measuring the gas should be carried on board: responsible members of the ship's crew should be trained in the use of the equipment.

Because self-heating is an oxidation process the oxygen concentration in the hold atmosphere under conditions of restricted ventilation is likely to become depleted: it should also be noted that carbon monoxide is toxic.

Coal emits a flammable gas, methane, particularly when newly worked or freshly broken. This gas, when mixed with certain proportions of air, will explode if brought into contact with a suitable ignition source such as an electrical or frictional spark, flame, or heated surface. Again, given the wide variability in the properties of coal, some cargoes will be particularly prone to this. For cargoes which are described as "liable to emit methane", therefore, the greater hazard is that associated with an explosion and precautions must be taken to avoid the accumulation of a flammable atmosphere. Adequate surface ventilation of the cargo will prevent the build up of a flammable atmosphere in the holds although excessive ventilation, particularly into the body of the coal, could promote self-heating. As with coals that are liable to self-heat, therefore, the cargo should be trimmed as flat as is reasonably possible, and equipment capable of measuring the concentrations of methane and oxygen in the holds, without requiring entry to the cargo spaces, should be employed to monitor the hold atmospheres. Whilst specific safety arrangements will vary with each ship, the following general precautions should be observed:–

1. Warning notices against smoking and the use of naked lights should be posted at the entrance to cargo compartments and adjoining spaces where flammable gases may accumulate.
2. Electrical circuits in cargo compartments, mast houses, deck houses, hatch trunks or other spaces where gas may accumulate should be isolated. These circuits should not be reconnected until the space has been adequately ventilated and checked to ensure there is no danger.
3. Torches specifically designed for safe use in potentially flammable atmospheres should be carried on board.

Where cargoes have both a tendency to self-heat and emit methane the more immediate hazard should be considered to be that due to methane and the precautions appropriate to that risk should be adopted.

Some coals, particularly those with a high sulphur content, may react with water to produce solutions which are corrosive to the ship's structure: this risk can be monitored by measuring the pH value (a measure of acidity/alkalinity) of cargo hold bilge samples.

The Master of a ship carrying coal should ensure that the requirements of the relevant Authority are complied with.

If the Master is not satisfied that he has been provided with sufficient information concerning the properties of the cargo, or has reason for concern about the safe carriage of the cargo, he should seek expert advice.

Commodity	*Characteristics*	*M3 per Tonne*	*Packaging*

Explosives are not permitted to be carried in a hold carrying coal in any compartment over that in which coal is carried, nor in one that is not separated from a coal compartment by a steel bulkhead.

Vessels carrying part cargoes of coal, separated from other cargo by bulkheads, should have the latter in a perfect dust-tight condition, otherwise claims for coal dust damage will likely result. Cases have occurred where vessels have been declared "unseaworthy" on it being proven that bulkheads were not dust-proof and of proper construction at the time of loading the cargo.

Coal shipped in a wet condition will turn out about 3% less in weight. B/Ls should adequately cover the ship against any claims arising from short delivery from this cause.

"Pond Coal" is the term used to describe coal which has been reclaimed after having been abandoned and dumped into fresh water ponds. The moisture content is usually high, and it may also have a high sulphur content. This latter can lead to the release of sulphuric acid, with the attendant risk of damage to the carrying vessel.

The Stowage Factor of coal, depending as much as it does upon trimming, varies very considerably for the same class and port. In respect of 2 and 3 deck vessels it is considerably higher than for single deck vessels or the large bulk carrier.

Sample stowage factors are shown below, but these may be exceeded to a substantial degree:—

Origin	*M3/Tonne*	*Origin*	*M3/Tonne*
American	1.17/1.28	North Country U.K.	1.31/1.39
Japanese	1.20/1.31	(small)	1.25/1.34
Lancashire	1.20/1.28	Scottish	1.25/1.34
New South Wales	1.08/1.25	(small)	1.20/1.25
Queensland	1.08/1.20	Welsh	1.20/1.23
S. Africa	1.19	(small)	1.11/1.14

ISO containers may be successfully used for the carriage of bagged coal. Specially treated coal, in multi-ply paper sacks for domestic use, has little risk off heating or sifting. If Open-top, Half-height, or even Closed Box containers are used for bulk (e.g. when otherwise empty containers would be moved), the containers should be properly lined against abrasive damage (by lining ends and walls with plywood sheets, hardboard, etc.) and against dirt — lining the whole container with plastic sheet or similar.

COBALT   1.60 Drums
0.50 Bulk

Is a mineral ore from which is obtained a reddish white, highly magnetic metal. Various kinds of this ore exist which are known as cobalt glance, black oxide, cobalt bloom, etc. See Part 2, Ores. See IMDG Code.

COCA  2.23 Bags

The leaves of a shrub native to Bolivia and Peru, valued for their medicinal properties.

Being very lightly packed, it constitutes exceptionally light cargo. Stow in a dry, well-ventilated place.

COCA COLA (concentrate) Drums. Cartons.

High value. May have alcohol base — See IMDG Code. Susceptible to high temperature damage. May be carried under temperature control + 2/7 degrees C.

Discharging bagged cocoa beans.

Note ventilation channels.

Commodity	*Characteristics*	*M3 per Tonne*	*Packaging*
COCHINEAL		1.67/1.81	Tin Lined Cases

A small insect (plant bug) used for dyeing purposes. It is a valuable cargo and peculiarly liable to deterioration through heat or moisture. Stow in strong room or other special stowage but place must be cool and well ventilated.

COCKSFOOT SEED		5.74	Sacks

See Grass Seeds.

COCOA (Cacao)	  	2.0/2.15	Bags

Cocoa beans are the seed obtained from the pods of the cacao tree, a native of the tropics. Shipped in considerable quantities from Trinidad, Brazil, Venezuela, Mexico, Philippines, Africa, etc. Shipped in new or good bags, this cargo has a high value. Very apt to heat, stow in a dry place preferably 'tween decks. Good ventilation is essential. Shippers normally require a minimum of 20 air changes per hour basis empty hold. Protect from metal stanchions, hatch coamings, etc., with mats or similar. Cocoa is liable to take taint, arrange stowage accordingly. Stow on sound separations to assist in picking up loose cargo on completion discharge. Loses value if allowed to become too dry and brittle −7% moisture content acceptable, vapour given off may contain acetic acid which may cause corrosion. *ISO containers* — due to high moisture content is usually unsuitable for carriage in closed box containers — particularly if moisture content exceeds 7%. Super ventilated containers with double lining to channel excess moisture away from the cargo, are ideal and produce good results. Payload in 20 ft containers approximately 12.5 tonnes.

COCOA BUTTER	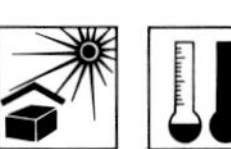		Cartons. Bags.

Protect from excessive heat. Protect ISO containers from direct sunlight. May travel under temperature control 10 degrees C (50 degrees F).

May be shipped in Polythene bags enclosed in cardboard cartons. Melting point of Cocoa Butter is 85–90 degrees F. Readily absorbs aroma taint odours. Cool dry stowage essential. Do not stow cartons more than 4-5 tiers high. Avoid exposure to heat or direct sunlight. Check temperature at loading and reject cartons with temperature in excess of 50 degrees C.

25 kg carton may measure 470 × 350 × 230 mm
380 × 290 × 280 mm
330 × 330 × 300 mm

Payload in 20 ft closed box container approximately 15 tonnes.

COCOA CAKE	 	2.26/2.46	Bags

Hygroscopic, may give off moisture. See IMDG Code. Usually palletised. Weight per bag 40.5 kg. Weight per pallet 1080/1095 kg.

COCONUTS		2.51/2.79	Bags
		2.65/2.79	Bulk

Usually shipped (with or without the husks) in nets or bags, but on occasions in bulk when, if B/L does not specify to the contrary, they are used as dunnage for light goods or for broken stowage.

Do not stow under or with dry sugar or delicate goods as the nuts heat and steam very considerably. Will give up moisture copiously to a drier ambient temperature and liable to heat. Notwithstanding the coarse nature of the cargo it requires gentle handling. *ISO containers* — by fitting a temporary wooden bulkhead in way of the container doors (see section 2, Bulk in Containers) loose coconuts may be suitably carried in closed box containers.

Commodity	*Characteristics*	*M3 per Tonne*	*Packaging*
COCONUT (Desicated)		1.95/2.23	Cases

The kernel of the coconut, specially dried and prepared, the better class being usually packed in tin lined cases. Similar carriage requirements to sugar confectionery.

Frequently shipped lined multiwall paper bags which may be liable to damage and weakening by moisture. The oily nature of the contents sometimes results in external staining of bags. This staining may not necessarily denote deterioration of product.

COCONUT CAKE		1.40/2.10	Bulk

See Seed Cake.
(See IMO Code of Safe Practice for Bulk Carriers).

COCONUT FIBRE (Coir)		2.79 4.74/5.30	Bales Dholls

The fibrous outer covering of coconuts used in the manufacture of mats, brushes, ropes, etc. Made up for shipment in bales and dholls. If the oily yarn is shipped it should be stowed clear of goods liable to be damaged by oil. See Coconut Oil.

COCONUT OIL	 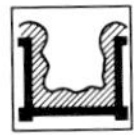	1.53/1.81 1.81/1.95 1.05/1.11	Drums Tins Bulk

Coconut oil is produced by crushing copra, the dried 'meat' of the coconut. The principal producer and exporter of coconut oil is the Philippines, which exports large quantities of the oil to North America and Europe. Smaller quantities of the oil are also exported by Indonesia as well as other countries. Coconut oil is in many respects similar to palm kernel oil, the two oils being the most important members of the oils referred to collectively as the 'lauric' oils. Coconut oil is semi-solid at ambient temperature, is very light in colour and has a density of approximately 890 kg/m^3 at 60 degrees C (920 kg/m^3 at 15 degrees C). The oil is normally exported in the form of the crude oil with a maximum free fatty acid content of 5%, calculated as lauric acid.

The recommended temperature range for the storage and bulk shipment of coconut oil is 27-32 degrees C, and a temperature range of 40-45 degrees C is recommended for loading and discharge of the oil.

Coconut oil is prized for its sharp-melting properties and is widely used as a confectionery fat in North America, Europe and Japan. It has also found numerous applications in the oleochemicals industry. See Part 2, Oils and Fats.

COCONUT OIL PRODUCTS			

As the result of splitting coconut oil into two parts High Acid (Acid Oil or Fatty Acid) and Glycerine are obtained. Coconut oil splits into 90 per cent High Acid and 10 per cent Glycerine both of which commodities are shipped from the Philippines and other copra producing countries. This weighs approximately 890 kg per M3 at 37.8 degrees C being about 3 per cent lighter than coconut oil and is practically black in colour.

Commodity	*Characteristics*	*M3 per Tonne*	*Packaging*
(High Acid)		1.16	Bulk

Owing to the corrosive effect of acid it is essential that the tank in which it has been carried be thoroughly cleaned and all traces of the acid removed immediately after completion of discharge.

See Part 2, Oils and Fats.

COCONUT SHELL (Broken)		1.67/1.81	Bags

Clean, dry stowage.

CODFISH: STOCKFISH	Salted	1.39/1.56	Bales
		1.90/2.00	Barrels
		1.67/1.73	Cases
		1.95/2.01	Kegs & Drums
		1.56/1.62	Bulk
	Cod Roe	1.39/1.45	Barrels

The fish is prepared for salting by removing the head, backbone and interiors. It is then kept under salt for 10 to 15 days, when it is well washed in clean water and afterwards stacked up in airy sheds to a height of about 6 ft, each layer of fish being kept well covered again with salt and left so for about two weeks. So treated, it is known as Wet Salted Codfish.

The fish is then ready for drying, which is done by laying it out in single layers on a stone lined floor, where it is exposed to the action of the sun and air. It is collected and stacked every evening and covered with waterproof cloths to protect it from dew and rain; this process being continued until the drying is completed when it becomes variously known as Salted Codfish or Sundried Codfish.

The fish as it comes out of the trawler, after removal of head, backbone and interiors and well salted, is known as Fresh Salted Codfish. This kind is carried in bulk.

Wet Salted·Codfish in bales, care should be taken to avoid pressure damage.

Sundried Salted Fish — the holds and bilges must be thoroughly cleaned and dried; sides, bulkheads and floors overlaid with thin wooden battens about a foot apart, secured in position by nailing to ceiling and cargo battens. The floor is then covered with sawdust to a depth of about 4 inches and then with clean separation cloths; the sides, bulkheads, pillars, etc., being also covered with separation cloths, as the fish quickly deteriorates if permitted to come into contact with metal. The 'tween decks are prepared in a similar manner.

Bales or "parcels" of fish are stowed, without wood ventilators, one over another. The drainage from "Sundried" is comparatively light.

Fresh Salted Codfish — the ground is covered with a layer of salt and stowage proceeds, each layer of fish being covered with salt, stowing as compactly as possible, approximately one ton of salt for every five tons of fish. Check for infestation.

Frozen Codfish — See Fish.

COD ROE  

See Codfish. Wet stowage and away from delicate and odorous goods.

CODILLA			Bales

The coarse part of Hemp and Flax. For stowage, see Hemp and Flax.

Metallurgical coke in stow, section of Australian hold ladders in background.

Commodity	*Characteristics*	*M3 per Tonne*	*Packaging*
COFFEE		1.72/1.81 1.81/2.09	Bags (Brazil) Bags

Green bean coffee is a sensitive product which is prone to moisture absorption during transport and storage. Care must be taken that the cargo is received in sound condition. Coffee may be carried in break-bulk mode or pre-slung but nowadays most frequently in containers. Except for short voyages these should be of the ventilated type, as coffee is vulnerable to moisture damage. The bags, normally 60 kg, may also be susceptible to taint, contamination and infestation. Cleanliness of the cargo space or container is essential.

With particular regard to moisture damage, it must be appreciated that normally the coffee is loaded in warm and moist conditions and transported to Europe where it may be winter. Ventilation of the cargo may not be possible for the last few days of the voyage due to rain and stormy conditions. During this time the sea temperature and air temperature may fall dramatically. It is essential, where practical, to apply prudent ventilation in order to avoid drip damage due to condensation forming on the internal surfaces of ships holds and the walls and ceilings of the containers.

Coffee beans are readily susceptible to taint. This may be due to 'field damage' caused by the burning of husks or improperly applied chemicals in the form of fungicides or insecticides.

Risk of contamination may be minimised by covering adjacent metallic parts of the ship or the internal surfaces of containers with paper.

Provided all normal cleaning of spaces and containers are properly observed, infestation is normally of pre-shipment origin. Coffee may be stowed for many days in infested warehouses. These may be purged from time to time with insecticide and care should be taken that the fumigant is fully dissipated before the coffee is loaded on board the ship or into containers.

COFFEE, INSTANT  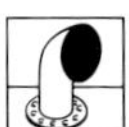

Dried instant coffee normally shipped in large plastic bag within cardboard liners, and strapped onto pallets. Contents are very hygroscopic and good ventilation is essential. Spillage should be cleaned off as soon as possible.

COILS Steel Sheeting. See Part 2, Iron and Steel Products.

COILS Wire Rods. See Part 2, Iron and Steel Products.

COIR **(Coir Yarn)**	 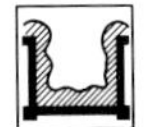	2.79 4.74/2.30	Bales Dholls

See Coconut Fibre. When coir yarn is shipped in the oily condition it should not be stowed on top or over dry goods. See IMDG Code.

COIR ROPES		2.51/2.79	Coils

Oiled coir rope should be stowed apart from and not above textile or fine goods or fibres such as cotton, jute, etc. See IMDG Code.

COKE **(Coal Origin)**		1.25/2.80	Bulk

Used as fuel for furnace work, etc. See also Petroleum Coke.

Commodity	*Characteristics*	*M3 per Tonne*	*Packaging*
COKE BREEZE	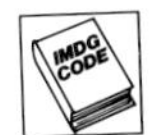	1.80	Bulk

May liquefy if shipped wet. Check origin and history of storage and moisture content prior to loading.

COLEMANITE		0.61	Bulk

A natural hydrated calcium borate. Used in boric acid and sodium borate. Light grey lumps. Angle of repose 47 degrees.

COLOMBITE ORE		0.56/0.61	Bags

Dusty, but non-odorous cargo. Valuable cargo. Only clean dunnage to be used.

COLOCYNTH		6.97/8.36	Pressed Bales

Ground plant used as a purgative drug. Dry, cool stowage.

COLOMBO ROOT	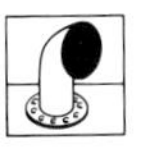 	2.79/3.07	Bales Bags

A root valued for its medicinal properties. Dry well ventilated stowage away from tea and edible goods.

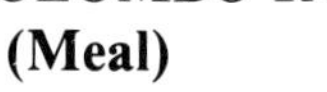

COLOMBO ROOT **(Meal)**	 	2.79/3.07 1.95/2.09	Bags Bags

Bitter taste, used medicinally. Stow as for Bark, which see, Dusty.

COLZA OIL		1.67/1.73	Barrels/Drums

Colza, otherwise rape oil, which is pale yellow in colour is obtained from the seeds of the cole and rape plants (see Rape Seed), the unrefined oil being known as brown rape oil, while "colza oil" denotes the most refined product of these seeds.

COLZA SEED		See Rape Seed	

See Vegetable Oil. See Part 2, Oils and Fats.

Commodity	*Characteristics*	*M3 per Tonne*	*Packaging*

COMPRESSED GASES    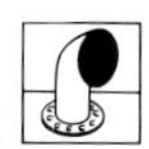

Compressed gases, which are inflammable and otherwise dangerous in varying degrees, include the following:—

Ammonia Anhydrous	Oxygen
Blaugas	Coal Gas
Carbon Acid Gas (Carbon Dioxide)	Hydrocarbon
Nitrous Oxide (Dental Gas)	Pintsch or Iolite
Aragon	Acetylene
Hydrogen	Chlorine
Nitrogen	Sulphur Dioxide

These and other gases of like properties are, by law, to be shipped in metal cylinders, protected by rope or matting in accordance with approved specifications, and tested periodically. The valves fixed to the cylinders must be properly protected by metal caps that cannot become detached by the rolling of the cylinders, and strong enough to protect the valves from damage.

They should be stowed in a cool, well ventilated place, and, if their nature calls for it, on deck, well away from living quarters and steam pipes, and effectively protected from the sun's rays. *ISO containers* — particular care must be exercised in securing cylinders within the container to prevent any movement during transit, and avoid cylinders tumbling out when the doors are opened. See IMDG Code.

CONCENTRATES 

Partially washed or concentrated ores from flotation process. Copper, lead, and zinc concentrates are shipped in bulk. Usually concentrates, which are powdery in form or character, carry a moisture content, however this is normally controlled by legislation of the exporting country.
May liquify. Relevant physical properties including moisture content should be furnished by the shipper. Trim below angle of repose.

See Part 2, Bulk Cargoes, also Sludge, Ore, Slurry Coal and Ores. See IMDG Code. (See IMO Code of Safe Practice for Solid Bulk Cargoes).

CONDENSED MILK  1.25/1.28 Tins in Cases

Fine cargo very liable to pilferage. Must be stowed in cool place remote from boilers and machinery space, otherwise it quickly deteriorates. Because of the high density of the contents, is susceptible to condensation damage particularly when loaded in a cool climate and transported to a warmer one. In these circumstances the practice of loading the cargo in a warm condition has proved successful on occasion, and in such circumstances no ventilation should be required.

CONFECTIONERY    2.34 & var. Cases. Cartons

Fine cargo very liable to pilferage. Most kinds are very liable to melt and mass if stowed in a warm place. Cool dry stowage and apart from odorous goods indispensable. Chocolate confectionery may suffer from "bloom" as a result of varying temperature or condensation. While this "bloom" does not alter the taste, the change in appearance affects the value. At 24 degrees C (75 degrees F) chocolate will melt and run. Sugar confectionery is less susceptible to temperature damage. Can be carried under refrigeration at a temperature of 4.4-7.2 degrees C (40-45 degrees F), but beware condensation problems when refrigerated goods are finally exposed to warm humid air. *ISO containers* — if carrying confectionery highly susceptible to temperature change (e.g. chocolate filled or covered) in closed box GP containers, better carriage conditions may be achieved by putting such sensitive a consignment in the centre and low down, surrounding it with less sensitive cargo (e.g. sugar confectionery, etc.).

Commodity	*Characteristics*	*M3 per Tonne*	*Packaging*

CONGOLEUM

A class of linoleum, shipped in rolls with wooden cores. See Linoleum.

CONTAINERS See Part 2.

COODIE  See Rice

COPAIBA  See Balsam Copivi

COPAL, GUM		2.23	Cases

A semi-fossilised resin used for varnish making, etc. A clean dry cargo, usually packed in cases.

Must be stowed in a cool place, well removed from boilers and from goods liable to heat or throw off moisture such as jelatong, gambier, areca nuts, etc., etc. See Gums.

COPPER		0.28/0.33	Loose Ingots or Strapped
		0.39/0.42	In Barrels
		0.84	Coils

This should be carefully tallied and watched to prevent pilferage. Given the opportunity, unscrupulous stevedores will drop ingots overboard and recover same after vessel has left. Anodes, Cathodes, Slabs, and Blisters may be 1200 × 900 × 200 mm. 350-920 kg stowing at 0.34 M3 per tonne. This cargo should be overstowed with some other goods immediately after loading in order to minimise risk of loss. Should not be stowed adjacent to zinc.

Intake tally should be taken in the hold if possible — never on the lighter or wharf. Avoid working this cargo at night if possible. Stow on ceiling or deck and, if receiving a large amount, not too many heights. For stowage in refrigerated compartments see Refrigerated Cargo. *ISO containers* — spread load as evenly over the container floor as possible. Any local concentrations of weight should be positioned adjacent to the walls where greatest floor strength prevails. Essential to provide sufficient chocking off (nailing timber dunnage to the floor, etc.) to prevent even the smallest movement during transit.

COPPER CONCENTRATES	 	0.39/0.50	Bulk
		0.61/0.78	Bags
		0.33	Slabs

See Concentrates, also Ores. Maybe self heating. Moisture content is critical. May liquify. Relevant physical properties including moisture content must be checked prior to loading. Cases have occurred where very considerable damage was caused to cereal and to coffee cargoes stowed in the same compartment as a parcel of copper concentrates, which, it was alleged, generated great heat causing discolouration, sweat and fume (taint) damage. See IMDG Code. (See IMO Code of Safe Practice for Solid Bulk Cargoes). See Appendix 13.

COPPER CUTTINGS	 	3.07/3.34	Bales

Liable to spontaneous combustion. Stow in or near hatchway. See IMDG Code.

Commodity	*Characteristics*	*M3 per Tonne*	*Packaging*

COPPER GRANULES 0.22/0.25 Bulk

Sphere shaped pebbles. Odourless. Angle of repose 25 to 30 degrees.

COPPER MATTE  0.42/0.53 Bulk
0.56 Bags

The crude black copper ore (native metal). See Ores.

COPPER ORE 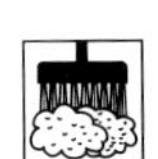 0.39/0.56

Do not overstow with oils or other cargo of any oily nature. See Part 2, Ores.

COPPER PYRITES
YELLOW SULPHUR 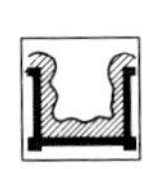  0.56/0.61 Bulk

A mineral sulphide from which is obtained about 10 per cent of copper. Very dusty and dirty. Becomes free flowing when wet. See Ores, also Pyrites.

COPPER SULPHATE
BLUE COPPERAS
BLUE VITRIOL      1.23/1.28 Drums

A highly corrosive, poisonous substance soluble in water. The fumes given off if wetted or damp attack other articles in the vicinity and corrode the hoops binding the packaging. Dry stowage remote from edible goods. See IMDG Code.

COPPERAS (Iron Sulphate)   1.53 Drums

Sulphate of iron or green vitriol, largely used in the manufacture of ink and black dyes. Must be stowed in a dry place, away from moist or wet goods and from perishable cargo. B/L should embody a clause to the effect that recooperage of packages be to the account of consignees.

COPRA    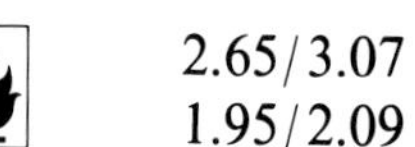 2.65/3.07 Bags
1.95/2.09 Bulk

Is the dried kernel of the coconut, having an oil content as high as 66 per cent. The oil obtained from the copra (coconut oil) is extensively used in the manufacture of butter, cream and milk substitutes, soap, candles, etc.

Shipped in large quantities — both in bulk and in bags — from the Philippines, East Indies, South Sea Islands, India, etc.

The rank, oily odours thrown off by copra are very penetrating, this and other properties rendering it very unsuitable for stowage with, or even near, fine and edible goods, especially tea.

Copra, should not be received into a vessel carrying tea unless stowage can be arranged for it whereby it will be effectively separated from the tea.

Commodity	*Characteristics*	*M3 per Tonne*	*Packaging*

Two kinds of copra are shipped, i.e. white copra (sun dried) and black copra (kiln dried) these being divided into 25/30 grades or classes. The degree to which copra is dried varies very considerably; that shipped on the Malabar Coast which has an average shrinkage due to evaporation of about 3 per cent, being regarded as the best dried, while the shrinkage of copra shipped from other countries varies between 3 per cent and 7 per cent, even touching 10 per cent in some cases from the East Indies and Philippines.

Copra being very liable to heat, special care should be exercised to allow the moisture, as well as the odours thus thrown off, to escape.

Imperfectly dried copra, or that which is shipped wet, is liable to deteriorate in the course of the voyage when Carbon Dioxide gases are generated — cases having occurred where loss of life has resulted from the presence of "copra fumes" in ships' holds.

Smoking and the use of naked lights in holds or near open hatches should be strictly prohibited.

Opinions differ as to whether copra is liable to spontaneous combustion, but owing to the high oil content (about 66 per cent) it certainly burns fiercely when once it "gets away". Some are inclined to attribute the origin of fires to the jute material, of which the bagging is made, jute being very liable to fire when in the wet or oily condition. Copra dust readily ignites when brought into contact with flame. A certificate of origin, temperature, and moisture content should be supplied by competent authority. See IMDG Code. (See IMO Code of Safe Practice for Solid Bulk Cargoes).

Copra Bugs — these little insects are bred in myriads in copra cargoes and this fact should be taken into consideration when arranging the stowage of copra. They are most troublesome to human beings.

These bugs readily attack any other cargo in its vicinity and, though they do not always destroy the cargo by eating it, they lay their eggs therein, round which cocoons form. Subsequently the larvae emerge from the cocoons leaving behind the cocoons which are very difficult to remove. In such a case it would be very difficult to resist a claim for bad stowage.

Receiving — wet copra, wet or damp bags should be peremptorily refused and a careful watch kept for same.

B/L should be claused "Not responsible for loss of weight through evaporation or for deterioration through inherent vice or properties".

If in bulk — ideally, suitable precautions should be taken to avoid contact with steel, but in practice much of the copra is simply bulk stowed.

If in bags — suitable methods should be adopted to keep the bags from working between and so touching the sides, it being preferable not to mat sides in this case as the circulation of air through the cargo is made easier. Copra should not be allowed to come into contact with iron.

Dunnage particularly well on box beams, stringers and cement caps, and on no account should the contents of torn bags be allowed to get inside spar ceiling.

Ventilation — constant attention to ventilation of holds is absolutely necessary to ensure the proper carriage of copra and immunity from damage to other fine cargoes that may be carried. The ventilators should be assisted in clearing the holds of copra smells, moisture and fumes, by keeping some of the hatch covers off at each end of the hatches when ever weather conditions permit, at which time all the ventilators should be kept on the wind to assist in expelling the fumes, etc. Great advantage in this direction follows the fitting of thick rough gratings along and inside the hatch coamings, extending slightly below the lower edge of the same — this serving as an easy escape for the moisture charged and heated air, fumes, etc. All ventilators should be fitted with wire mesh guards. Temperature should be regularly taken and recorded.

Stowage factors — these vary very considerably owing to the different degrees to which copra is dried and fineness of chopping:—

Well dried copra should stow—
in bulk in about 1.95 M3 tonne hold and 2.09-2.15 M3 tonne T.D. space.
In bags in about 2.09 M3 tonne hold and 2.37 M3 tonne T.D. space.

It is not unusual, however, for these figures to be exceeded by 20 per cent.

Commodity *Characteristics* *M3 per Tonne* *Packaging*

ISO containers. Do not stuff in same container as other cargoes. Container may require fumigation after stripping. Limited quantities have been carried in ISO containers with due regard to stowage positions on board to reduce possibilities of self-heating and permit access in case of fire. About 12 tonnes can be loaded into a 20 ft closed box container — if IMDG regulations permit such carriage.

COPRA CAKE   1.67/1.81 Bags
See Seed Cake Bulk

(See IMO Code of Safe Practice for Solid Bulk Cargoes).

COPRA EXPELLER PELLETS   1.67/1.81 Bulk
See Seed Cake

(See IMO Code of Safe Practice for Solid Bulk Cargoes).

COPRA MEAL   2.23/2.37 Bags
See Seed Cake Bulk

(See IMO Code of Safe Practice for Solid Bulk Cargoes).

COQUILLA NUTS  1.67/1.81 Bags

A vegetable ivory; the fruit of the Brazilian palm coquilhos, which yields a brown substance used in the manufacture of buttons, door knobs, etc. Dry stowage and away from moist goods.

CORAL   2.23/2.79 Bags

Of which there are many varieties, red, black or white in colour. Very fragile, requires handling carefully and must not be overstowed except with light goods. Owing to the smell given off, stow in well ventilated space clear of edible goods. Coral ornaments should be stowed in special cargo locker to prevent pilferage.

CORESTOCK 3.90 Crates

Similar to plywood.

CORIANDER SEED  3.43/3.62 Bags

The highly aromatic seeds of a tropical plant extensively used for culinary purposes, Never shipped in large quantities. Dry stowage away from tea and other delicate goods.

CORK   8.36/11.71 Bales
4.18 Bags

The outer bark of trees native to Spain, Portugal, Algeria, etc. Can be discoloured by sweat. Becomes brittle when cold. Due to the variety of uses to which the end product is put, it should not be stowed in the same compartment or container as highly odorous cargoes. Dry cargo. Do not permit slinging by the bands. Cork board in cartons stows at 9.75 M3/tonne. Cork blocks, Cork discs and cork (agglomerated), all in cartons, stow at 4.18 M3/tonne.

Commodity	*Characteristics*	*M3 per Tonne*	*Packaging*
CORK SHAVINGS		7.80/8.36	Bales
GRANULATED		7.80/8.36	Bags

Dry cargo. Do not permit slinging by the bands.

CORN 

In the U.K. this term is usually applied to all kinds of grain, but in the U.S.A. it is applied to Maize, which see.

CORNFLOUR		1.25/1.39	Bags

A farinaceous food. Dry stowage, remote from odorous cargo.

COROZO NUTS		1.67/1.81	Bags

A vegetable ivory; the fruit of a South American dwarf palm, largely used by turners in the manufacture of various articles.

Dry stowage and away from moist goods.

CORPSES 

The coffin should be packed in lead lined cases and properly sealed. A sanitary certificate should be obtained from the Health Authorities, endorsed by the proper Consular official, before shipment. May also be carried under refrigeration.

CORRUGATED IRON 

See Galvanised Iron.

CORUNDUM		1.11/1.25	Bags
(ORE)		0.67/0.61	Bulk

An oxide of aluminium being, next to the diamond, the hardest of minerals — used in the manufacture of emery, etc. Shipped from India, etc. See Part 2, Ores.

COSTUS ROOT		3.07/3.34	Bags

An aromatic root possessing medicinal properties. Dry stowage remote from edible and odorous goods.

COTTON 

Is the delicate fibre which clothes the seeds of the cotton herb, an annual, and the seeds of the cotton shrub and tree which live for 2 to 3 and 6 to 10 years respectively. Extensively cultivated in the Southern United States, India, China, Egypt, and to a lesser degree in Brazil, Peru, West Indies, Turkey, Australia, etc.

Commodity *Characteristics* *M3 per Tonne* *Packaging*

Cotton is packed in bales, pressed to varying densities, the highest density bales being shipped from India and China. The bales are covered with coarse gummy cloths and are bound by steel, wire or rope bands.

The following are the average sizes of bales and stowage factors for the principal cotton exporting countries:—

		Cu. M. per Tonne
Australia,	Pressed	3.62-3.76
China,	High Density	1.39-1.67
India,	High Density	1.53-1.67
Turkey,	Pressed	3.07-3.62
U.S.A.,	Standard Bale = 22½ lbs per cu. ft.	
	Standard Hand stowed, 29-30 cu. ft per bale	3.62-3.76
	Standard Screwed 27 cu. ft per bale	3.34-3.43
	Webb or high density = 34½ lbs per cu. ft	2.37-2.45
	Hand stowed 221-22 cu. ft	
	Screwed 18 cu. ft	2.17-2.23

Owing to the great risk of fire which is ever present with cotton cargoes, the handling, stowage, and carriage of same calls for the greatest care and attention.

The National Cargo Bureau have issued Rules which govern the loading of cotton in all U.S.A. ports, and these may, with advantage and confidence, be adopted when loading cotton in ports where no specific rules are enforced.

Fire Extinguishers — before commencing to receive cotton, the ships fire fighting system should be thoroughly examined. Fire extinguishers should be placed adjacent to working hatch. Stevedores forklift truck should be battery driven and checked that arresters are in proper and good condition.

While loading or discharging cotton, the fire hoses should be connected and ready for use. "No Smoking" notices conspicuously displayed, and the order rigidly enforced, "spark arresters" fitted on main and galley funnels and the use of forges or naked lights prohibited.

Wet cotton, if stowed in a confined space, will heat and deteriorate but no danger of spontaneous ignition is to be expected. Wet and dry cotton should not be stowed together.

On the other hand, cotton which is or has been in contact with oil or grease is very liable to spontaneous combustion, for which reason holds, and especially spar ceiling, should not be painted shortly before loading cotton unless it is certain that there is sufficient time for the paint to harden before cotton is stowed up against it.

Receiving — bales of cotton must be carefully examined on wharf or lighter, and all wet, damp or oil stained bales, also those with wrappers torn off and marks missing, or, with burst or missing bands should be rejected. Loose cotton should never be received on board.

When bales are wrapped in hessian wrappers, as usually is the case, many countries (Canada especially) insist on bales being fumigated before entry is permitted where second-hand material has been used for wrapping. It is therefore advisable for a certificate to be obtained from Shippers certifying the wrapper used to be new material and that such be visaed by the consul of the country of destination.

Where bales are partly opened up for sampling and grading only new material should be used to replace parts of wrappers cut off for that purpose.

Stowing — the bales stow on flat, edge or end according to which arrangement will ensure the greatest number being carried in the compartment, the particular arrangement depending on dimensions of bales and depth of compartment.

Cotton fires — in the event of a fire breaking out in a cotton cargo at sea, prompt measures are necessary:—

Batten down and close all apertures by which air may find its way into the holds — turn on the CO_2, according to instructions. See IMDG Code.

Commodity	*Characteristics*	*M3 per Tonne*	*Packaging*
COTTON PIECE GOODS		1.68/7.0	Cases. Cartons. Bales

Top stow. Do not stow in same compartment with goods liable to leak or stain. Use no hooks.

COTTON SEED	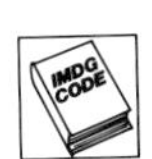	2.23/2.51	Bags

Shipped from all cotton growing countries, the stowage factor varying considerably.
See IMDG Code.

COTTON SEED CAKE (See Seed Cake)	   	1.34/1.53 1.56/1.67	Bales Bulk

Is readily damaged by moisture and susceptible to tainting damage if stowed near or with odorous goods. Particular care should be observed if loading in Southern U.S.A. ports to keep this cargo clear of turpentine fumes. Usually put up in bales covered with gunny or in bags, but at times shipped in the loose condition. Requires to be well dunnaged and well ventilated.

Shipped in bulk may heat and ignite if excessive moisture content. Commodity should be properly aged prior to shipment in order to reduce oxygen content and certificates of moisture and oil content should be furnished prior to shipment.

(See IMO Code of Safe Practice for Solid Bulk Cargoes).

COTTON SEED OIL		1.09	Bulk

This oil is obtained from the seed of the cotton plant, whose oil yield is about 25 per cent and is shipped from most of the cotton growing countries — sometimes in bulk (see Vegetable Oil in Bulk) when it sometimes (according to season) is necessary for heating coils to be fitted, as the oil solidifies at between 10 and 11 degrees C (50 and 54 degrees F) but care should be exercised to ensure that the temperature does not approach boiling point otherwise the oil will deteriorate greatly. Pumping temperature about 18.3 degrees C (65 degrees F).

See Vegetable Oil Constants. See Part 2, Oils and Fats.

COTTON WASTE	 	3.90/4.46	Bales

Is liable to spontaneous combustion, depending upon the percentage of oil — vegetable or mineral — that it contains.

If certified to be free of oil, ordinary stowage. Waste containing not more than 5 per cent of oil — properly baled — stow in open bridge deck, poop space, or where it is readily accessible and remote from Cotton. Reject all wet bales. See IMDG Code.

COWGRASS SEED		1.73	Sacks

See Grass Seeds.

COWRIE SHELLS		1.25/1.50	Bags

Small white glossy shells abundant on certain African and Asiatic shores.

Commodity	*Characteristics*	*M3 per Tonne*	*Packaging*

COWTAIL HAIR 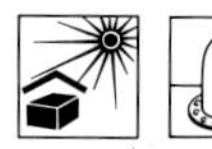 1.95/2.09 Bales

Stow in a dry well ventilated space and apart from all heat throwing goods. See Hair, Animal.

CRATES See Part 2, Techniques

CRAYFISH 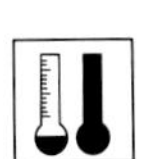 1.53/2.70 Boxes. Cartons.

Careful stowage. Liable to pillage and therefore should be given locker stowage whenever possible. Properly prepared frozen crustaceans will keep for 6 months at −29 degrees C (−20 degrees F) and 3 months at −23 degrees C (−10 degrees F).

Pre-packing hygiene and treatment is most important to the successful stowage of these products. Carriage and temperature to shippers instructions. See Fish.

CREAM  1.39/1.53 Cartons

Stow as for cartons. −15/−18 degrees C.

CREAM OF TARTAR (Argal)  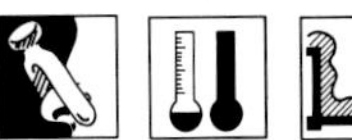  1.62/1.67 Drums; 1.81 Casks

A white crystalline compound, susceptible to damage by taint or very high temperatures. *ISO containers* — Protect containers from direct sunlight.

CREOSOTE 

 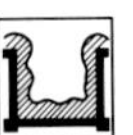 1.67/1.87 Drums; Bulk

A tar oil — is colourless — and inflammable with a highly pungent smell. Often shipped in second-hand drums which are prone to leak. A very undesirable cargo, and should only be received on a vessel carrying fine and clean cargo if fumes cannot come into contact with cargo, ships' stores, or find their way into living quarters. It should be handled and stowed very carefully to avoid leakage, and the greatest care should be taken to remove all traces of creosote stains and smells after discharge before shipping the next cargo. Lower holds with wooden ceiling should be completely lined with impervious material (e.g. plastic). See Sleepers, Creosoted. *ISO containers* — containers should be lined (floors, and well up the walls) with impervious material (e.g. plastic). Preferably not in containers subsequently to be used for foodstuffs or fine goods. In any case never in insulated or refrigerated containers. Fumes/spillage will almost certainly damage the fabric of container tilts (open sided, open topped containers).

When carried in bulk, the oil must be carried in completely self contained tanks to avoid contamination of the creosote. In bulk it is normally carried at a temperature of between 32.2 and 37.8 degrees C (90 and 100 degrees F) throughout the voyage. See IMDG Code.

CRUDE OIL 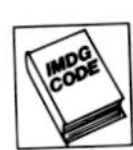

0.8/0.9 Specific Gravity Bulk

See Part 2 Petroleum.

Commodity | *Characteristics* | *M3 per Tonne* | *Packaging*

CRYOLITE  0.70 Bulk

A fluoride of sodium and aluminium used in production of aluminium and for ceramic glazes. Grey pellets. Slightly pungent odour. Prolonged contact may cause serious damage to the skin and nervous system. Angle of repose 45 degrees.

CUBEBS
Tailed Pepper  2.23 Bags

Oily berries having a pungent smell. Not to be stowed in the vicinity of fine cargo such as tea, coffee, cigars, nor in the vicinity of tapioca, sago. Should be well ventilated.

CUBE GAMBIER   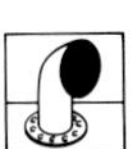 3.07/3.34 Package

A substance prepared from the leaves of certain East Indian shrubs used in dyeing and tanning. In the cube form it has no objectionable properties and may be stowed with fine cargo provided it is in a cool place. Good ventilation, liable to damage from damp heat. Should not be confused with Gambier which see.

CUDBEAR  2.37/2.51 Bales

A weed from which a dye similar to Orchil is obtained. Stow away from any damp cargo, as it will absorb moisture. Dry stowage.

CUMMIN SEED   2.23/2.93 Bags

The aromatic seeds of an annual plant which have a bitterish warm taste, used for medicinal purposes and as a substitute for pepper. Dry stowage and away from fine and delicate goods.

CURIOS 4.18/5.02 Cases

Shipped from the Far East, mostly in large cases fairly well constructed. This cargo loses a lot of space in stowage unless a lower hold can be allocated for it but unless plenty of bottom weight is assured this is difficult to arrange owing to the extreme lightness of this class of cargo.

CURRANTS  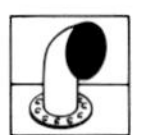 1.25/1.50 Cases/Bags

Shipped in considerable quantities from Greece and other ports of the Mediterranean. Must be well ventilated to avoid heating and fermenting.

CURRENCY NOTES See Bullion, also Part 2, "Specials"

CUSPARIA BARK  Bales

See Barks.

Commodity	*Characteristics*	*M3 per Tonne*	*Packaging*
CUTCH		1.45/1.56 1.67/1.81	Cases Bags

The resin of a thorny tree grown in India and Malaysia, used for dyeing and tanning purposes. Is a form of refined gambier. Odorous. Put up in cases and bags, also made up into small square bales covered with gunny. It melts at 32-33 degrees C (90-100 degrees F) during the tropical part of the voyage and sets very hard when the ship gets into lower tenmperature. Stow apart from perishable goods, bulk latex and cargoes likely to throw off moisture such as jelatong (may become mouldy if exposed to damp). Give floor space stowage, well dunnaged and with a bed of sawdust. Also sprinkle sawdust between each tier. Adjacent cargoes should be well dunnaged. Good ventilation.

CUTTLEFISH  4.18 Cases

A mollusc with an internal soft shell called cuttlebone used in polishing fine metals and for making dental powders. Ordinary dry stowage.

CYANIDE   Drums

All cyanides are high poisonous, the packing and stowage of which are prescribed by Official Regulations. See IMDG Code.

DAMAR   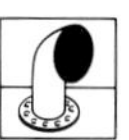 1.81/1.95 Cases; 1.91 Bags

A resin obtained from several kinds of East Indian pine, also from the Kauri pine, the largest tree of New Zealand. Used for varnish making.

Damar is very liable to melt in hot weather, setting hard when cold weather is reached, and to get into a "matted" condition or even to "block", the latter very seriously affecting its value.

Stow in a cool, well ventilated 'tween deck compartment or, as top stowage in the lower hold, avoiding pressure as damar gives off heat, etc., when under pressure. Dunnage well, and for bag damar, separate each tier with matting leaving about 150 mm air space at ends, if possible, to assist ventilation and keep the damar cool. Avoid stowing with or near goods that give off heat or moisture. See Gums.

DANGEROUS GOODS  See Part 2, Techniques

DARI JOWAREE  1.48/1.56 Bags

An Indian grain used as a rice substitute or poultry food. Also known as Indian millet, Turkish millet and Guinea corn. See Seeds.

DATES 

M3 per Tonne	Packaging
1.14/1.20	Cases Wet
1.25/1.31	Cases Dry
1.23/1.39	Baskets Wet

The fruit of the date palm, exported in considerable quantities from Tunis, Persian Gulf ports, Far East, etc., both in the wet and dry condition. When in the wet condition from which there is considerable drainage, treat and stow as wet cargo. Dry dates, ordinary stowage for edibles.

DEAL: DEAL ENDS See Part 2, Timber

Commodity	*Characteristics*	*M3 per Tonne*	*Packaging*

DECK CARGO — See Part 2, Techniques

DEEP FROZEN PRODUCTS, see Part 2, Techniques. Also individual commodities.

DETERGENTS  1.25 Bulk

A variety of detergents are now carried in bulk liquid form. Most of these are of a non-hazardous nature, although many have an offensive odour. It is essential that tanks are thoroughly cleaned prior to loading and it is advisable to call survey of space before the cargo is accepted. Suitable for carriage in bulk bags in ISO Closed Box containers. Detergent concentrate has a specific gravity of 0.98.

If carried in plastic containers or cartons, should be stowed on own ground to avoid affecting other cargoes if leakage occurs.

DEVILS DRUG  See Assafoetida

DHALL, DAL 1.39/1.53 Bags
1.95/2.09 Bags

A pea grown in India and E. Africa. Clean stowage.

DIAMMONIUM PHOSPHATE  1.20 Bulk

Fertilizer. Slightly pungent odour. Angle of repose 30 degrees. See also Fertilizers.

DIAMONDS — See Part 2, Specials

DICHLOROPHENOL   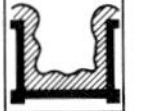 1.11 Drums

Dichlorophenol is a chemical which has an extremely potent odour. A minute quantity of which is capable of contaminating a vessel's entire hold. Contamination in foodstuffs can be detected with only one part of this chemical in 200 million.

Must on no account be carried adjacent to any foodstuffs nor stowed in insulated compartments. All drums should be thoroughly examined before acceptance.

ISO containers — extreme care to ensure no leaking or potentially leaking drums. Container floor and walls should be lined with a single piece of impervious material (e.g. plastic). The slightest spillage can render the container unsuitable for subsequent carriage of foodstuffs or fine goods. Under no circumstances should this commodity be carried in insulated or reefer containers. Destroy any dunnage or timber after unpacking container.

See IMDG Code. See also Part 2, Obnoxious cargoes.

DIRECT REDUCED IRON    0.5 Bulk

It is essential that DRI is properly and accurately described prior to shipment. The Code of Safe Practice for Solid Bulk Cargoes must be consulted and preshipment requirements and carrying procedures strictly observed.

Cold moulded briquettes, lumps or pellets are the most hazardous. Hot moulded DRI is less so having a greater density. In both types a certificate attesting to the suitability of the DRI for shipment must be issued by the competent authority.

Commodity	*Characteristics*	*M3 per Tonne*	*Packaging*
DIVIDIVI	  	3.21/3.24	

Extract from pod of Divi Divi tree. Used in tanning and dyeing. Damp reduces its value. May be infested. Do not stow in same compartment as foodstuffs.

DJARAK		See Castor Seed	

DODDER		2.79/3.07	Bags

A leafless parasitic plant known also as "Hell-weed" used as a Manure and is to be stowed as such, which see.

DOGS DROPPINGS	 	1.81/1.95	Bags

Used in the dyeing industry and manure works. Shipped from the Arabian Gulf in bags containing 68 kg to 90.7 kg weight. Should be protected by tarpaulins from rain and spray, as moisture and heat causes it to throw off very offensive smells.

On deck only.

DOGSKINS		2.51/3.34	Bales

Like all other skins, is a very odorous cargo. Put up in bales, hand or machine pressed. Dry stowage, clear of tea and delicate goods. See Skins.

DOGTAIL SEED		2.45	Sacks

See Grass Seeds.

DOLOMITE		0.56/0.65	Bulk

A carbonate of calcium and magnesium. Used for refractory purposes, road construction and as a fertilizer compound. Odourless. Angle of repose 35 to 42 degrees.

DOM NUTS		2.79/3.34	E. Africa Bags
		1.53/1.67	Red Sea Bags

Shipped mostly from E. Africa. Non-odorous but very dusty cargo. Dry stowage.

DRAGONS BLOOD	 	2.37	Cases

A resin obtained from several varieties of E. and W. Indian trees and large rattans used for colouring. For Stowage, see Gums. Liable to taint. Inflammable.

DRIED BLOOD	 	1.11/1.67	Bags

An animal manure — packed in bags generally double, not very dusty, neither does it give off offensive smell if kept dry. May be infested. *ISO containers* — containers may require fumigation after emptying. Insulated and reefer containers may be used (if properly lined with plastic or similar) for "general" leg of voyage.

Commodity	*Characteristics*	*M3 per Tonne*	*Packaging*
DRIED MEAT		See Carne Seca	
DRIPPING	 	1.95/2.09	Casks

Cool dry stowage.

DROSS			Slabs Pigs

Metallurgical refuse. Stowage factor varies according to nature.
Check specification, may be IMDG Code cargo.

DRUMS		See Part 2, Techniques	
DUNNAGE		See Part 2, Techniques	
DUR-DUR	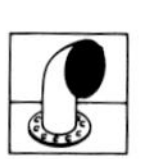	2.37/2.51	Bags

An astringent vegetable pod used for dyeing and tanning. Well ventilated stowage remote from greases and oils.

DURRA **Turkish Millet** **Negro Corn**		1.48/1.67	Bags

A species of small seed, same as millet, used in some countries as a rice substitute, also for poultry food. Grown in E. & W. Indies, U.S.A., etc.

DYES	 		Drums Bags

See Aniline Dyes.
Many vegetable commodities used for dye making are referred to herein, such as cochineal, fustic, lac, logwood, madder, myrabolams, turmeric, etc. Stow away from foodstuffs. May be powder, liquid, paste. May sift or stain. See IMDG Code.

DYNAMITE		See Part 2, Dangerous Goods	
EARTHENWARE		1.81/1.95 1.48/1.67 2.79/3.34 5.57	Cases Pipes Crates Unpacked

This cargo is shipped both in bulk, either protected by straw binding or totally unprotected, and in large crates. Loss through breakage is almost unavoidable, but unless very carefully handled the loss is apt to be serious and productive of claims.

Earthenware pipes are usually nested in pairs when the nature of shipment permits such.

When fairly large shipments of earthenware pipes and like ware are made, it is not unusual for shippers to have their representative in attendance to direct the stowage and in some ports it is the practice for freight to be based on actual space occupied by the ware, measured, after stowing.

Should always be stowed in 'tween deck spaces, and if these are deep it is best to stow over a tier or two of firm cargo laid on deck in order to reduce the top weight on bottom tier of earthenware as much as possible. Straw is usually used for stowing this cargo (which should be dry to avoid self heating). If light case goods are available they should be used for beam fillings and so act to secure top tier from movement when vessel rolls.

B/L should be claused so as to protect vessel from claims for breakage of this class of goods.

EBONY 1.53/1.67 Bulk

Is the black heart wood of a tree grown in Madagascar, Mauritius, Sri Lanka, West and East Indies, etc. Is a comparatively valuable commodity and should be carefully tallied in and out.

Shipped loose, in short lengths sometimes used for dunnaging dry cargo. Ends are sometimes painted or tarred.

EGG ALBUMEN   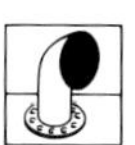 2.65/3.07 Cases
Crates

See Albumen. Eggs are not only fragile but a very delicate cargo liable to be rendered worthless by tainting if stowed near scented or strong smelling goods, such as spices, sugar, onions, cheese, oranges, apples or other green fruit, hay, newly sawn timber, etc.

EGGS   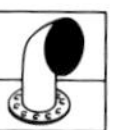 2.65/4.46 Cases/Crates
2.93/3.48 Boxes

Eggs in the shell are packed in crates or cases with bran, odourless wood shavings, fine straw or husks and shaped plastic trays. Stowage factor can vary greatly.

If not carried under temperature control, stow in a cool dry well ventilated space, well removed from boiler or machinery spaces, and from odorous goods. A separate compartment should be set aside for the stowage of eggs.

Suggested carriage temperature for chilled eggs −0.5 to +0.5 degrees C. Liquid eggs, without their shells (usually in cans or packs) are hard frozen.

Handle gently. Packages should always be handled to the hatchway from the wings and ends of the compartment, and not dragged or hove out. Cases and crates should be stowed on a firm platform on close laid dunnage. Packages should be stowed fore and aft in line with the air flow, on their flat (never on edge). If vessel is equipped with brine grid system only, then intermediate lathes should be laid between each tier to assist air circulation.

Bran, husks or odourless wood shavings may be used for packing. It is important that such material is not allowed to become damp, or damaging heat may be evolved.

Treat as cargo liable to leak, and do not stow over goods liable to take damage from leakage of egg contents.

See Part 2, Refrigerated Cargoes.

EGGS DESICCATED 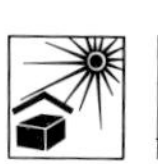  2.03/2.09 Cases

From Far East, etc. Stow in cool place clear of odorous cargo if not carried in refrigerated chambers.

Commodity	*Characteristics*	*M3 per Tonne*	*Packaging*
EGGS LIQUID	  	1.11/1.25	Cases & Packs

Hard frozen. Suitable for stowage with clean frozen meat of compatible temperature.
See Part 2, Techniques, Refrigeration.

EGG YOLK POWDER	 	1.95/2.23	Tins in Cases

From Far East. Dry cargo. Stow in a cool place clear of odorous cargo if not carried in refrigerated chambers.

ELEPHANTS

On deck, at shippers risk, B/L to clause "Ship not responsible for mortality". A full grown animal weighs 3 tonnes and over and varies from 2.28 m to 2.75 m in height. Special slings must be used for lifting these animals. Allow 120 Litres of water and 280 kg of green foodstuff per day to each full grown animal.

ELEPHANTS TEETH		See Ivory	
EMERY		0.56/0.84	Bags

The coarser form of corundum.

EMPTIES		8.36/9.75	Drums

Returned empties, i.e. wine, spirit and beer casks, drums, etc., are usually shipped as deck cargo, but on occasions are stowed below. Empty soft drink cases may stow at approximately 7 M3/tonne.

As the packages may have lain some time in the open, it not infrequently happens that water is retained, or that a small residue of former content is present, either of which will damage fine or dry cargo if overstowed with empties. For this reason a consignment of empties has potentialities for trouble, so that care should be exercised in disposing of same. Stow with bunghole up. Empty drums, cans, barrels, etc., which have contained inflammable spirit should be thoroughly drained and screw plugs firmly fitted — guard against risk of gas explosion. Gas cylinders may still be under pressure. Any empties which have carried dangerous goods must be handled and stored in accordance with the requirements of the dangerous goods previously carried. This rule may only be relaxed when the empty is certified free of all previous contents.

EMULSIFIED ASPHALT		See Asphalt	
ERNODI		See Podophyllum	
ESPARTO GRASS **Alfa**	 	3.62/4.74	Bales

A strong grass fibre grown in N. Africa and Southern Europe, used in the manufacture of paper, cordage, etc.

Packed in bales and said to be liable to spontaneous combustion from which, or some other cause, many serious fires have occurred on ships carrying this commodity. Guard against fire from sparks (galley and main funnels should be fitted with spark arresters) smoking, etc.

The bales are frequently bound either with grass rope or iron bands and vary greatly, as between one port of shipment and another, as to the density to which they are pressed.

It is the practice in some ports to open a certain percentage of bales so as the better to fill up broken stowage, a most reprehensible practice which should never be permitted. In other ports 5 per cent, loose grass is shipped for filling up broken stowage. Loose grass, like loose hay, adds to the risk of fire.

All bales which are not properly bound up should be rejected, thus saving a lot of trouble at discharging ports.

B/L should adequately protect the ship for any loss of weight as this cargo suffers such to a considerable degree in the course of a long voyage. Wet bales should be rejected.

When ore is carried under the grass the vessel has to level same down. See IMDG Code.

ESSENTIAL OILS

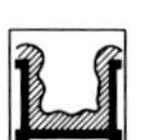

These are all of plant origin, are charged with a concentration of the odour of the particular plant from which they are obtained, are highly volatile and inflammable, have an acrid taste and extremely pungent smells. A very small package of essential oil, stowed in a compartment containing tea, edible goods, tobacco, etc., would, if broken damage the whole contents of the compartment. Proposed loading should consider whether leaking oil could penetrate into another compartment through common hatchways, scuppers, or ventilation.

Whilst all varieties of essential oils are valuable, some are extremely so. The oils are variously packed — according to their value — in hermetically sealed drums, tin lined containers crated or boxed up, etc., some of the more valuable kinds being delivered on board in sealed containers.

The packages should be carefully examined and checked, any showing signs of leakage to be rejected on the spot. Stow in a dry, well ventilated peak — never in holds — where it can be safely locked up. Dunnage and chock off well. Dunnage used for this cargo should not be used a second time, in order to avoid the risk of damaging fine cargo.

The following is a list of the most important Essential Oils of Commerce:—

Aniseed	Cinnamon	Lime	Peppermint
Almond	Citron	Marjoram	Rose
Attar of Roses	Cloves	Mint	Rosemary
	Copaiba	Mustard	Rosewood
Bay	Eucalyptus	Myrrh	Sandalwood
Bergamot	Fennel	Myrtle	Sassafras
Cajaput	Garlic	Nutmeg	Spearmint
Calamus	Ginger	Oolang Ooland	Thyme
Caraway	Jasmine		Valerian
Cassia	Lavender	Orange	Verbena
Cedarwood	Lemon	Patchouli	Wormwood

Note — a number of the above are specially referred to in the alphabetical section of this book.

ISO containers — as well as the extreme care against leakage mentioned above, the container floors and walls should be lined with a single piece of impervious material (e.g. plastic). Spillage will render the container totally unfit for the carriage of foodstuffs and fine goods. On no account should insulated or reefer containers be used. Destroy any dunnage or shoring timber after unpacking containers.

ETHANOL (Ethyl Alcohol)

See Chemical Cargoes. Part 2.

Commodity	*Characteristics*	*M3 per Tonne*	*Packaging*
ETHYL ACETATE	 	1.50/1.78	Drums

Flammable liquid. See IMDG Code.

ETHYL CHLORIDE	 	3.62	Drums

Flammable liquid. See IMDG Code.

ETHYL DICHLORIDE (EDC) — See Part 2, Chemical Cargoes

ETHYL HEXANOL			Bulk

Non-hazardous cargo with a flash point of about 74.2 degrees C.

Absolute cleanliness of the tanks is essential prior to loading, Tanks should be steamed, washed out in caustic soda, hosed down and dried. It is important to remove any kind of organic matter such as grain from the tank. No kerosene or similar cargoes should have been contained in the tanks for at least three voyages prior to loading Ethyl Hexanol.

It has an objectionable odour and tanks should be well ventilated after discharge, before men are allowed to enter.

EUCALYPTUS OIL **White Wood Oil** **Kavu Putch**		1.23/1.39	Drums

A resinous oil obtained from the Australian "blue gum" tree, mainly used for medicinal purposes. It is not of the same class, as regards value, as essential oils (which see) but owing to its very pungent smell it should be stowed in the same manner as such.

EUPHORBIUM			Cases

The inspissated sap of an East Indian plant having an exceedingly sharp and bitter taste, valued for its medicinal properties. Exported from Morocco, East Africa, etc. For stowage see Gums.

EXPLOSIVES  — See Part 2, Dangerous Goods

EXTRACTS

Many varieties of extracts are shipped in liquid paste and powder form. Such have to be dealt with according to their nature and properties. Valuable commodity.

FARINA	 	1.53/1.67 1.59/1.67	Bags Cases

A starchy substance — very susceptible to damage if stowed near odorous or wet goods. Stow as for Flour.

FATTY ACID

May be in second hand drums. Do not stow with fine goods. See Coconut Oil Products.

Commodity	*Characteristics*	*M3 per Tonne*	*Packaging*
FEATHERS (ORDINARY)	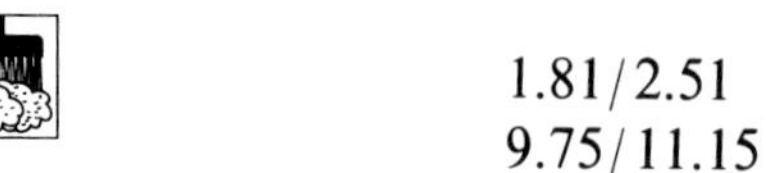	1.81/2.51 9.75/11.15	Pressed Bales Hand Packed Sacks

The commoner kinds of feathers used for bedding and upholstery are shipped in considerable quantities from the Far East, Russia, etc., in bales, pressed to various densities or unpressed in large sacks. In most cases the production of a Sanitary Certificate obtained from the country of origin and endorsed by Consular officer is necessary in respect of this commodity.

FEATHERS (VALUABLE)

Feathers are variously packed in cases, bags or bales, according to quality and value.

Certain classes of feathers constitute very valuable cargo, amongst which are the following: Argus, Pheasant, Adjutant, Bird of Paradise, Egret, Ostrich, Peacock and White Heron.

Valuable feathers are shipped in sealed boxes or cases.

Packages and seals should be carefully examined and tallied. Any packages showing signs of having been tampered with should be rejected. Should be stowed in strong room or special cargo locker and well guarded against pilferage.

FEEDSTUFF	Fodder Pellets	1.39/1.67	Bulk

The minimum quantity of feedstuff to be carried on vessels carrying livestock is laid down in Regulations appropriate to the country of shipment; generally speaking not more than two days supply is permitted to be carried on the weather deck (except in the case of containers). The feedstuff mostly carried is the following:—

Barley	1 tonne in bags = 1.95 M3
Bran	1 tonne in bags = 3.62
Chaff	1 tonne in bags = 3.36
Gluten	1 tonne in bags = 1.87
Hominy	1 tonne in bags = 1.81
Oats	1 tonne in bags = 2.17
Rye	1 tonne in bags = 2.12
Hay	1 tonne in bags = 3.62/4.46

FELDSPAR **FELSPAR**	 	0.98/1.11 0.60	Bags Bulk

Crystalline white or red mineral. Used in ceramics and enamelling. Angle of repose 35-45 degrees. Cool, dry stowage.

FELT		Variable	Rolls Bales

A rough fabric or cloth made by rolling and pressure, from wool, hair, etc. Ordinary dry stowage. (Do not confuse with Inodorous Felt, a dangerous cargo, which see). Stowage factor varies over a very wide range according to class of material, etc., being in some cases 1.67 and 4.18 M3/tonne and more.

Commodity	*Characteristics*	*M3 per Tonne*	*Packaging*
FELT, INODOROUS	 	Variable	Rolls Bales

Owing to its liability to spontaneous combustion, this commodity is classed as a Dangerous Cargo. It is made from flax, jute or similar refuse treated with rosin previously moistened by mixing with oils of various kinds.

Most shipping companies only receive this for carriage on deck. If stowed below, stow in a cool place near the hatch where it is easily accessible. The stowage factor varies very considerably. See IMDG Code.

FENNEL SEED		2.51/2.79	Bags

The seeds of an aromatic hay-like plant used for medicinal purposes.

FERROCHROME		0.18/0.26	Bulk

Raw material of iron mixed with chrome. Angle of repose 45 degrees.

FERROCHROME **exothermic**		0.18/0.26	Bulk

Alloy of iron and chromium. No welding or hot work should be permitted in the vicinity. Angle of repose 45 degrees.

FERRO MANGANESE		See Manganese Ferro	
FERROPHOSPHORUS	  	0.20	Bulk

If mixed with water may produce toxic and inflammable gases. Stow only in mechanically ventilated space.

(See IMO Code of Safe Practice for Solid Bulk Cargoes).

FERROSILICON (See Ferrophosphorus)	  	0.47/0.71	Bulk

Mixed with water and air may produce explosive mixture. Pay strict attention to description of cargo. Preloading certificate must be provided by manufacturer. Stow only in mechanically ventilated space. Exhaust gases must not reach living accommodation. Detectors for presence of phosphine and arsine gas should be carried.

(See IMO Code of Safe Practice for Solid Bulk Cargoes).

FERROUS METAL	 		

Iron Swarf, Steel Swarf, Turning or Cuttings. Liable to heat and ignite spontaneously particularly if oil or other impurities are present.

Careful monitoring of temperatures prior to and during loading essential. Continue monitoring temperatures on passage.

(See IMO Code of Safe Practice for Solid Bulk Cargoes).

Commodity	*Characteristics*	*M3 per Tonne*	*Packaging*
FERTILISERS AND CHEMICALS FOR SOIL TREATMENT	IMDG CODE	1.39/1.67	Bags Bulk

See IMDG Code. (See IMO Code of Safe Practice for Solid Bulk Cargoes).

These may be classified as natural materials which comprise, guano, chalk, lime, bone meal, rock phosphate, gypsum, dried blood and rarely organic manures, peat etc., and by-products such as basic slag and artificial fertilisers. The latter materials are produced chemically and constitute the principal quantity of fertilisers carried on ships.

Organic materials, guano dried blood and bone meal often have a strong unpleasant odour and should not be stowed in the same holds as foodstuffs, tobacco, paper and other sensitive products.

Fertilisers are designed to either control the acidity of soil; such products being lime, basic slag and to a lesser extent some other calcium compounds or to provide nutrients necessary for plant growth; these are principally

1. Nitrogen (abbreviation N) which is provided by guano, nitrates, ammonium compounds and urea
2. Phosphorus (abbreviation P) which is provided by dried blood, basic slag, phosphates, super phosphates and triple super phosphate
3. Potassium (abbreviation K) which is provided by potassium chloride (also known as muriate of potash) and potassium sulphate.
4. Other nutrients, principally sulphur (S) but also certain trace elements which may be added to artificial fertilisers.

It is common to supply mixed fertilisers under the designation NPK. The numbers following the term NPK indicate the proportion of each nutrient present on the basis of a trade terminology. They do not indicate the actual chemical composition of the particular fertiliser.

Most artificial fertilisers are shipped as bulk cargoes and are bagged off subsequent to discharge for distribution to end users. Modern practice is to produce such fertilisers in prill form to facilitate distribution in fields by means of mechanical spreaders. If such fertilisers are wet or caked they cannot be used in mechanical spreaders. All artificial fertilisers are hygroscopic to varying degrees. Furthermore if moist fertiliser dries out caking occurs. There is no necessity to ventilate these products apart from the requirement to remove traces of ammonia gas which are sometimes produced during a voyage before either the ship's crew or stevedores enter cargo spaces.

Fertilisers incorporating ammonium salts or urea must not come into contact with alkaline materials such as basic slag or lime. As most NPK fertilisers contain ammonium salts these must not be allowed to come into contact with alkaline materials. Most NPK fertilisers also undergo self-sustaining decomposition with the release of toxic gases when significantly heated. They are classified in the IMDG Code because of this property. Hence great care must be taken to ensure that such materials cannot come into contact with steam pipes, cargo lamps and any other hot surfaces. Under no circumstances should hot work be conducted in the vicinity of holds containing artificial fertilisers.

Commodity *Characteristics* *M3 per Tonne* *Packaging*

Table of Compatibility of Stowage of Fertilisers

	Basic Slag	Tri Basic Slag	Nitro Plus	Sulphate of Ammonia	Calnitro 20%	Sodium Sulphate	Super Phosphate	Triple Super Phosphate	Nitro Phoska	Nitro Chalk	Muriate of Potash
Basic Slag	—	Yes	No	No	No	Yes	Yes	Yes	No	No	Yes
Tri Basic Slag	Yes	—	No	No	No	Yes	Yes	Yes	No	No	Yes
Nitro Plus	No	No	—	Yes	Yes	Yes	Yes	Yes	Yes	Yes	Yes
Sulphate of Ammonia	No	No	Yes	—	Yes	Yes	Yes	Yes	Yes	Yes	Yes
Calnitro 20%	No	No	Yes	Yes	—	Yes	Yes	Yes	Yes	Yes	Yes
Sodium Sulphate	Yes	Yes	Yes	Yes	Yes	—	Yes	Yes	Yes	Yes	Yes
Super Phosphate	Yes	Yes	Yes	Yes	Yes	Yes	—	Yes	Yes	Yes	Yes
Triple Super Phosphate	Yes	Yes	Yes	Yes	Yes	Yes	Yes	—	Yes	Yes	Yes
Nitro Phoska	No	No	Yes	Yes	Yes	Yes	Yes	Yes	—	Yes	Yes
Nitro Chalk	No	No	Yes	Yes	Yes	Yes	Yes	Yes	Yes	—	Yes
Muriate of Potash	Yes	Yes	Yes	Yes	Yes	Yes	Yes	Yes	Yes	Yes	—

FESCUE SEED 3.62 Sacks

See Grass Seeds.

FIBRE  2.79/3.34

Fibre is the raw material used in the textile industries for weaving various classes of cloths, etc., a large variety of which is shipped from most parts of the world.

The principal fibres of commerce are:—

Asbestos	Cotton	Jute
Bristles	Flax	Kapok
China Grass	Hemp	Ramie
Coconut	Horn	Rhea
Coir	Istle	Rice

all of which are dealt with herein. See IMDG Code.

Commodity	*Characteristics*	*M3 per Tonne*	*Packaging*
FIBRE ROOT		5.02/5.30	Bales

Shipped from Far East ports.

FIGS		1.25/1.39	Cases
		1.34/1.45	Baskets
		1.25/1.39	Bags

Shipped in boxes and baskets from Mediterranean ports. Care should be exercised not to overstow dry cargo with figs unless satisfied that the fruit is thoroughly dried. Should be kept apart from odorous goods. May be infested. May require refrigeration for long passages (green figs) where suitable temperature might be −1 degree C (30 degrees F). See Part 2, Techniques, Refrigeration).

FILMS   

Some cinema and picture films, being readily inflammable, may be classified as Dangerous Goods. Susceptible to damage by heat or moisture.

Stow in a dry cool place, removed from engine and boiler rooms and near the hatchway.

X-ray film will "fog" if subjected to high temperature, i.e. in excess of 18.3 degrees C (65 degrees F). May require temperature control. Typical carriage temperature being +5 degrees C (41 degrees F).

FIRE CRACKERS		2.51/2.79	Cases

Shipped in considerable quantities from Hong Kong and other Eastern ports. Usually are packed in very fragile cases covered with matting which are quite unsuitable for overstowing with other cargo; come in very handy for beam filling over dry cargo, but not for broken stowage. Should not be slung with other cargo to avoid crushing.

Requires careful watching to avoid broaching and pilferage.

See IMDG Code.

FIREWOOD

Firewood C/Ps usually contain the following provision:—

"The cargo to be loaded and discharged in tiers of lengths, as customary in the firewood trade".

Cases have been brought before the Courts where this requirement of the C/P has not been respected. Consignees as a result, having been involved in expense in sorting out the cargo, the claim including loss of despatch money in some instances.

FISH	 		
SHELLFISH		2.28	Crates. Cartons.
CRUSTACEANS		2.34	Crates. Cartons.
SALTED		1.95/2.23	Bales, Crates, Casks
FROZEN		2.50	Boxes, Cartons, Loose

Fish for refrigerated stowage should be received in the hard frozen condition and unless under deep freeze, stowed in a compartment by itself to avoid tainting damage to other cargoes. Where fish is carried under deep freeze conditions. i.e. −18 degrees/−15 degrees C, there is little danger of taint to other cargoes provided they are suitably wrapped in watertight cellophane bags. It will be necessary to deodorize the compartment and where applicable lower hold bilges after discharge.

Commodity	*Characteristics*	*M3 per Tonne*	*Packaging*

Dried fish which may be carried without refrigeration for limited periods, depending on ambient temperatures, etc., may have a strong odour and must not be stowed adjacent to cargo susceptible to damage by taint. Suggested carriage temperature + 4 to + 7 degrees C.

Salted fish, suggested carriage temperature −1 to + 1 degree C, has a high moisture content, is very odorous and may leak brine. If packed in bales, intermediate horizontal dunnage (50 mm × 50 mm) will be required every second tier to allow cooling air full access.

Shellfish — scallops, oysters, etc., are hard frozen or deep frozen. Suggested carriage temperature −15 to −18 degrees C. May be stowed in the same compartment or container as other hard frozen fish or clean meat if carriage temperatures are compatible. Must be hard frozen at, or below, carriage temperature at time of loading. The minimum time must elapse between presenting the cargo and putting it under refrigeration.

Fish fingers — cartons deep frozen. Due to bread and batter content they are susceptible to damage by taint. Also the quality may be adversely affected by fluctuations in temperature during transport.

Fish white — hard or deep frozen. The lower the temperature the longer the shelf life. Must be at the correct carriage temperature at time of loading/stuffing. May be carried in the same compartment/container as clean meat if temperatures are compatible. Very susceptible to damage by taint.

Fish with high fat content, e.g. Mackerel, herrings, etc. Very susceptible to damage by taint, also provide a source of taint. They may be susceptible to fat migration manifesting itself in the form of brown rust spores on the outer flesh if exposed to wide temperature deviation above −18 degrees C. Indications of loss of quality are haemorrhaging from the gills and opaque glazing in the eyes. Stow away from other frozen produce that may take taint or be a source of taint.

Crustaceans are normally carried between −18 degrees C and −25 degrees C being wet block glazed or spray glazed to prevent oxidisation and dehydration. The deep water red shrimp caught off Greenland and Northern Canada are handled in a different manner to their warm water counterparts. They are normally frozen at sea and spray glazed after initial freezing prior to packing into retail type cartons and thence into market cartons. If the post freezing glaze technique is improperly carried out starvation of glaze to product block can lead to dehydration and white spots under the skin after two to three months in storage. Further, by applying a cold water glaze to a frozen block at −20 degrees C will cause the overall block temperature to rise due to temperature equalisation, often as high as −4 to −5 degrees C necessitating the needs to further blast freeze. Failure to do so by placement in conventional cold chamber will result in slow freezing, changing the structure of the glaze and tissue ice crystals. This will allow tissue to break down which is irreversible and will shorten the product finite storage life.

In most Far East exporting countries, the shrimp are normally finger/block packed and frozen totally immersed in a wet block. By doing so, dehydration and oxidisation is reduced, minor variations in temperature during commercial breaks in the cold chain during handling are less likely to have detrimental effect on product tissue.

Spray glazed products requires a storage/carrying temperature of 3 to 5 degrees C below 18 degrees C due to time voids between the product; not present in a solid wet glaze block.

FISHMEAL	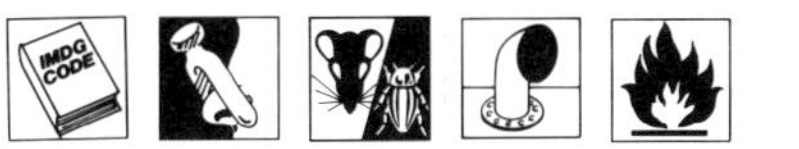    	1.73/1.81 1.34	Bags Bulk

The residue of fish after oil extraction and drying. Used as a valuable nutrient in feeds for poultry, fish (fish farming) and other animals. A source of taint, and possibly infested. May be liable to spontaneous combustion. See IMDG Code. Ship must be fitted with hold CO_2 fire fighting system.

May be susceptible to damage by moisture migration or by condensation. The moisture content should be in the range of 6 to 12%. An acceptable mean of 8% will probably give no condensation problems. Existing mould spores will not germinate and grow provided moisture content is below 12%.

Fishmeal in stow.

Discharging sound bags. Fishmeal fire in background.

Commodity	*Characteristics*	*M3 per Tonne*	*Packaging*

Care should be taken when handling or slinging paper bags to prevent tearing or breakages. The B/L may require that second-hand Jute bags may not be used.

Holds should be clean, dry and free of residues from previous cargoes. Any permanent dunnage or spar ceiling should be complete and in place.

Bagged fishmeal should not be allowed to come into contact with bare metal surfaces.

There are three types of stow that can be adapted for bagged fishmeal IMO Class 9.

1. Compact Stow — which incorporates limited channelling.
2. Conventional Stow — which incorporates extensive athwartships channelling and may also be known as a "Double Strip Stow". (This is the stow recommended by IMO and non anti-oxidant treated Class 4.2 which has been properly aged before shipment).
3. Block Stow (i.e. no internal chanelling) — which is recommended by IMO for Class 9 fishmeal.

Provided the fishmeal has been *properly treated* by an anti-oxidant agent there is no reason why the Block Stow should not be adopted. The tank tops and decks should be adequately dunnaged and space left at the shipside and topstow to allow the air to circulate freely round the block.

A certificate from the competent authority should contain moisture content, fat content, details of anti-oxidant treatment for meals older than six months. Anti-oxidant at the time of shipment which must exceed 100 mg/kg (ppm), packing-number of bags and total mass of consignment, temperature of fishmeal at the time of despatch. Factory date of production, wet or damaged bags should be rejected for shipment. At the time of loading the temperature should not exceed 35 degrees C, or 5 degrees C above ambient temperature whichever is the higher.

Temperature should be taken three times daily during voyage. If temperature exceeds 55 degrees and heating continues, ventilation should be excluded, the holds and spaces sealed and CO_2 released.

Shipments in bulk may be in granular or pelletised form. Serious heating in this stow has also occurred in the past. However it is suspected that this was due to ineffective treatment of anti-oxidant at the time of production. It cannot be over emphasised the effective treatment of the fishmeal by the anti-oxidant is vital in precluding over heating of the cargo. Bagged fishmeal may be carried in containers provided relevant certification is produced. It is recommended that the containers be carried on deck. Internal surfaces must be clean and dry. Stow must be tight with minimum air space. Doors and ventilators to be sealed after stuffing. In case of heating container should be cooled with water. If combustion occurs top of container should be punctured and contents flooded with water.

FISH OIL	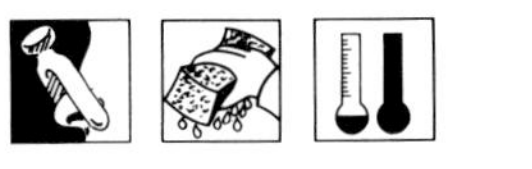  	1.39/1.48 1.09	 Bulk

The oil generally referred to as fish oil may in fact be obtained by the processing of one of a variety of fish, the principal types of fish harvested for oil recovery being capel, anchovy, herring, sardine and menhaden. This range of types of fish harvested for oil recovery means that the chemical composition of fish oil can straddle a wide range, but it is generally the case that fish oils are highly unsaturated and therefore easily oxidised oils. The high level of unsaturation means that the oils are predominantly liquid, though in some cases solid fat may be present in the oil in small quantities and may settle to the bottom of storage tanks. The fish oils are generally dark in colour.

Most of the fish processed for oil are harvested in the North Atlantic and the north and south Pacific regions. South American exports of fish oils are substantial, but Japanese exports have tended to decline in recent years. Fish oil is always shipped as a crude oil, with the free fatty acid content generally required to be not more than 2%. It is refined and processed further for use as an edible fat.

		1.09	Bulk

Seal oil, shipped from Greenland, Newfoundland, Aleutian Islands, etc., is a very odorous oil, of light yellow to brown in colour; used principally for soap making.

Commodity	*Characteristics*	*M3 per Tonne*	*Packaging*
		1.13	Bulk

Sperm oil — of two kinds, i.e., the Southern sperm and the Arctic sperm, the oil of the former being the higher priced. Both are pale yellow in colour, with strong fishy smell. This oil is of a particularly light specific gravity.

Most bulk fish oils are carried at a temperature of 20-25 degrees C. Pumping temperature 30-35 degrees C.

		1.09/1.11	Bulk

Whale oil differs considerably in colour, some being dark brown, and has a very strong smell.

Whale oil is also called “train oil”.

Whale oil carried in bulk, should not be allowed to fall below a temperature of 20 degrees C. If the temperature falls below this figure, there is the risk of separation of solids which are difficult to absorb by heating.

Carrying temperatures should be at about 23.9 degrees C or as indicated by shippers. Where it is required to heat the oil to a requisite temperature for pumping out, the temperature should be raised at the maximum rate of 0.5 degrees C per day.

The most effective way of cleaning the tank after discharge is to first cover the bottom coils in a solution of water and detergent, close the tank down and heat the water to a temperature of at least 48.9 degrees C and leave for about 12 hours. After, wash down the tank with the detergent treated water and afterwards wash down with fresh water. After pumping out, again pass heat through the coils to dry out the tank. It is then advisable to coat the tank with a mixture of boiled and raw linseed oil.

See Part 2, Oils and Fats.

FLAVINE		1.23/1.34	Cases

A vegetable extract from which is obtained an olive yellow dye. Stow apart from edible goods.

FLAX		2.93/3.34	Bales
		4.18/4.32	Baltic Bales
		2.37/3.62	Bales

Is the fibre of the inner layer of an annual herb from which linen is made. Grown and shipped from Russia, Italy, India, Argentina, Australia, New Zealand, etc. Flax is not pressed to a great density to avoid damaging the fibre. The size and weight of bales differ for various ports and countries, but generally they average 181 to 204 kg per bale. Should be well dunnaged and matted and kept well ventilated. See IMDG Code.

FLAX SEED		1.39/1.50	Bulk
		1.59/1.67	Bags

The seed of the flax plant, more generally known as linseed, from which linseed oil is obtained, the seed residue being utilised for cattle feed, i.e. linseed cake.

FLAX SEED OIL

See Linseed Oil, also Vegetable Oils, also Part 2 Oils and Fats.

Commodity	*Characteristics*	*M3 per Tonne*	*Packaging*
FLOUR	   	1.39/1.59	Bags

A delicate cargo, very liable to damage by tainting if stowed near to or in the same compartment as odorous goods, also readily damaged by moisture, so that flour should always be stowed apart from odorous, wet goods, or such as are liable to heat and throw off moisture, and should never be stowed with or on newly sawn timbers.

Large claims have had to be met for flour damage as the result of having been stowed over maize and other cargoes liable to heat and throw off moisture; such stowage should be avoided at all costs.

Flour is particularly susceptible to damage by turpentine and spirit fumes, and should not be received into a vessel carrying such, unless stowage can be so arranged that the turpentine, etc., is separated from the flour by the engine and boiler room spaces.

If required by the trade, bagged flour should be suitably dunnaged and kept from contact with bulkheads, pillars, etc. The use of hooks for handling bagged flour should not be permitted.

In some trades shipments of full cargoes of bagged flour are loaded by being dump stowed sometimes called random stow. Open hatch vessels are suitable for this method of stowing. Bags are lifted aboard on nets or in tubs and rolled out into the hold to find their own position. Some cubic is of course lost but there are considerable savings in loading costs and the loading operation is much faster than placing individual bags. Paper is used to protect the cargo from ships' sides, bare metal, etc., where necessary. The stowage factor for this form of stow using an open hatch ship would be 1.73/1.87 M3/tonne.

FLOWERS (FRESH)  

Carried on, or under deck for short sea passages. Can be carried for lengthier periods under refrigeration. Carrying temperature + 5 degrees C (41 degrees F).

FLOWERS (PLASTIC)  

From Hong Kong and Far East. May be infused with artificial scent. Examine before stowing with taintable foodstuffs. Stow away from heated areas, protect from standing in direct sunlight (ISO containers).

FLUORINE		2.37/2.93	Cases

Obtained from fluorspar and used in the manufacture of glass. Compressed Gas.

FLUORITE OR FLUORSPAR	 	0.92/0.98 0.33 0.55/0.71	Baskets or Mats, Blocks Bulk

One of the most beautiful of minerals formed of cubical crystals of a variety of colours, being a compound of calcium and fluorine. It is used to a large extent in the manufacture of glass and hydrofluoric acid is obtained from it. Shipped from Far East, etc., usually in bags or in solid blocks. Ordinary dry stowage — guarding against damage to other goods through siftage of fluorite, dust or powder from packaging. Maybe harmful if inhaled.

FLY ASH	 	1.26	Bulk

Light finely divided powder. Residual ash from coal-fired power stations. Used in commercial products. Angle of repose 40 degrees. See Calcined Pyrites. See IMDG Code. (See IMO Code of Safe Practice for Solid Bulk Cargoes).

Commodity	*Characteristics*	*M3 per Tonne*	*Packaging*
FODDER		See Feedstuff	
FOGGRASS SEED		4.74	Sacks

See Grass Seeds.

FOODSTUFFS			Cases. Cartons.

Do not stow with odorous or objectionable cargoes, particularly skins, bones, hides or animal bristles on account of Anthrax risk. Chinese foodstuffs are usually in frail packages, which are easily damaged. May be infested. See also Chinese Groceries.

FORMALDEHYDE	 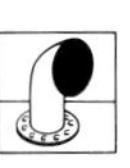 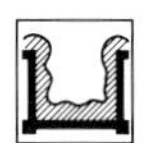 	1.53/1.67 3.07/3.34	Liquid Cases, Powder

A strong disinfectant, shipped both in the liquid and powder form.

In the powder form, ordinary dry stowage away from delicate foodstuffs.

Any leakage of the liquid releases strong pungent fumes by which men are apt to be overcome or be made sick, for which reason it should be stowed near the ventilators in spaces well removed from crew quarters, where good air circulation is assured. *ISO containers* — liquid should never be stowed in insulated or refrigerated containers. Doors should be opened with care prior to unpacking, and a few minutes allowed for any fumes to disperse before entering the container.

FORMIC ACID	  		

Pungent and corrosive. Stow away from foodstuffs. Check *ISO container* carefully after use for any damage to fabric or structure. See IMDG Code.

FOSSIL WAX		1.81/1.90	Drums

An infammable substance of a wax-like nature (Cerasine) used in textile trades. See Part 2, Petroleum.

FRESH WATER		See Appendix 7, Fluids Weights and Capacities	
FRANKINCENSE		1.67/1.81	Drums

A resinous juice obtained from certain trees which, when burnt, emits a strongly fragrant smell. A valuable cargo, should be carefully tallied and watched. Stow in a special cargo locker or strong room.

FROZEN MEAT		See Part 2, Refrigerated Cargoes Also under individual categories	
FRUIT, DRIED	  	1.95/2.09 1.42	Cases Cartons

Clean, dry stowage, away from any odorous cargoes such as sheepskins, hides, casings, etc., as the fruit is susceptible to taint.

Stow away from flour, as the flour will become tainted. Avoid extremes of temperature. Adequate ventilation at all times. Under warm conditions liable to develop grub.

Commodity *Characteristics* *M3 per Tonne* *Packaging*

FRUIT, GREEN 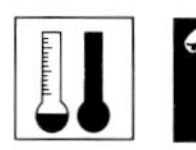 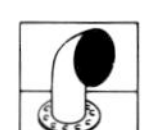

Fresh fruit respirate. This process absorbs oxygen and emits carbon dioxide and other trace elements back into the atmosphere. This process produces heat which, if unchecked, will eventually cause the fruit to decompose and rot. It is therefore essential, if a long passage or shelf-life is contemplated, that the heating process is minimised or arrested. This is done normally through refrigeration. Shippers loading and carrying instructions should be strictly observed, with particular regard to temperature. If the fruit is not received precooled, it should be brought down to carrying temperature as soon as practical. Check pulp temperature prior to loading.

For relatively short voyages, through temperate latitudes, green fruits, if picked at the right time, carry, with natural ventilation, without undue loss. Apples from N. America, oranges, lemons, grapes and peaches from the Mediterranean and N. Atlantic Islands form the bulk of fruit carried under such conditions. Even tropical inter-island voyages are acceptable if transport time is of only a few days duration.

Thorough and uninterrupted ventilation is indispensable; inattention to or defective ventilation, even for a few hours, may spoil the entire cargo. See Part 2, Ventilation.

When a mechanical system of ventilation is not provided, it is both necessary and customary to form large vertical air shafts by means of boards and cases, extending from the hatchways and from the underside of ventilators to the bottom of compartments, where they connect with gutters or air passages, formed of cases, leading to the sides. These, in turn, connect with similar air passages leading fore and aft through the cargo — all being designed to ensure the best possible circulation of air through the mass of the cargo, by which means only can it be kept cool and the heated air and gases, which the fruit throws off discharged, this being necessary in order to retard the natural process of ripening.

In some trades it is customary to arrange side and fore and aft air passages at more than one level.

Further, to assist in the ventilation, the cases of fruit are loosely, and not compactly stowed — air spaces of 200 mm to 250 mm clear being left at and across the bulkheads, the stowage stopped 200 mm or 250 mm short of deck beams and laths laid athwart between the tiers. Owing to the difficulty of adequately ventilating deep lower holds, it is seldom that such are utilised for more than a limited quantity of green fruit.

Fruit compartments, bilges, etc., should be thoroughly cleaned and sweetened and 'tween deck scuppers cleared.

Green fruit should not be stowed with or over any cargo that is odorous, moist or liable to heat, as it is liable to receive damage from such; while, on the other hand, edible and delicate goods such as tea, coffee, eggs, vermicelli, macaroni, dried fish, flour, etc., are readily damaged, if stowed with or near green fruit.

Mechanical means of ventilating fruit cargoes are now generally fitted, especially in large vessels carrying large consignments of apples, etc. This, usually, consists of extractor fans, fitted to each compartment, by which means the heated air and gases that ascend to the upper part of the compartment are withdrawn and expelled. Better results are obtained when controlled humidity air is forced by fans through specially constructed air trunks to and through the bottom tiers of cargo, thus circulating cool air through the mass of the stowage and not through or over the uppermost tiers.

The ventilating system for Refrigerated Fruit Cargoes is referred to in Part 2, Refrigeration, which see.

The hatch covers should be kept off whenever weather conditions permit. Sometimes booby hatches are fitted over hatch coamings so that the latches can remain uncovered and, at the same time, ensure that rain or spray does not wet the cargo.

Decomposed fruit throws off poisonous fumes and a number of fatal accidents have resulted owing to men having entered ill-ventilated compartments or recesses containing, or which recently had contained, decomposed fruit.

ISO containers — except in very limited circumstances involving hardy products and short voyage times closed box general purpose containers should never be used. If circumstances prevail where this

Commodity | *Characteristics* | *M3 per Tonne* | *Packaging*

type of container has to be used, then limited ventilation may be introduced by lashing the doors open, stowing the cargo with pallets or dunnage on the floors and lining the sides. CO_2 (given off by the fruit), being heavier than air, may then be induced to "pour" out of the doors. Special attention is, of course, needed to the securing of the cargo face in way of the open doors.

Open sided containers with tilts lifted (dropped if exposed to rain) are better. Reefer/insulated containers or containers with forced air ventilation facilities are best. Open top containers may also be used, but more attention will be required to covering up when exposed to rain, and dunnaging as suggested above for general purpose closed box containers.

N.B. It is doubtful if closed box containers with small ventilation grilles in each side can be properly effective with this type of cargo.

The stowage factors of green fruit vary considerably as for different ports and countries. As a result of difference in style and size of packages, method of packing, the degree of compactness of stowage, number and sizes of air shafts, etc.

FUEL OILS 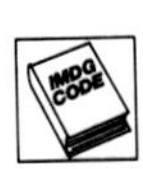

See Part 2, Petroleum.

FUEL, PATENT 

  0.92/1.00

A mixture of small coal, tar and various other combustible materials made up into brick shape or brickets.

It is a very dusty and strong smelling cargo.

Stow away from foodstuffs and delicate goods and protect adjacent cargo from the dust which is always present where this commodity is stowed and handled. Trays, tubs or like means should be used for loading and discharging patent fuel. See IMDG Code.

FULLER'S EARTH  1.11 Bags

A clay-like substance used in the preparation of cosmetics and in the clothing trade, etc. Reject all torn or rotten bags. Dry stowage, remote from timber and goods, liable to be defaced or stained by the earth leaking from the bags.

FURNITURE  1.1/2.2 Packages

Usually in fairly large packages. Stowage similar to case cars, etc. Containing personal property as they do, especial care is needed to ensure good outturn. Delay in delivery can cause considerable inconvenience.

ISO containers — vibration of container over roads may cause abrasion and chaffe. Use adequate packing materials and tight stowage to guard against this.

May require in conainer fumigation prior to shipping to certain ports. Check current destination government requirements.

FURS   2.79/4.18 Tin Lined Cases
Boxes or Bales

A large variety of furs are shipped from Russia, Far East, Canada, etc., a number of which constitute very valuable cargo, the following being amongst the higher priced furs:—

Commodity	*Characteristics*	*M3 per Tonne*	*Packaging*

Badger	Otter
Bear	Polecat
Chinchilla	Racoon
Ermine	Sable
Fox	Seal
Marten	Skunk
Mink	Squirrel
Muskrat or Musquash	Weasel
Mutria or Coypu Rat	

and are shipped in tin-lined cases, boxes (sometimes containing a disinfectant powder) or are made up into well protected bales. Bales may be canvas wrapped 1 or 2 layers. Furs packed in pairs, fur side to fur side, then folded. Naptha crystals sprinkled between skins.

Fur slip destroys commercial value and bald patches most likely to occur along the fold. Preparation — scraping, drying, curing — influences safe carriage.

Fur packages should be carefully tallied, marks checked, any packages showing signs of having been tampered with rejected, and should be stowed in strong room or special cargo locker, which is free of rats and rat-proof.

Amongst the cheaper kinds of furs are the rabbit, hare, cat and the skins of certain types of dogs, which usually are made up into bales.

All furs should be stowed in a well ventilated compartment, which contains no moist or wet goods or any that are susceptible to damage by odours from the skins.

In many cases a Sanitary Certificate from country of origin suitably endorsed by the appropriate Consular Officer is necessary.

Pelts may be carried wet, chemically treated or pickled. Wet blue hides have little smell, but give off moisture. Unlikely to deteriorate in transit. See also Hides.

See Skins for the coarser grades.

High quality furs are also carried under refrigeration, to preserve their condition. Carrying temperature 2 degrees C to 5 degrees C (35-40 degrees F).

FUSTIAN

A coarse twilled cotton fabric, including moleskin, corduroy and velveteen.

FUSTIC		1.95/2.51	Bulk

The wood of the species of West Indian mulberry tree, which yields yellow, green, brown and olive dyes. Usually shipped in lengths of from 1.22 m to 1.52 m, mostly very crooked, as a result of which its stowage factor varies very considerably, as shown, and for that reason it is not very suitable for dunnaging purposes, though at times it is used for such purpose with dry cargoes, but it is very useful for broken stowage.

GALANGAL		2.65	Well Packed Bales
		4.18	Loose Packing

The aromatic roots of an Asiatic plant, which has a hot spicy taste not unlike ginger. Ordinary dry stowage.

GALBAN OR GALBANUM

A gum used for medicinal purposes also in the manufacture of varnish. See Gums.

Commodity	*Characteristics*	*M3 per Tonne*	*Packaging*
GALENA (Lead Ore)		0.45	Bags
		0.36/0.39	Bulk
(Lead Sulphite)		1.95	Baskets

Ore shipped in bulk from Spain and elsewhere. When very soft, Galena can oxidise and give off toxic Sulphur Dioxide Gas. May liquefy if shipped wet. See Part 2, Ores. (See Code of Safe Practice for Solid Bulk Cargoes).

GALLNUT EXTRACT		1.11/1.25	Tin Lined Cases
		1.20/1.34	Cases of Powder

Stow in a cool place.

GALLS OR GALLNUTS		2.17/2.23	Cases
		2.23/2.37	Bags

A hard excrescence deposited by insects on the leaves and twigs of various plants growing in Far East, etc., from which gallic and tannic acids are obtained. Dry cargo, but one that is liable to heat, which property should be considered when arranging the stowage, which should be in a cool, well ventilated space.

GALVANISED IRON		0.67/0.78	Packages
		0.84	Coils

Sheet iron coated with zinc deposit, the bulk of which is corrugated, though plain flat sheets are not uncommon. Is made up into unprotected packages held together by iron bands or clips or in crates, which, not infrequently, are of a construction that necessitates very careful handling to avoid their collapse and/or damage to the sheets. Claims in respect of this cargo are usually due to the buckling or rusting of sheets and re-cooperage. To guard against the former, the use of chain slings on packages that lack stout protection must be prohibited; stow on the flat wherever possible on a firm flat surface, with ends free.

If overstowing sheet iron, protect by overlaying with stout boards. To avoid rust claims, dunnage well (with boards), if stowing on ground floor; do not receive goods in wet condition, a condition frequently experienced when loading direct from railways trucks; stow in a position remote from moist goods as well as from sulphates, salt and other bagged chemicals; do not overstow with acid or wet goods and avoid chafing and scratching exposed sheets by dragging other cargo over them, as the protective zinc coating is thereby damaged and rust quickly forms.

Galvanised goods should not be stowed in the same compartment as fertilisers. Fertilisers coming into contact with galvanised goods will cause corrosion of the protective zinc coating. If stowage of the two commodities has to be effected, adequate separations must be provided. See also Part 2, Steel.

ISO containers — the size of sheet — and resultant crate — may make the package awkward and uneconomical to load into a closed box container. Open-sided containers and flat racks often allow a greater number of crates to be carried, with side access suitable for fork-lift truck operations. Appropriate weather protection (tarpaulin or plastic sheets) must be provided for flat rack stowage. Particular care must be exercised in securing, particularly in the absence of side wall restraint. Lashing equipment must be set up tight in way of bearers or skids to avoid loosening when cargo flexes and vibrates. Non-stretch material (e.g. wire or metal strapping) should be used. Where appropriate and possible, timber shoring to be used, with wire nails to secure battens, etc., on the floor.

See also Part 2, Neobulk.

Commodity	*Characteristics*	*M3 per Tonne*	*Packaging*
GAMBIER		2.65/2.79	Bales
		3.07	Cases
		3.34	Baskets

An astringent substance extracted from the leaves of certain East Indian shrubs, used for tanning and dyeing purposes.

It is one of the most objectionable cargoes of the Far East and should not be stowed over anything except coal, nor against any commodity except jetalong. The drainage from gambier (a thick viscuous fluid) is great in hot weather, which, when cold weather is experienced, sets very hard, rendering it exceedingly difficult to detach the packages from deck, bulkheads, stringers, etc., and, unless a liberal supply of sawdust is introduced between tiers, the packages become almost a solid mass.

The greatest care should be exercised to keep gambier drainage from fouling bilge suction pipes, strum boxes, 'tween deck scuppers pipes, etc., as, if the fluid gets into such, it will necessitate the whole pipe line being disconnected and, in all probability, the removal of such, as that will be cheaper than clearing them.

If possible, stow in bridge, poop or shelter deck space; if in the latter, airtight bulkheads, the lower part of same to be made watertight, should be erected to protect other cargo from damage by the heated moisture and odours emitted by or drainage from gambier.

Stow on 50 mm or more of sawdust and cover each layer with sawdust to the depth of 25 mm or more, also at ships' side if sparring is not fitted.

Gambier generates humid heat, so that every care is necessary to protect delicate cargo in the vicinity from the effect of the fumes emitted. Ample ventilation of compartments containing gambier should be provided for this purpose, which should include the removal of hatch covers whenever conditions permit.

Dry or delicate cargo should not be stowed in the same compartment or in one that is connected by ventilators to that in which gambier is stowed.

If overstowing with other coarse and suitable cargo, the gambier should be boarded over and well matted to prevent the top cargo sinking into the gambier. Cutch, being of the same nature as gambier, is suitable for overstowing, if the need arises, but, if possible, overstowing should be avoided.

GAMBOGE	 	1.84/1.95	Cases

A gum resin obtained from trees growing wild on the coasts of Far East, Malabar, Sri Lanka, etc., from which a yellow water colour pigment is obtained, It is a strong poison with a bitter taste. Stow in a cool place away from foodstuffs. See IMDG Code.

GANGA

Used in lieu of tobacco and as a narcotic. A product of the hemp plant. Special stowage, very liable to pillage.

GAPLEK		1.53/1.67	Bags
		2.37	Chips
		2.09/2.23	Roots
		1.95	Roots Bulk

See Tapioca.

Commodity	*Characteristics*	*M3 per Tonne*	*Packaging*
GARLIC	 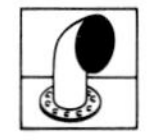	2.65	Bags

The bulbous root of a plant, not unlike the onion; has a highly pungent taste and very strong smell. Perishable cargo, stow same as onions. May be carried under refrigeration. Suggested temperature 0 degrees C. Full air freshening to reduce build up of CO_2 and other gases. Observe shippers loading and carrying instructions.

ISO containers — not suitable for closed box containers, except for short journeys when doors are best left open to assist ventilation. May carry somewhat better in open-sided containers, tilts rolled up and dunnage (or pallets) on the floor to improve ventilation. Care must be exercised in the Terminal and on board to ensure adequate weather protection, e.g. tilts rolled down if exposed to rain.

See also Fruit — Green.

GARMENTS, HANGING		var.	Unpacked

In closed box *ISO containers* on specially constructed racks that have to be dismantled and returned complete to shippers. Garments may be polythene-wrapped and partially exhausted of air at time of stuffing into the container, to reduce moisture content and increase number of garments in the limited space. Very important that they are not unpacked for inspection purposes except by expert personnel, since there is little hope of replacing all the garments temporarily removed. May have dessicants in the container to absorb migrating moisture.

GARNET ORE		0.50/0.56	Bags

Also known as Garent, Rubble Ore, a dry dusty cargo. See Ores.

GASES	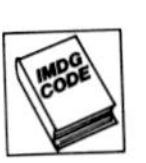	See Compressed Gases	
GASOLINE	 	1.39/1.45	Cases Drums Bulk

A highly inflammable spirit with low flash point. See Part 2, Petroleum Cargoes.

Shipped in 5 gallon cans, two in a case, also in drums of varying capacity and in bulk. See Case Oil for Stowage. See IMDG Code.

GAUGAO			Bags and Cartons

Starch used in the preparation of clothing materials. Keep dry.

GELATINE	  	1.67/1.95 1.53/1.81	Cases Bags

An animal product obtained from the bones and resembles jelly. May also be shipped as powder or sheet. Susceptible to damage by moisture (causes powder to lump) and heat causes melting.

ISO containers — stow and store away from direct sunlight. Cool dry stowage, free from taint.

GEMS			See Precious Stones Also Part 2, Specials

Commodity	*Characteristics*	*M3 per Tonne*	*Packaging*
GENEVA (Hollands, Schiedam)		1.62/1.73	Cases

A spirit distilled from grain and flavoured with juniper berries. Should be carefully watched to avoid broaching. Stow in special cargo locker or overstow immediately with other cargo. Also carried in tanks. See IMDG Code.

GENTIAN ROOT		3.34/3.48	Bales

A dried herb root with bitter taste, its tonic properties making it valuable for medicinal purposes. Ordinary dry stowage away from oily, moist or odorous goods.

GHATTI		An Indian gum. See Gums.	

GHEE	 	1.39/1.53	Tins in Cases Tanks

An Indian clarified butter, mostly prepared from buffalo milk and largely used in the Far East. Often shipped in second-hand receptacles, treat as wet cargo liable to leakage and stow well clear of odorous goods.

ISO containers — may be carried in closed box containers. Probable payload 17.5 tonnes for 20 ft. Closed box containers.

GILSONITE		See Pitch	
GIN		See Geneva.	
GINGELLY (Sesame)		1.62/1.67	Bags

The seeds of an annual, cultivated in large quantities in East Mediterranean countries, India, China, Japan, etc., from which is obtained "Gingili" oil. See Seeds for Stowage.

GINGER 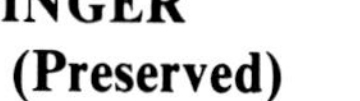 **(Preserved)**	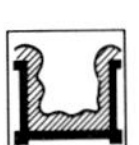	1.95/2.09 1.58/1.81	Cases Casks

The ginger root preserved in a syrup. Wet cargo, very liable to leakage; the cases require careful handling and should be watched to avoid pilferage. See Chinese Groceries.

GINGER ROOT (DRIED)	 	2.23/2.65	Bags

The dried stem-like root of a reed-like plant grown in the East and West Indies, valued for its hot and spicy properties. Shipped in bags and comes in handy for filling broken stowage amongst clean cargoes. Stow away from tea and other goods liable to taint.

Commodity	*Characteristics*	*M3 per Tonne*	*Packaging*
GINGER ROOT (GREEN)	  	2.23	Crates

Succulent aromatic root from sub-tropical countries. Used in oriental cooking. May be damaged by drying and shrivelling (too much ventilation), mildew and rot (too high humidity). Roots from Singapore and Hong Kong have the reputation of being unsuitable for long storage.

Pre-shipment condition is important. Roots should be dry on the outside with little or no earth adhering. There should be no evidence of mildew, green shoots, or excessive drying (shrivelling).

May be best suited to carriage under refrigeration. Suggested carrying temperature + 10 degrees C (50 degrees F).

ISO containers — open-sided containers with tilts raised to allow maximum ventilation, with due regard to the prevention of moisture (rain, etc.) entering the container.

GINSENG		3.07/3.34	Baskets

A dried plant root grown in the Far East, which is very popular amongst the Chinese as a restorative medicine. Ordinary dry stowage, away from moist, oily or odorous goods. Care against pilferage at time of loading.

GLASS		1.11/1.39	Crates
		1.39	Plate Common
		1.39/1.53	Good Grade

In the plate or sheet form this is a very fragile cargo requiring great care in handling and stowing to avoid breakage. Certain sheet glass may "cloud", if it becomes damp. Plate and window glass is packed in strong crates or cases which should be devoid of battens on the outside edges to enable them to rest on deck, etc., for their length.

They should be stowed on firm ground, on the 'tween deck preferably, with extra large and heavy packages in square of hatch for ease of handling. On no account should glass be stowed on top of any cargo which is liable to settle such as coke, bagged stuff, etc. The Courts have held that the stowage of glass on sand which was not properly levelled and boarded over was "improper Stowage".

Dunnaging should not be resorted to, as it is preferable for the package to be supported along its entire length.

Crates and cases of glass should always be stowed On Edge and in all cases plate glass should be stowed Athwart. The general run of window glass will stow satisfactorily fore and aft if desired.

It is essential that the crates, etc., be well chocked off and all broken stowage filled with suitable material in order to reduce to a minimum any movement in a seaway.

Slings of glass should be made up in such a manner that the deeper packages are central and the smaller on the outside, grading upwards towards the centre, thus avoiding the rope slings straining the crates, etc., at their upper edges with disastrous results to their contents.

The use of trays for handling glass has many and decided advantages.

Crates of glass bottles stow at 2.23/2.79 M3/tonne. Crates of glassware vary in stowage over a very wide range.

Glass beads should be stowed in metal drums. There is a case on record of glass beads shipped in hessian bags; some bags were torn, as a result of which some of the contents spilled on to a flour stowage, resulting in some of the flour shipment being condemned.

GLAXO		2.23/2.65	Cartons. Bags.

A milk by-product shipped in large quantities from Australia and New Zealand. See Milk Powder.

Commodity	*Characteristics*	*M3 per Tonne*	*Packaging*
GLUCOSE		1.17/1.25	Cases
		1.28/1.34	Barrels
		1.11	Drums

A sugar, obtained from the juice of grapes and other fruits, also by the action of sulphuric acid upon starch which is largely used in brewing. Do not stow with cargoes that are susceptible to wet damage.

Commodity	*Characteristics*	*M3 per Tonne*	*Packaging*
GLUE (Animal)		4.18/5.57	Bales
		3.34	Drums
		3.07/3.34	Casks Liquid
		1.81/2.09	In Cases
		3.34/3.62	Bags

A viscid substance obtained by boiling the hides and hoofs of animals. Glue pieces are liable to spontaneous combustion. Stow in poop or bridge space or in square of hatch. Shipped both in the liquid and dry condition. See IMDG Code.

Commodity	*Characteristics*	*M3 per Tonne*	*Packaging*
GLUE REFUSE	 	1.25/1.29	Casks

An animal manure composed of hide trimmings, etc., after the gelatine has been extracted. On account of its offensive odour, suitable for on deck stowage remote from crews quarters. See IMDG Code.

Commodity	*Characteristics*	*M3 per Tonne*	*Packaging*
GLUTEN		2.23	Bags

Product of flour. Treat as Flour.

Commodity	*Characteristics*	*M3 per Tonne*	*Packaging*
GLYCERINE		1.11/1.39	Cases
		1.25/1.53	Drums
		0.81/0.84	Bulk

A colourless viscid sweet tasting fluid soluble in water and in alcohol, obtained from animal fats, oils, etc. Packed in a variety of ways. Wet stowage. Drums from the Far East stow as high as 2.23 M3/tonne (80 cu. ft/ton). Weighs approximately 1250 kg per M3 at 37.8 degrees C (78 lbs per cu. ft at 100 degrees F) and is shipped in the form of a solution of 90 per cent glycerine and 10 per cent water. Salt content varies between 0.5% and 7% or more.

While the caustic alkali content of crude glycerine will dissolve any ferric oxide (rust) present on the steel of the tank, it has no harmful effect on the steel itself.

This commodity does not solidify, but being very heavy may require slight heating (which, it is said, causes it no harm) to facilitate pumping during cold weather. Pumping temperatures 35-40 degrees C. (95-105 degrees F).

While this cargo is not in any sense dangerous, it has very penetrating qualities, which makes it very difficult to avoid leakage even when tanks are known and certified to be oil tight. In order to protect plate joints, points of rivets, etc., from corrosion, it is the practice to protect them with a coating of paraffin wax. Wax does not contaminate glycerine.

Shipment is in steel drums or in bulk when B/L should be claused "Not responsible for leakage, all glycerine remaining in tank to be delivered whereupon vessel's responsibility shall cease". See Tank Cleaning, etc., under Vegetable Oils. See also Coconut Oil Products.

Commodity	*Characteristics*	*M3 per Tonne*	*Packaging*
GOATHAIR		See Mohair	
GOAT MEAT		2.09/2.23	Cartons

Goat meat is shipped in carcase, cartons and, occasionally, bags. The meat is extremely lean and, as much of this is shipped to West Indian and Mediterranean ports where it is discharged during hot weather, it should be stowed in the most accessible place and discharged as quickly as possible.

Suggested carrying temperature −12/−9 degrees C.

See Part 2, Techniques, Refrigeration.

GOATS		See Part 2, Techniques, Livestock.	
GOATSKINS		See Skins	
GOLD, BULLION OR COIN		See Bullion. See also Part 2, Specials.	
GOLD SLAG		0.42	Bags

A heavy metal dross. Dry cargo.

GOORA NUT		See Kola Nut	
GRAM	  	1.53/1.67	Bags

An Indian pulse or bean, known also as Chick Pea, used as an article of food in India and in considerable demand as a cattle food. It contains a high percentage of oxalic acid and is very liable to sweat, though not giving to fermenting. Cool stowage away from moist goods and those liable to damage from sweat. Dunnage well.

GRAIN			Bags, Bulk

The successful transportation of grain in bulk is dependent upon the moisture content. The moisture content of wheat should not exceed 13 per cent and that of maize, 14 per cent. For voyages of long duration the moisture content should be less. If the moisture content varies in different parts of the cargo it may well be that pockets of high moisture content will ferment and generate heat. This heat travels upwards and the moisture at the top of the stow may sometimes be mistaken for sweat damage. Flaking is normally a sign of excessive moisture content.

Most self trimming bulk carriers are designed so that the hatch coamings become a feeder which, in turn, minimises the space in the hold wings and ends largely between the top of the stow and the underlying deckhead. Ventilation is therefore ineffective. Ventilation may remove warm air from the surface of a heating cargo, but will not eliminate the source of internal heating as most systems are not designed to force air through the bulk of the cargo. It is therefore felt that the efficiency of ventilating a bulk cargo is minimal. Further unavoidable but small damage to the cargo may occur when the warm grain is loaded to a ship destined for discharge in a cold climate. Shipside sweat will occur and condensation from the underside of deckheads may drift back onto the surface of the cargo. Where there is surface damage this should be carefully removed, preferably manually, in order to effectively separate the good from bad.

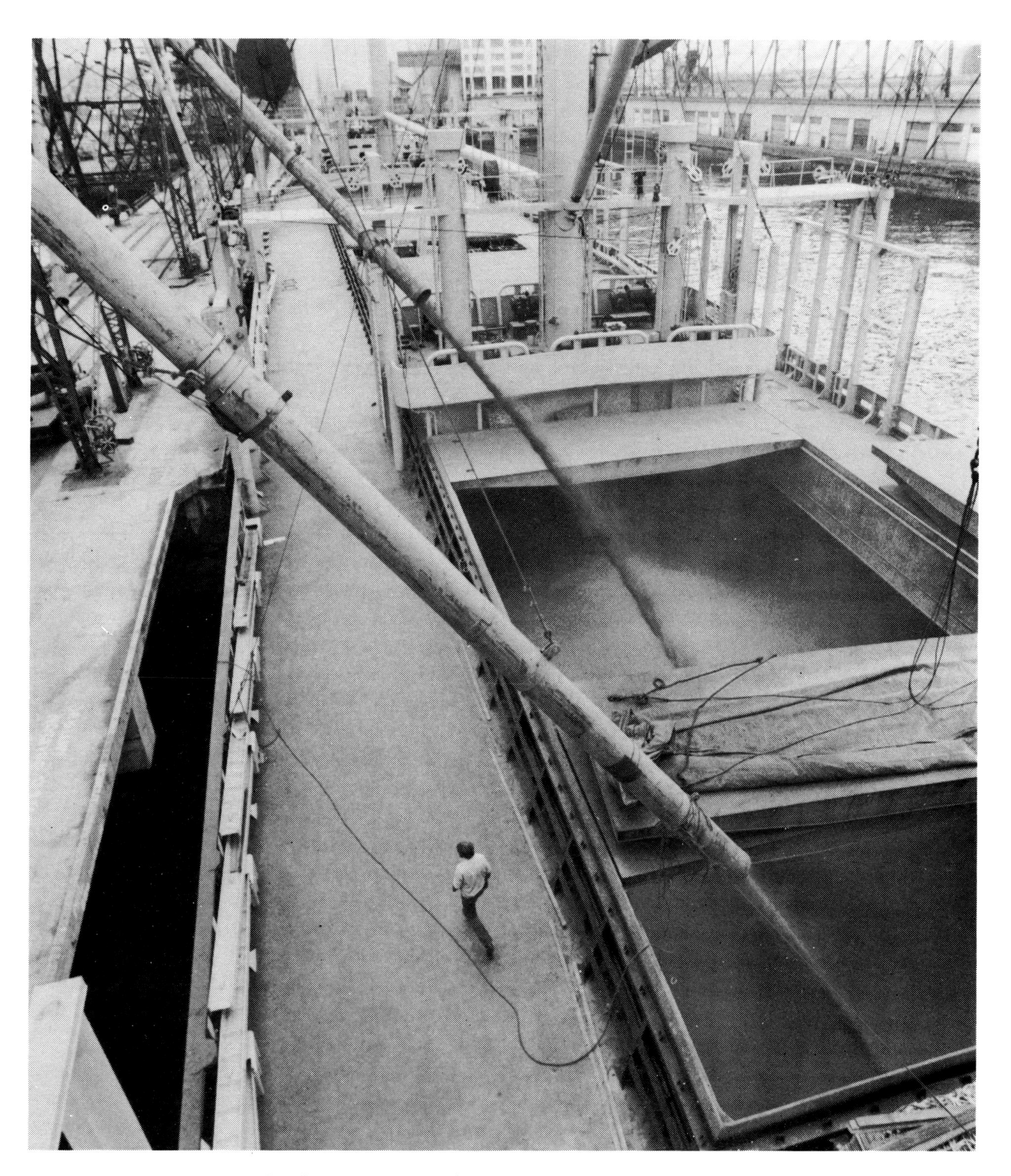

Loading grain on bulk carrier fitted for carrying logs/packages timber.

Commodity	*Characteristics*	*M3 per Tonne*	*Packaging*

Care should be taken to ensure that the heat applied to double-bottom fueltanks is not excessive. Many claims arise where the grain adjacent to the tanktops are toasted from this source. Soyabean meal is particularly susceptible to this problem.

It is virtually impossible to detect visually the presence of excessive moisture content in the cargo. However, if it is being loaded from a country known for this problem, it may be prudent to take random samples and seal them in jars or bottles for future analysis.

There are many parts of the world where discharge of grain is almost always routinely followed by a claim for shortage. Normally this is based upon weights obtained from a weighbridge and truck discharge or, alternatively, from an elevator where the ship has patently no control over the discharged tonnage measurements. In this case it may be prudent for the shipowner to arrange for a sealing of the hatches upon completion of loading and to invite the receiver to be present at the unsealing prior to commencement of discharge. Upon completion of unloading, empty hold certificates should be issued by an independent surveyor. These measures may assist an owner in defending a claim for shortage.

Grain	*M3/tonne bags*	*M3/tonne bulk*
Barley	1.67/1.81	1.45/1.67
Buckwheat	1.67/1.81	1.39/1.53
Maize	1.39/1.67	1.34/1.48
Oats, clipped	1.95/2.09	1.78/1.89
Oats, unclipped	2.34/2.40	2.01/2.17
Rye	1.48/1.53	1.34/1.45
Wheat	1.34/1.48	1.25/1.39

See Part 2, Grain Cargoes in Bulk.

ISO containers — bagged grain is suitable for closed box containers. Preferably not to stow bags with their length parallel to the long axis of the container, or settling bags may deflect the container sides. Bulk grain may be loaded into dry bulk containers direct by overhead hoppers. Trimming is usually required to achieve the payload — or a quick "drive round the block" to settle the grain. Care should be taken at time of loading to ensure that the roof hatch seals are not obstructed by loose grain. Also that the hatches are well tightened down onto the seals to prevent any ingress of water. (In some makes of containers the drain holes from the hatch wells may also need to be checked clear to prevent a build-up of rain water). Closed box containers may usually be used in conjunction with a temporary bulkhead in way of the doors and throwing or blowing equipment to load the grain. See Part 2, Containers. In either case (i.e. bulk container or ordinary container), labels, locks and seals should be used to prevent doors being opened by accident.

GRANITE SLABS		0.45/0.50	Large Dressed
		0.50/0.56	
		0.64/0.70	Macadam (broken)
		0.81/0.84	Chippings

Ground stowage for large dressed slabs or blocks. Macadam (broken) sets if stacked. Chippings may be liable to shift.

ISO containers — bulk granite chips are not suitable for container stowage on account of their abrasive qualities.

GRANULATED SLAG		0.90	Bulk

Residue of blast furnaces in granulated form. Detrimental if loaded too hot. Angle of repose 40 degrees.

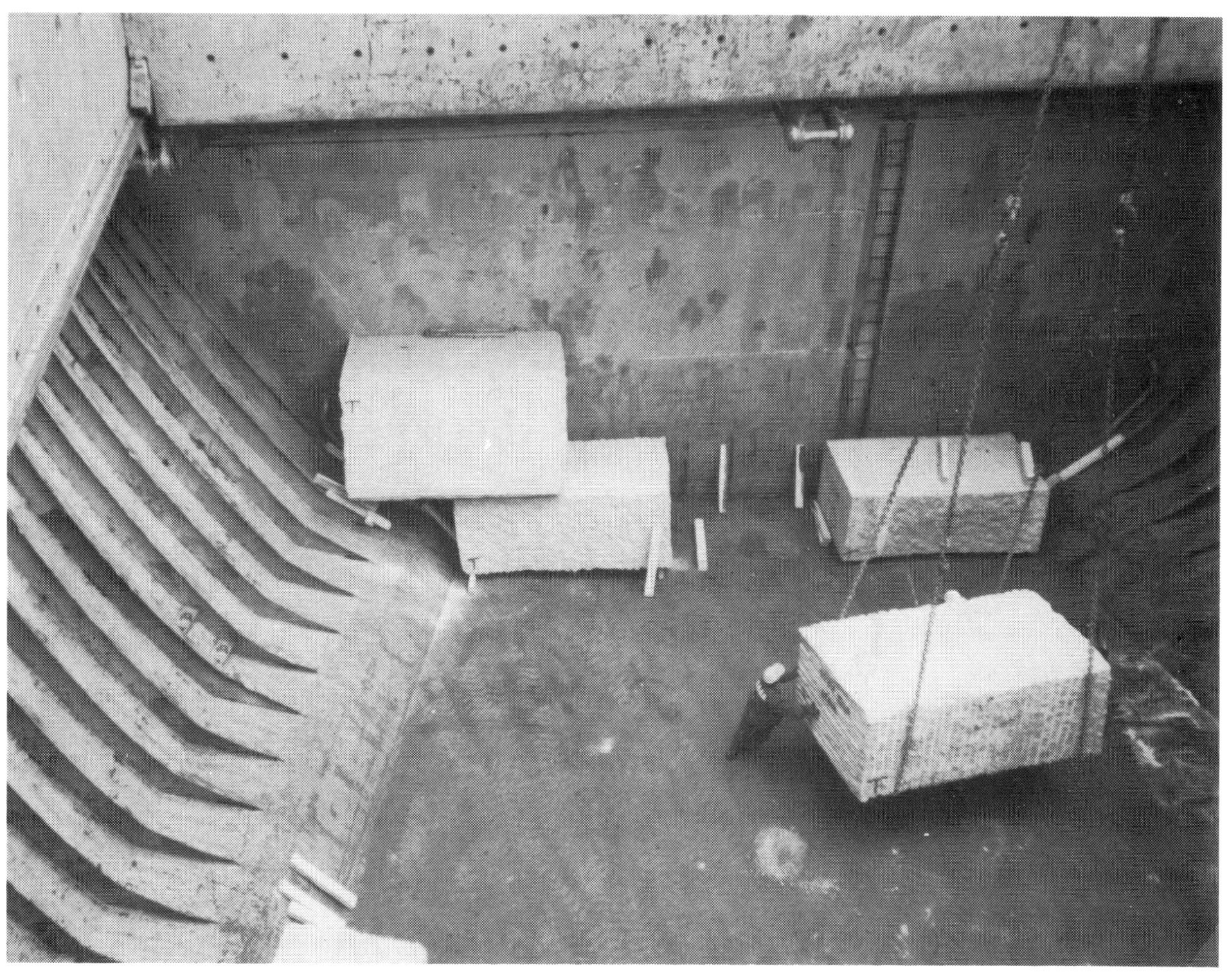

Granite blocks — adequate dunnaging essential.

Granite blocks. Completion of loading.

Commodity	*Characteristics*	*M3 per Tonne*	*Packaging*
GRAPEFRUIT		See Citrus	
GRAPES		3.29/4.18	Cases, Cartons

Shipped in cartons, trays or cases. Stow preferably in 'tween decks. Floor dunnage should be suitable battens laid in line with the air flow, with a covering of 150×25 mm boards to ensure full support of the packages. Damage noted on out-turn may be of pre-shipment origin and not readily detected at the time of loading. Table grapes do not ripen after cutting and should be chilled as soon as practical after they are cut.

If damage is noted on discharge, a competent surveyor should be instructed to investigate. Samples should be taken in order to analyse the degree of infection or deterioration. If palletised on 2 way entry pallets, ensure stowage is such as to promote cold air passage through and around each pallet. Normally refrigerated. Observe shippers' carrying instructions.

See Fruits, Green, also Part 2, Techniques, Refrigeration. Carrying temperature −1 to 0 degrees C.

Commodity	*Characteristics*	*M3 per Tonne*	*Packaging*
GRAPHITE **Plumbago**	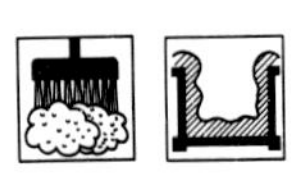	1.34/1.39 1.25/1.34 1.11/1.25	Paper Lined Kegs Cases Double Bags

An iron black or steel grey mineral — known also as plumbago or black lead — shipped in considerable quantities from Siberia, Canada, India, etc., in kegs and cases.

Unless the containers are well made and lined with stout paper a great deal of siftage of contents occurs, so that all precautions should be observed to protect other cargo in vicinity of graphite from damage by the powder. Dry stowage.

Commodity	*Characteristics*	*M3 per Tonne*	*Packaging*
GRASS CLOTH		2.79/3.07	Cases

Clean dry cargo.

Commodity	*Characteristics*	*M3 per Tonne*	*Packaging*
GRASS RAFFIA		3.34/3.62	Pressed Bales

A fibre root. Ordinary dry stowage.

Commodity	*Characteristics*	*M3 per Tonne*	*Packaging*
GRASS SEEDS	   	1.39/4.18	

Many varieties, all more or less valuable, are shipped. Generally they are shipped in bags. Sometimes double bags are used to prevent loss of contents, as the seeds are, mostly, very fine and small and easily pass through ordinary sacking in course of handling.

The higher priced of these seeds are, not infrequently, packed in boxes.

Must be stowed in a cool dry place where no moisture is apt to reach them. The commoner and coarser class of seeds are referred to under Grain, which see. In some cases may be carried under temperature controlled conditions. Suggested carriage temperature + 10 degrees C.

Different varieties have different stowage factors:—

Name	*M3/tonne*	*Package*
Browntop	3.09	Bags
Clover	2.20	Bags
Cocksfoot	3.73	Bags
Cowgrass	2.09	Bags
Dogstail	2.68	Bags
Fescue	4.18	Bags
Rape	2.42	Bags
Ryegrass	3.09	Bags
Yarrow	2.50	Bags

Commodity	*Characteristics*	*M3 per Tonne*	*Packaging*
GREASES	 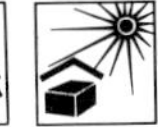	1.53/1.81	Drums

All greases are solidified oil and liable to melt if subjected to heat. Stow in a cool place well removed from engine and boilers. Treat as Wet Cargo liable to leakage. See IMDG Code.

GREEN FRUIT  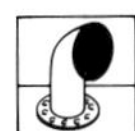

See Fruit, Green.

GREY AMBER See Ambergris

GROUNDNUTS		1.67/1.81	Bags
PEANUTS		3.34/4.18	Shelled
ARACHIDES		2.79/2.93	Unshelled
MONKEY NUTS		1.53/1.67	Bulk
KERNEL		1.67/1.81	Bags
		2.51	Bulk

An edible earth nut which when new contains a large amount of moisture, shipped from Southern U.S., Argentina and China, and valued for its oil yield which averages 45 per cent of its weight. Also used for human consumption.

Groundnuts are normally in single hessian bags and may be sound in appearance at both loading and discharge port. However, claims for mould damage on out-turn may reach epidemic proportions.

This mould growth may be influenced by one or all of the following:—

1. The cargo shipped onboard the ship may be the crop from the previous season.
2. The temperature of the groundnuts allied to the moisture content.
3. The quality of the groundnuts: foreign matter, broken kernels, etc.
4. Embryonic mould growth at the time of shipment.

Preloading history, quality of content and moisture content, may have to be accepted at face value and it is appreciated that it is virtually impossible for the master to question the veracity of the certificates.

Holds and decks prior to shipment of cargo must be clean and dry. Discontinue loading during rain. Ventilation channels should be left in the stow and the bags protected from direct contact with metallic surfaces by dunnaging.

The Master and ship's staff should pay particular attention to the condition of the cargo prior to loading. As many bags as possible should be inspected and those showing any signs of mold growth or excessive moisture should be rejected. Meticulous attention should be paid to ventilation and an accurate record kept. It is important that the temperature of the groundnuts is measured during loading. If the ambient air temperature is 2 to 3 degrees celsius below the temperature of the cargo at the time of loading, then ventilation should commence.

In view of the potential severity of claims arising from the carriage of this cargo, it may well be that the prudent shipowner may wish to appoint a surveyor to monitor, sample and report upon the out-turn of the cargo. See Part 2, Techniques, Bag Cargo, also Nuts.

GROUNDNUT CAKE	 	1.40/2.10	Bulk

See Seed Cake. See IMDG Code. (See IMO Code of Safe Practice for Bulk Cargoes).

Commodity *Characteristics* *M3 per Tonne* *Packaging*

GROUNDNUT (PEANUT) OIL

Groundnut oil which, when produced from good-quality groundnuts, is a light-coloured oil, is produced for export mainly in West Africa, although production of the oil in India is on a far larger scale. Indian production is, however, mainly for domestic consumption. The oil contains a significant proportion of saturated material which will cause the oil to partially crystallise when exposed to low temperatures. The oil as shipped should have a low free fatty acid content, contract limits normally being no higher than 1.5%. It should be stored at ambient temperature when shipped, but recommended loading and discharge temperatures are 20-25 degrees C.

The oil has a density at 25 degrees C of approximately 910 kg/m^3. In the refined state it is used as an edible oil and can also be used as a fat if processed further (hardening) after refining. The high cost of the oil limits its use in Europe and North America.

See Part 2, Oils and Fats. See Appendix Table No. 11.

GUANO  1.11 Bulk
1.17/1.23 Bag

Is the more or less fossilised dung of seabirds, mainly collected from islands off the West Coast of S. America and the Pacific Ocean.

It is a valuable manure containing a high percentage of lime phosphate and ammonia and is quite unsuitable for carriage in a vessel carrying foodstuffs and other delicate goods.

Must be kept apart from nitrate of soda, which is also shipped from S. American ports, and carefully protected from contact with salt water, but rain water does not adversely affect it.

Charterers supply bagging, etc., when required.

GUINEA CORN 1.50/1.39 Bags

See Durra.

GUM  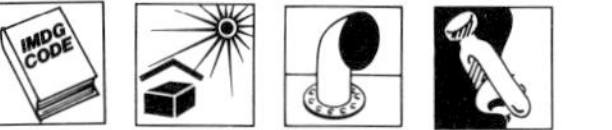  1.39/1.81 Bags
Gum Resin

Gum is the generic term applied to a large variety of substances composed of the viscous exudation of numerous trees and plants — mostly tropical.

The true gum is soluble in cold water, while the other varieties, which contain resin and essential oil (the gum resin kind), are not so, but are soluble in spirits. Some of the latter kind will, however, absorb moisture more or less freely and form into an adhesive paste.

Gums are used for a variety of purposes, by food, pharmaceutical, and industrial companies, to glaze and stiffen the fabric, for making mucilage, varnish, etc., as well as for medicinal purposes, some of the true gums being classed valuable cargo.

All are inflammable and may be damaged if heated, so that dry, cool stowage, well dunnaged in a well ventilated space is necessary. They should also be stowed apart from oils and greases, and the finer kinds from strong smelling goods.

Arabic	Dragon's Blood	Locust Bean Gum
Assafoetida	Euphorbium	Mastic
Bdellium	FossilGums	Mouri
Benjamin or Benzoine	Frankincense or Bdellium	Myrrh
Caboge or Gamboge	Galban or Galbanum	Olibanumor Bdellium
Camphor	Ghatti	Persian
Cerasin or Cherry	Herol	Senegal
Chicle	Karage	Thus
Churrah	Kalera	Tragacanth
Copal	Kauri	Wattle
Dammar	Lac	Yacca

Commodity	*Characteristics*	*M3 per Tonne*	*Packaging*

Gums are variously packed for shipment, sacks, chests, cases, etc., being used. They form a dry, clean cargo. Stow in a cool place.

ISO containers — may be carried in closed box containers. Keep away from heat and direct sunlight when stowed on board or in the stack ashore. See IMDG Code.

GUM OLIBANUM **BDELLIUM**	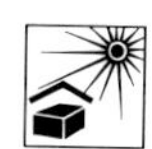	1.90 1.84	Cases Bales

Obtained from trees in the Far East. Stow in a cool place. See Gums.

GUNGA, GUNJA **GANJA**			Cases

A narcotic drug obtained from the hemp plant. In a certain form it is mixed with tobacco leaf and smoked. Some countries prohibit or restrict the importation of this drug. See Hashish.

GUNNIES	 	1.38/1.81	Bales

A coarse jute cloth woven with single threads, while hessian is of double threads, shipped in large quantities from Calcutta, etc. Clean cargo, but care, especially during the S.W. Monsoon, must be exercised to see that damp or wet bales are not shipped and particularly if receiving from lighters. When bales are wetted, the lightermen, given the time, are able to dry the outside folds and escape detection that way, leaving the ship to fight claims for rotted bales. Stow apart from oils, greases, moist and sweat producing cargo and do not permit the use of hooks for handling.

GUNPOWDER		See Part 2, Dangerous Goods	
GUR		See Jaggery	
GUTTA PERCHA **GUTTA**		2.37 2.23	Bags Cases

Is the dried resin of a very large tree grown in Malay Archipelago and is used largely for insulating electric cables, etc.

Has no objectionable properties.

Stow as for Rubber, which see.

GYPSUM		1.20/1.28 1.06	Bags Bulk

A soft mineral, a common form of which is alabaster, another form is known as monstone or selenite, which is transparent. On being dried it crumbles into powder and is then called plaster of paris.

Do not store near goods liable to be damaged by moisture or dust. Liable to cake if wetted.

HAEMATIN			Cases

Logwood crystals used for dyeing purposes.

Commodity *Characteristics* *M3 per Tonne* *Packaging*

HAIR, ANIMAL 1.53/1.95 Pressed Bales
3.34/3.90 Unpressed

Animal hair is usually shipped in bales, compressed to various densities, or in bags and, in some instances, a Sanitary Certificate from country of origin endorsed by the appropriate Consular officer has to be produced at destination.

HAIR, HUMAN 1.67/2.23 Cases
1.53/1.95 Bales

This is a valuable cargo. Cases should be carefully examined before receiving for any indications of tampering. The weighing of each case is advisable as a check against pilferage. Some cases contain a disinfectant powder and in some instances a Sanitary Certificate as for animal hair has to be produced. Dry cargo stow in special cargo locker or in holds, provided it can be overstowed quickly with other cargo for the same destination, but not in contact with other cargoes.

HANGING GARMENTS See Garments, Hanging

HARDBOARD  1.67 As for Plywood

HARES  Crates

See Rabbits.

HASHEESH
HASHISH
BANQUE

A narcotic drug obtained from hemp, the importation of which is only permitted to most countries under very strict regulations. See also Opium.

HAY 

 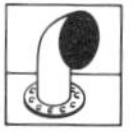  3.34/4.46 Bales

Loose, wet or damp hay should not be received on board under any circumstances; all broken bales should be rejected. It is pressed into bales of varying densities.

If stowing large quantity, air shafts should be introduced to ensure air circulating through the stowage, where no mechanical ventilation is available, as proper ventilation is essential to avoid heating and spontaneous combustion.

Every protection against an outbreak of fire should be taken. See Feedstuff. See IMDG Code.

HEAVY INDIVISIBLE LOADS See Part 2, Techniques

HEAVY SPAR  Bags

Chemical powder. Odourless but dirty.

HEAVY WEIGHTS/LIFTS See Part 2, Techniques.

Discharge of heavy lift. Essential that shipper's instructions regarding slinging and point loading are strictly observed.

Commodity	*Characteristics*	*M3 per Tonne*	*Packaging*

HECOGINON — Drums

A liquid extract of sisal. A valuable commodity which should be given lock up stowage.

HEMP	 	2.56/2.93	Bales Pressed
(Abaca)		1.39/1.67	High Density
		6.13/6.97	Unpressed
		2.51/2.79	N.Z. Bales dumped

The fibre of a variety of plants grown in Russia, the East, etc. It is longer, coarser and stronger than flax, to which it is allied, and is used in the manufacture of cordage, paper, sailcloth, etc.

It is a dry, clean cargo, devoid of any objectionable properties, but must be kept clear of greases, oils, etc., as it is liable to spontaneous combustion if it has been in contact with such.

Manila bales run 8 to the ton of 2240 lbs, stow at about 0.37 cu. m per bale. A certain class of China hemp is pressed to a very high density, stowing at about 1.39 cu. m per tonne. See IMDG Code.

ISO containers — traditional, loosely compacted 125 kg (275.6 lbs) bales can stow 60 to the 20 ft container. New high density bales (125 kg) can stow 80 to the container. However, the Trade deals in 25 bale units, so 75 bales per TEU has become a normal load.

HEMP SEED		1.95	Bags
		1.62	Bulk

The seed of the hemp plant. See Seeds.

HENEQUEN		2.51/3.26	Bales
SISAL HEMP			

The fibre of the agave or maguey growing in Mexico and other central American countries, also in the Bahamas, Philippines, etc. Otherwise known as sisal hemp, used in the manufacture of cordage, for which it is well adapted, as this fibre resists the action of salt water better than other fibres.

HERBS	  	5.30/5.57	Cases. Bags.

The dried leaf, stalk or seeds of certain plants used for culinary purposes and in preparing medicines. Shipped in relatively small quantities packed in cases or well-made bags. Most kinds are scented and may taint other cargoes. Stow in a dry cool place.

HERRING		See Fish	

HESSIANS		1.39/1.81	Bales

A cloth woven from jute, shipped from Scotland and India, etc., in large quantities, generally referred to as gunnies, which see, or burlap amongst nautical men.

Shipped in rolls or made up into bags or sacks which are hard pressed into bales, varying in size and contents for different destinations.

Is a clean dry cargo — ordinary stowage. See Gunnies. The use of hooks for handling should be strictly forbidden. Must not be contaminated with vegetable oils.

Commodity	*Characteristics*	*M3 per Tonne*	*Packaging*
HIDES	  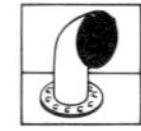 		

Hides are shipped in the dry and wet conditions.

They should be very carefully tallied and Sanitary Certificates, endorsed by the appropriate Consul or other Authority, should be obtained almost in all cases.

Dry Hides — see Skins. Should not be stowed near wet hides.

Wet Hides — (Salted or pickled). Various methods are adopted in shipping wet hides, i.e.

In casks, stowing at about 1.67 M3/tonne.
In barrels, stowing at about 1.53 M3/tonne.
In bales or bundles — with the hairy side out — 1.39/1.67 M3/tonne wet.
In bags, wet, 1.81/1.95 M3/tonne.
In bags, dry, 2.09/2.23 M3/tonne.
In the loose condition about 1.95 M3/tonne wet 2.79/4.18 M3/tonne dry.
Unitised — compressed, polythene wrapped and strapped to disposable pallets.

When packed in casks or barrels, treat as wet cargo, which throws off a very strong smell, and stow away from dry, fine and other goods liable to damage from moisture or tainting.

Wet hides put in bundles or bales or when shipped loose are very fruitful of claims and, unless the greatest care is observed in preparing holds for their reception also in stowing and in ventilating them, the vessel is liable to be held responsible for damage which, likely, is attributable to conditions existing prior to shipment.

Much of the damage noted after discharge is due to the hides being old and/or improperly cured or to the hides having lain for a long period in leaky lighters or on earth or damp floors of badly ventilated warehouses. It is a most difficult matter, however, for this incipient form of damage to be directed by tally men or ships' officers at the time of shipment and, as clean receipts are usually demanded, the only defence which can be set up against a claim for damage arising from the above or like conditions is that of having properly prepared holds and that the hides have been stowed and ventilated in the best possible manner.

Whether wet hides are shipped in bales, bags, bundles or loose, it is essential to ensure sound delivery, so that they do not come into contact with iron, steel or any other metal or with oak timber.

Dunnage should be carefully examined and any pieces of oak rejected.

Reject all hides with the hair off, any that are damaged and any hides showing signs of Red Heat. This is a form of heating on the flesh side of the hides and can lead to advanced stages of damage on discharge.

The use of hooks in handling and stowing hides should be strictly forbidden.

When preparing holds, see that the bilges and strum boxes are properly cleaned so that the drainings — which will be very considerable — can be taken care of.

Bales, Bags or Bundles of Wet Hides can be stowed in 'tween or shelter deck spaces, but should be well removed and separated from any fine or dry cargo that is liable to be damaged by moisture or by the very strong and offensive odours thrown off.

Adequate floor dunnage should be laid under the hides and care should be taken to ensure that there is a clear passage for the seepage of brine to the scuppers. In order to ensure adequate ventilation to avoid overheating, it is recommended that flat dunnage be placed some 300-400 mm apart every six tiers high. The entire stow should be interspersed by channels for the full depth of the stow. These channels should be approximately 150 mm wide and should run in both fore and aft and athwartship directions, so dividing the cargo up into series of approximately 6 sacks of hides. The dunnage used must be clear of nails.

Wire bands should not be used to secure hides on pallets; there have been cases of rust damage through this practice.

Loose Wet Hides — Steer hides average about 27.2 kg each; about 36 hides to the tonne, cow hides average about 21.8/22.7 kg each; about 44 hides to the tonne; stowing about 1.25/1.34 M3/tonne.

Generally only received into lower holds or deep tanks — the latter for preference if the lids can be left off for ventilation.

The deck or tank top, stringers, box beams, gussets, bulkhead brackets, etc., must be well dunnaged and the whole, as well as the sides, bulkheads, tunnel, stanchions and ladders covered with mats, paper, etc.

Some trades demand the lining of sides and ends of deep tanks or holds intended for the carriage of loose salted hides with boards, secured edge to edge and butt to butt — not spaced like spar ceiling.

When stowing loose, each hide is to be laid perfectly flat, hairy side up. On no account are hides to be doubled, bent or folded to any extent, otherwise they will rot in way of the bend.

The stowage of wet hides should cease not less than 1220 mm below overhead deck and thus ensure plenty of room in which the men can work in spreading the hides out. A layer of salt 150 mm thick should be laid over he uppermost tier.

The pickle may be mixed with either fresh or salt water and is usually prepared in large receptacles conveniently placed on deck, the contents being led into the tank or hold by means of a flexible hose.

Notwithstanding the hides being wet, it is important that they be carefully protected from rain or salt water.

Ventilation — on account of the heat, moisture and strong odours thrown off, it is most important, in order to avoid deterioration of this and damage to other cargo, that hatch covers be taken off whenever conditions permit.

After Discharging, the compartment in which salted hides have been carried should be thoroughly washed out with fresh water, using plenty of force, so as to remove all traces of brine and salt crystals from behind frames, etc. This is necessary both in order to prevent corrosion of the steel work and to ensure the proper drying of the compartment to render it fit to receive and carry the next cargo. It is common to treat wet hides with Napthalene. This chemical has a very persistent odour and great care should be taken to ensure holds are odour free after cleaning.

ISO containers — palletised units of wet salted hides can be carried successfully in closed box general purpose containers. Such containers must be lined (floors and well up the walls) with a single piece of impervious material (e.g. plastic) to protect the container from corrosive brine and facilitate cleaning. On the floor a layer of sawdust (50 mm, 2″) deep, covered by sheets of hardboard or similar to allow mechanical handling equipment to enter. The containers, once loaded, should not be left standing in full sunlight and should be stowed on the ship away from sources of heat such as direct sunlight. Under certain circumstances temperature controlled carriage may be required (10 degrees C. 50 degrees F). Reefer containers so used should be lined with protective material in such a way that the circulating air can move unimpeded.

HIGH ACID  See Coconut Oil Products.

HOMINY CHOP 

  1.40/2.10 Bulk

See Seed Cake. See IMDG Code. (See IMO Code of Safe Practice for Solid Bulk Cargoes).

Is roughly crushed maize — an offal, the best part of the maize having been taken out for human food. Exported from S. Africa in large quantities as a cattle food, in two separate grades, viz. — yellow hominy chop, to which a very small percentage of ground maize cob is added, and white hominy chop which contains a higher percentage of cob. The higher the percentage of cob, the larger becomes the stowage factor. Stow and care for exactly as for maize which see.

Commodity	*Characteristics*	*M3 per Tonne*	*Packaging*
HONEY		1.25/1.34 1.45/1.50 0.99/1.11	Barrel Cases Drums Tanks

Wet cargo, very liable to leakage. May crystallize as a result of impurities, which may affect handling (tanks) or processing — blending.

Stow clear of dry goods, the bottom tier to be well bedded.

ISO containers — tank containers must be clean, dry and odour-free. May require heating at time of discharge to improve the flow properties. Any heating should be applied gradually.

HOPS	 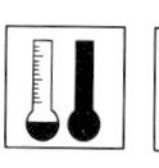 	5.02/6.97	Bales and Bags

A dried leaf exported from U.S.A., Australia, Holland, etc. Pungent odour, can take harm from cross taint. Moisture content may be up to 9½%. Stow in a dry cool place, well removed from moist cargoes. May require temperature controlled carriage; recommended temperature 1 degree C + 1 degree.

ISO containers — when not requiring refrigeration may be suitable for closed box containers. Payload at approximately 10.5 tonnes may be achieved for a 20 ft container.

HORNS **HORN SHAVINGS** **AND HOOF TIPS**	  	2.79 4.18/5.02 2.51/2.79	Loose Bags Bags

Horns are usually shipped in bulk, horn shavings and hoof tips in bags.

Hoof tips, if allowed to get wet in stowage, will crack and split when drying.

It is customary in all trades to use them as dunnage (see Hides, also Dunnage) and for filling broken stowage. When horns are used for that purpose between frames, back of spar ceiling, they should be arranged points up so that they will not catch and retain moisture.

As these are apt to give off strong smells, guard against tainting edible and other delicate goods.

In ports of India and Far East, horns are preferred to native wood as dunnage, as there is a danger of the destructive white ants being introduced into the vessel with the latter.

If different parcels are shipped, it is necessary that they are kept entirely apart to avoid mixing and claims on delivery.

If shipped loose, the B/L should never specify quantity, but should have "quantity unknown" clearly shown thereon. See IMDG Code.

HORSE HAIR		1.95/2.79	Bales

Dry stowage, well dunnaged. See Hair, Animal re Sanitary Certificate.

HORSES AND MULES

The transport of horses by sea is governed by Official Regulations prescribed by the Government of the exporting country, whose inspectors are held accountable for their observance. See also Part 2, Techniques.

HORSE MEAT (Chilled)		See Chilled Beef	

Commodity	*Characteristics*	*M3 per Tonne*	*Packaging*
HUMAN HAIR	 	See Hair	
ICE BLOCKS		1.11	Blocks Boxes

Relatively heavy stowage. Clean handling.
Suggested carrying temperature −18/−12 degrees C.

ICE CREAM			Cartons

Carton stowage. Can be carried at higher temperature at shippers request. At no stage should this commodity be allowed to thaw. Suggested carrying temperature −18/−15 degrees C.

ILLIPE NUTS	  	1.67/1.81	Bags

An oil bearing nut. From Malaysia, India, Sumatra. Size of harvest varies enormously from year to year. For manufacture of edible fats, known as chocolate fats. Gathering process may involve the nuts being submerged for several weeks to reduce infestation and to burst shells of nuts by germination. Liable to heat — see IMDG Code. If moisture content greater than 7.5% may require on deck stowage only.

For stowage see Nuts.

ILMENITE SAND		0.36	Bulk

Ilmenite is a very heavy sand, almost black in colour, to be found on the Coast of India, Malaya, Australia, etc. Holds must be to an exceptional state of cleanliness in order to pass shippers inspection for carriage of certain grades of this mineral.

On the coast of India this sand is carried from the interior mountains to the sea by the rivers and, later, by the action of the S.W. Monsoon, is thrown up on the beach.

The sand is sun dried, sifted and run on belts under strong magnets after which it appears jet black and resembles gunpowder in appearance. By further treatment two valuable minerals — Monazite and Zircon — are extracted.

From Ilmenite also is extracted Titanium Dioxide from which a white pigment is obtained. See Appendix 13.

INDIGO, POWDER **INDIGO, PASTE**		1.73 2.09/2.232	Cases Casks

A deep blue dye prepared from the leaves and stalks of the Indigo plant which grows in Indonesia, Philippines, India, Egypt, W. Indies, etc. Shipped both in powder and paste form, the former is usually packed in strong cases lined with suitable material to prevent siftage, and the latter in kegs.

It is a valuable commodity; packages should be carefully examined for any evidence of tampering and unsound packages promptly rejected. Stow in a special cargo locker.

INFUSORIAL EARTH **(KIESELGUHR)**		1.39/2.51	Bags

Used in building material, ceramics, sugar refining and the soap-making industry, etc. Cool, dry stowage. Stow away from foodstuffs to avoid contamination.

INDOROUS FELT		See Felt, indorous	
INTERMEDIATE BULK CONTAINERS		See Part 2, Techniques	

Commodity — *Characteristics* — *M3 per Tonne* — *Packaging*

I.P.A. (ISO — Propanol, ISO Propanol Alcohol)

See Part 2, Chemical Cargoes.

IPECACUANHA ROOT
(Ipecac Root)

Bags or Bales

The root of a plant grown in and exported from Central America, Brazil, etc.
Dry, cool stowage, remote from commodities likely to heat and sweat and from odorous and oily goods.

IRON, GALVANISED
(Pipes)

0.56

Bundles
Envelopes

See Part 2, Iron and Steel Products.
See Galvanised Iron sheet.

IRON ORE CONCENTRATE

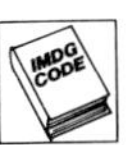

See Hematite and Ores. Also Part 2, Techniques, Bulk Cargoes.

May liquify, check physical properties prior to loading. See Appendix 13.
(See IMO Code of Safe Practice for Bulk Cargoes).

IRON OXIDE
(Spent)

0.45

Bulk

See Iron Sponge (Spent).

IRON, PIG

0.28/0.33

Bulk

See Steel.

IRON PYRITE

See Pyrites

Bulk

May liquify, check properties before loading. (See IMO Code of Safe Practice for Solid Bulk Cargoes). See Appendix 13.

IRON, SCRAP

See Scrap Iron

IRON SPONGE
(Spent)

0.45

Bulk

Derivative of coal gas purification. Liable to spontaneous combustion. Avoid contamination. May evolve toxic gases. Shipper certificate required. These cargoes are sometimes misinterpreted as D.R.I. not so, but may be used in further processes to produce D.R.I. The cargo is extremely dusty and the dust, if it becomes wet and allowed to lie will attack and penetrate paintwork.
(See IMO Code of Safe Practice for Bulk Cargoes).

Commodity	*Characteristics*	*M3 per Tonne*	*Packaging*

IRONSTONE — 0.39 — Bulk

Angle of repose 36 degrees. See also Part 2, Ores.

ISINGLASS  2.51/2.56 Casks; 2.68/2.79 Bales

A white gelatinous substance prepared from the sounds or air bladders of certain fresh water fish, used for culinary purposes and clarifying wines.

Dry stowage, special cargo locker.

ISOPHORENE   Drums

Chemical used in printing industry. Obnoxious. Combustible. See IMDG Code.

ISTLE
(Ixtle)  2.51/2.79 Bales

A fibre exported from Mexico — pressed into bales of various densities. See also Fibre and IMDG Code.

IVORY — 0.84/0.98 Loose or Bundles; 1.53/1.67 Cases

A very valuable commodity which includes elephant tusks, wolves, narwhal and hippopotamus teeth, etc. Generally shipped in the loose condition. Some countries now ban all imports of ivory.

Should be carefully tallied and scrutinised for damage and should be gently handled to avoid chipping and marking, must be kept clear of oils and greases. Stow in a special cargo locker. See Part 2, Specials.

IVORY NUTS  1.67/1.81 Bags

See Corozo Nuts.

IVORY SCRAP — 1.81/1.95 Barrels

Dry cargo.

JAGGERY  1.25/1.39 Bags

An exceedingly moist sugar obtained from a certain variety of palm tree grown in India. It quickly melts in hot weather and tends to become one viscous mass from which a thick syrup drains. The loss of weight sometimes amounts to 10 per cent or more after a voyage through the tropics.

If overstowing with other cargo, the jaggery should be well boarded over and heavily covered so as to preclude all possibility of top cargo getting into contact or mixing with jaggery.

Goods which are susceptible to damage from moisture should not be stowed in the same vicinity, neither should bag seed be stowed over same, if that can be avoided, owing to the danger of seed mixing with the jaggery. B/L should be adequately claused to protect vessel for any loss of weight.

Commodity	*Characteristics*	*M3 per Tonne*	*Packaging*
JAMS		1.11/1.39	Cartons

Stow as for canned goods. May be shipped in glass jars, in which case top weight must be limited.

JAPAN WAX		See Vegetable Wax	

JARRAH WOOD		0.84/1.060	

The most valuable timber found in Australia, sometimes called Australian mahogany, is extensively used for furniture and boat making and the coarser grade for railway sleepers and for paving. The unseasoned timber has a high moisture content. See Part 2, Techniques, Neobulk.

JARRY BEANS	 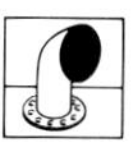	2.65/2.79	Bags

Stow in well-ventilated position, remote from odorous or heat-giving goods. See Beans.

JAVA COTTON		See Kapoc	

JELATONG		1.95/2.09	Cases

Is a form of wet rubber exported in considerable quantities from Malay Archipelago.

An objectionable cargo from which a great quantity of water drains. Sometimes packed in frail cases, which generally appear quite sound when shipped but which, from the action of their contents, become rotten before discharge and require a great deal of recoopering. B/L should be claused to the effect that cooperage charges be to account of consignees.

If available, stow in deep tank taking care that the moist and heated air ascending from the tank (or other compartment) has direct access on deck and does not find its way into compartments above.

Do not stow with any delicate goods including rubber, which becomes mouldy if stowed near jelatong; it stows all right with gambier, which see.

JERKED BEEF		1.62/1.73	See Carne Seca

JEWELLERY

For valuable jewellery see Precious Stones. See also Part 2, Techniques, "Specials".

JEWELLERY, CHEAP OR IMITATION

Considerable quantities of these are shipped from time to time and should be closely watched as they are very liable to pilferage, when claims are considerable.

Packages should be carefully inspected on receipt and doubtful packages rejected.

Stow in special cargo locker or strong room. If on delivery a clean cargo receipt is not forthcoming, the contents should be counted and package weighed in the presence of the consignee and, if necessary, of Customs officer — and results noted on receipt. See Part 2, Techniques, "Specials".

Commodity	*Characteristics*	*M3 per Tonne*	*Packaging*
JOWAREE DARI JOWAR		1.48/1.56	Bags

An Indian grain. See Seeds.

JUNIPER BERRIES		2.23/2.39	Bags

Small berry smelling strongly of turpentine which renders it most unsuitable for stowage with or near tea, flour, arrowroot, tapioca and like edible goods.

JUSI CLOTH

Native cloth of the Philippines.

JUTE		1.81/1.87 1.62/1.67	Bales per Tonne of 5 Bales

Is the fibre of an annual Indian plant used in the manufacture of Hessian (which see) gunnies, carpets, cordage, etc., also, when mixed with silk, of low grade satins.

Jute grows in water and the fibre is stripped by maceration in water. As a result of this form of preparation, also of exposure after packing to heavy monsoon rains and, in some instances, to deliberate wetting to ensure heavier weight, this cargo contains a considerable amount of moisture.

The moisture tends to rot the fibre — the heart of the bale more particularly in the former case — and when in stowage it evaporates and condenses on sides, bulkheads and overhead, thus producing sweat damage which often is quite impossible to avoid. The loss of weight by evaporation with certain jute cargoes is very considerable.

Bales which are noted to be wet or damp should be rejected. The dunnaging and matting for jute cargoes should be very thorough and such as to prevent the bales coming into contact with ships' sides, frames, beams, decks, bulkheads, stringers, pillars, brackets, etc., on which moisture will collect; contact with condensed moisture will cause the fibre to rot.

Jute should not be stowed in the same hold as seeds or goods, which are liable to heat and sweat, or with those which are likely to be damaged by moisture; contact with oils and greases should also be guarded against.

Every precaution against fire from sparks, smoking, etc., should be taken while jute is on board.

Jute, whilst not liable to spontaneous combustion unless impregnated with oil, is readily combustible. See IMDG Code.

JUTE BUTTS		1.62/1.67	Bales

Stow as for Jute.

JUTE CADDIES		4.46/4.65 2.37/2.51	Bales Pressed Bales

Jute Caddies are waste pieces of jute collected in and around jute mills, which, when baled, are shipped from India and Pakistan. One of their eventual uses is the making of felt.

Commodity	*Characteristics*	*M3 per Tonne*	*Packaging*

These pieces of waste jute contain vegetable and/or mineral oils to a varying degree. In consequence some caddies are liable to spontaneous combustion and should be given stowage accordingly.

They are believed to have been the cause of many fires both on board ship and ashore and should never be shipped without both a shipper's declaration Certificate of Cleanliness and also a Surveyor's Certificate.

It was usual for Surveyors in Calcutta to base their reports on the "Regulations of the Board of Underwriters of New York" insofar as they relate to the carriage of Cotton (a commodity with the same properties). These rules also apply in the case of Caddies shipped in American vessels and are quoted as under:—

"Cotton waste, mill sweepings, and similar materials free of oil can be stowed under deck, provided it is properly baled and covered with bagging.

Any of these commodities containing not more than 5 per cent by weight of animal, vegetable or mineral oils, if properly baled, may be stowed in any deck or hold providing ordinary cotton is not stowed in the same deck or hold.

These commodities may also be stowed in the forward end of a bridge deck or the after part of a shelter deck separated from ordinary cotton.

Cotton waste, mill and oily sweepings containing more than 5 per cent of mineral oil only requires on deck stowage.

Cotton waste, mill and oily sweepings containing more than 5 per cent of vegetable or animal oil is prohibited".

See IMDG Code.

JUTE CUTTINGS  1.53/1.59 Bales

Stow as for Jute.

KAINITE   1.06/1.11 Bag
0.98/1.03 Bulk

A mineral — hydrated compound of the chlorides and sulphates of magnesium and potassiumn — found in Germany and used as a fertiliser.

Shipped in large quantities from the Weser ports, etc., both in bulk and (in the finer form) in bags.

Dry stowage is essential, as, if Kainite gets wet or damp, it sets in a hard solid mass.

Kainite C/P makes it obligatory upon the ship to provide mats and boards for separations.

Holds should be thoroughly cleaned and dried on the discharge of Kainite to prevent dampness and corrosion.

KAOLIN See China Clay

KAPOC
KAPOK   3.34/4.74 Pressed Bales
1.67/1.81 Bales

A vegetable down — a silk-like fibre covering the seeds of a species of silk cotton tree grown in the East Indies, used for making life jackets and upholstery work. Ordinary dry stowage. Like all vegetable fibres there is a danger of combustion if wet or oil stained. See IMDG Code.

Commodity	*Characteristics*	*M3 per Tonne*	*Packaging*

KAPOC SEED 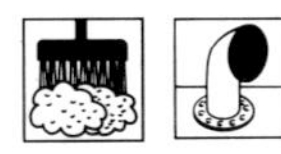 Bags, Bulk.

Bags should be adequately dunnaged. Good ventilation. This is a very fine cargo; take all precautions to avoid same penetrating into bilge suctions, etc.

KAPOC SEED OIL 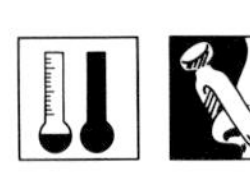 1.09 Bulk

Specific Gravity at 15.6 degrees C=.9154. Stowage factor 1.09 M3/tonne. Solidifies (partly) at about 12.8 degrees C.

An edible oil not unlike cottonseed oil in appearance, exported from Indonesia and other Malaysian ports. This oil is probably the most delicate of the vegetable oils which move in sufficient volume as to permit of bulk shipments. It is readily contaminated even by slight leakage of other vegetable oils from an adjoining tank, from which it follows that the tank must be scrupulously cleaned, properly tested and its unsuitability for the reception of the oil properly certified by a competent authority before loading commences. See Vegetable Oils in Bulk.

It is peculiarly susceptible to damage by scorching and consequent discolouration through overheating, the temperature to which it may be raised without being some 28 to 30 degrees C lower than that to which some other vegetable oils are heated to facilitate discharge by pumping.

That fact is of great importance when other oil is carried in an adjoining tank (not separated by a cofferdam). For instance, if Palm Oil which requires to be heated to a temperature of 52-55 degrees C for pumping is carried in a tank adjoining that which contains Kapok Seed Oil (whose discharging temperature is probably not to exceed 24 degrees C or 30 degrees C). It is obvious that the Palm Oil must not be heated above the prescribed discharging temperature of Kapok Seed Oil until the latter is discharged.

That, in turn, is likely to affect despatch as, especially during the winter months, it will take a considerable time to raise the temperature of the Palm Oil 25 to 28 degrees C after the Kapoc Seed Oil tank is empty.

Some shippers recommend that the temperature of this oil should not be permitted to get below 15.5 degrees C during the voyage gradually raising it, as the discharging port is approached, to just short of the prescribed discharging temperature.

When shippers prescribe the discharging temperature, it should be done so in writing, but, if such is left to the discretion of Consignees, time may be saved by the local Agents at discharging port obtaining the necessary particulars in writing and communicating same to the Master of the ship, the ship arriving in port with the temperature just a little below the prescribed temperature.

Further to avoid any risk of local scorching and discolouration, the heating of this oil (as indeed all vegetable oils) should be gradual and not forced.

Samples of oil taken at commencement of and during progress of discharge may come in very useful should a claim have to be considered.

See Part 2, Oils and Fats.

KAURI GUM  1.95 Bags
1.67/1.81 Cases

The semi-fossilised resin of the kauri pine (Dammara), the largest of New Zealand trees, used for varnish making. The resin is usually obtained by digging near standing trees or in ground previously occupied by the kauri pine. Dry cargo. See Gums.

KAVU PUTCH 

See Eucalyptus Oil, also Essential Oils.

Commodity	*Characteristics*	*M3 per Tonne*	*Packaging*
KEROSENE		1.39/1.45	Cases
		1.73/1.78	Drums
		1.76/1.78	Barrels

An inflammable illuminant oil. See Case Oil Petroleum. Shipped in 5 gallon cans, 2 in a case, weighing 36.74/37.19 kg more rarely in barrels and in drums, also in bulk. See IMDG Code.

Commodity	*Characteristics*	*M3 per Tonne*	*Packaging*
KIESELGUHR (Infusorial Earth)		1.39/1.48	Bags

See Infusorial Earth.

Commodity	*Characteristics*	*M3 per Tonne*	*Packaging*
KIWI FRUIT (Chinese Gooseberries)			Cartons
			Wooden Trays
			Palletised

Harvested in May/June with 3 to 4 months shelf life. Mostly grown in New Zealand. Carrying temperature must be in strict observation of shippers'/charterers' instructions. Careful handling to avoid bruising. Pallets from New Zealand usually 1016×889×2000 mm high, 870 kg.

ISO containers — 20 ft × 8 ft high insulated and integral 12 pallets of 174 trays. 7621 kg. 20 ft × 8 ft 6 in high integral 12 pallets of 186 trays, 8147 kg.

See Part 2, Techniques, Refrigeration.

Commodity	*Characteristics*	*M3 per Tonne*	*Packaging*
KOGASIN			Bulk

An oil made from coal and used in the soap industry. Non-corrosive with a flashpoint of 92 degrees C. No heating required. See Vegetable Oils.

Commodity	*Characteristics*	*M3 per Tonne*	*Packaging*
KOLA NUTS	 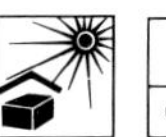 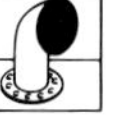	1.85/1.95	Bags

The fruit of an African tree possessing great stimulating properties. Liable to heat, otherwise has no objectionable properties. Dry stowage in a cool, well-ventilated space. See Nuts.

Commodity	*Characteristics*	*M3 per Tonne*	*Packaging*
KOLAI		1.56/1.62	Bags

An Indian edible grain. See Grain for stowage.

Commodity	*Characteristics*	*M3 per Tonne*	*Packaging*
KRAFT LINERBOARD		1.53/1.67	Reels

Used for the manufacture of cartons. Usually about 1200 mm in diameter, but heights vary considerably. Stow as for paper rolls. See also Part 2, Techniques, Neobulk.

Commodity	*Characteristics*	*M3 per Tonne*	*Packaging*
KYANITE		0.50/0.61	Bulk
		0.84/0.98	Bags

Kyanite ore must be kept dry as it solidifies if wet.

See Part 2, Ores.

Commodity	*Characteristics*		*M3 per Tonne*	*Packaging*
LAC		Stick Lac	1.95/2.09	Bags
STICK LAC		Lac Dye	2.09/2.23	Cases
SHELLAC, ETC.		Shellac	2.23/2.37	Lined Cases
		Grain Lac	2.15/2.31	Cases
		Garnet Lac	2.15/2.23	Bags

A resinous exudation found on the bark of the Indian fig and the Banyan fig trees where punctured by the lac insect, which is shipped under various names according to the processes through which it has been put.

Stick Lac is the crude substance in which is embedded the insects and their eggs, small portions of twigs, etc., which, when boiled, becomes known as Seed Lac.

Lac Dye is precipitated from Stick Lac during boiling.

Shellac is that into which Seed Lac is resolved by melting, refining and pressing into thin flakes, and is used in the manufacture of sealing wax and special varnishes, etc.

Grain Lac and Button Lac are different grades of Shellac differing from one another in the proportions of wax and resin content.

Stow cases on edge, not on flat, to prevent massing of flakes.

Garnet Lac is obtained by treating Seed Lac with alcohol.

All the foregoing Lac products require cool dry stowage, well removed from boilers.

LAC DYE

A dye obtained from Lac, which see.

LACTOSE
(Sugar Milk)

Bags, Drums

Sugar used in the preparation of food and medicine. Will deteriorate if damp.

LAMB, MUTTON
HOGGET OR TEGS

3.04/3.12 Hard Frozen — Carcases

Lamb and mutton is shipped in bulk in the hard frozen condition, the carcases being wrapped in white gauzy material.

Gross weight and measurement vary considerably at each port and according to the time of year. It is, therefore, necessary to obtain particulars of current stowage factors of all classes of meat from the local agent or stevedore.

For the purposes of classing, Lamb is meat derived from sheep of either sex under 12 months of age at date of slaughter. Most lamb carcases exported from New Zealand would be from animals aged between four and six months. Hogget is meat derived from sheep. Either maiden ewes or wethers, sheep with no more than two permanent incisor teeth; the weight of these dressed carcases is not permitted to exceed 56 lbs. Wether mutton is a class of meat derived from sheep which are either wethers or maiden ewes. Ewe mutton is meat derived from female sheep which have lambed.

Carcases should be carefully examined for softness and any found in that condition rejected; also for any sign of mould, paying special attention to throats and necks. Mouldy packages should be rejected.

Dunnage battens, horizontal and vertical.

With mixed cargo, lamb and mutton should be stowed in 'tween deck spaces, reserving the holds for the heavier goods. Mutton and lamb stow well with beef and butter, but lamb should not be overstowed with any other goods, neither should mutton except with lamb, to avoid crushing and mis-shaping the carcases, which is easily done.

Commodity	*Characteristics*	*M3 per Tonne*	*Packaging*

Stow carcases fore and aft, heads and tails, lower tier back to battens, then belly to belly and so on.

At ends of sides, care should be exercised to avoid shanks, protruding into and interlaying brine pipes, which, if permitted, will bring strain to bear on the pipes, as weight is superimposed and the natural settling of cargo occurs, neither should carcases be allowed to contact with pipes or insulation.

Mutton and lamb shanks in the hard frozen condition are very brittle and easily broken so that they require careful handling to avoid damage.

Stow to reaching height from each end towards the square of hatchway, after which the carcases in way of hatchway should be covered with a clean canvas screen and a platform laid over same on which to land incoming cargo, repeating as necessary. Walking boards should be provided and used to prevent damage, disfigurement and soiling by men walking on the carcases.

In order to achieve rapid discharge it is essential to obtain a loading which gives the stevedore at discharge port the best "run of parcels" together with producing as much information as possible relating to stowage dispositions. The necessity of planned stowage and detailed preparation of meat plans is essential.

ISO containers — carcase lamb. Gives poor payload approximately 7-10 tonnes.

See Part 2, Refrigerated Cargoes.

LAMB **MUTTON CHILLED**		4.18 1.81	Carcases Cartons

When packed in cartons, the lamb and mutton should be laid on their flat, usually under the hanging quarters of chilled beef. Clean close spaced 150×25 mm boards should be used as floor dunnage. If the cartons are block stacked, the stowage should be broken every two tiers high with 150×25 mm dunnage.

When shipped in carcase form, the shanks are secured to a bar which in turn is secured to the chain or hanging hook. Care must be taken to ensure that the meat is stowed as closely as possible to avoid chafage.

See Chilled Beef.

See Part 2, Refrigeration.

LAMPASOS			Bags

Coconut husks used for scrubbing floors. Usually inter-island carriage only.

LAMP BLACK **(Carbon Black)** **Carbon non-activated** **Charcoal non-acctivated**	 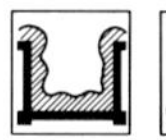 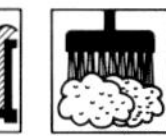  	3.21/3.34 3.07/3.21 3.01/3.12	Kegs Bags Paper Cartons

A sooty substance liable to spontaneous combustion, packed in barrels and in bags. From hydrocarbon sources, for fillers and reinforcing agents in rubber and plastic products. Very dirty. May sift. Stow near hatchway and protect other cargo from damage by sifting of the lamp black, particularly greasy or oily materials. See IMDG Code.

LANOLINE			Drums

Often in second-hand drums. May leak. Gives off odour. Stow away from foodstuffs and fine goods.

Commodity	*Characteristics*	*M3 per Tonne*	*Packaging*
LARD		1.53/1.61 2.01/2.69 1.73/1.81	Cases Pails Tierces

The melted fat of the hog, etc. Liable to melt if subjected to heat.

Stow apart from scented goods, such as turpentine, and away from delicate edible goods, this commodity being of a delicate nature itself and liable to taint other delicate goods.

LARD OIL		1.67/1.78	Barrels/Drums

An animal oil, primarily, though certain oils of vegetable origin (notably rape seed oil) are referred to under this name.

Can also be carried under refrigeration. Carrying temperature 4.5 degrees C. If lower temperatures are required these should be confirmed in writing.

Wet stowage, see Barrels.

LATEX		See Rubber Latex	
LATHS		See Timber	
LEAD CHROMATE	 	See Chrome Ore	
LEAD CONCENTRATES	 	0.33/0.39	Bulk

May liquify. Check physical properties before loading. See Ores. Also Part 2, Bulk Cargoes.

LEAD DROSS			

Is classed as Dangerous Goods, which see.

LEAD NITRATE	  		Bulk

Avoid dust inhalation. Toxic. Aggravated and intense fire if mixed with combustible materials.
See IMO Code of Safe Practice for Solid Bulk Cargoes.

LEAD ORE		0.24/0.65	Bulk

May liquify. Check physical properties before loading. See Galena. See Part 2 Ores.
See IMO Code of Safe Practice for Solid Bulk Cargoes.

LEAD, PIG		0.22/0.31 0.28/0.33 0.20	Pigs Ingots Slabs

Pig lead should be well distributed over tank top or deck and should not be tiered in a high block.
See Refrigerated Cargo.

Commodity	*Characteristics*	*M3 per Tonne*	*Packaging*

When lead is shipped in one-ton or two-ton ingots or strapped, it is usually the practice to load and discharge the lead with the aid of fork lifts. If shipped in vessels prior to loading refrigerated cargoes and in this case care should be taken to ensure that there is no damage, through excess weight, to the insulated tank tops. If necessary, plates should be laid under the track of the fork lifts to obviate damage. 'Tween deck stowage is not advised for heavy ingots. Soft lead blocks and strapped bundles may average 1 tonne weight.

ISO containers — it is most important that high density cargo of this nature is well distributed over the strongest part of the floor, i.e. adjacent to the walls and ends. Particular attention must be paid to securing against the slightest movement — it is not possible for ships' staff to see inside a closed box container as it is loaded on board.

See Steel and Iron.

LEAD, BLACK  See Graphite

LEAD PIPING — Crated Coils

Avoid crushing by overstowing with heavy goods.

LEAD SHEET

M3 per Tonne	Packaging
0.56/0.61	Cases
0.45/0.56	Rolls

Stow on the flat. Protected by straw roping or burlap.

LEAD, WHITE — 0.67/0.70 — Kegs/Drums

In paste.

LEATHER 

M3 per Tonne	Packaging
1.95/2.79	Bales
5.57	Rolls
1.67/2.23	Cases

Many kinds of leathers are shipped either in rolls, bales, protected or otherwise, and in cases.

Some are valuable and should be stowed in a special cargo room. Stow in a dry place, well removed from greasy or oily goods, from acids and from goods liable to transmit taint.

LEMONS 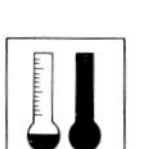

See Fruit, Green, also Citrus.

LEMON GRASS OIL  Drums

May corrode the drums. Very strong smell. Obnoxious. See Essential Oils.

LENTILS  

M3 per Tonne	Packaging
1.25	Bulk
1.38/1.53	Bags

The seed of a leguminous plant grown in Ecuador, Chile, North America and the Mediterranean countries. May be infested. Dry stowage, well ventilated.

Commodity	*Characteristics*	*M3 per Tonne*	*Packaging*
LICORICE		2.62/2.73	Bales (Alexandria)
	Liquid Extract	1.67/1.81	Barrels
	Paste Extract	1.25/1.39	Cases

The dried root of a perennial plant grown in Spain, Sicily, Italy, Asia Minor, etc., from which a sweet juice of medicinal value is obtained.

Bales from Alexandria = 0.34 M3 = 160 kos; from Beirut = 0.34 M3 = 125 kos. Dry stowage required.

Liquid Extract and Paste Extract both require wet stowage.

LIGNITE 

Brown coal, mined in Germany, Canada, New Zealand. See Coal.

LIGNUM VITAE 0.70/0.84

The hardest and heaviest wood grown in Jamaica, etc.
See Part 2, Techniques, Neobulk.

LILY BULBS  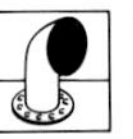  1.39/1.53 Cases/Cartons

Lily bulbs are exported from Japan and certain European countries in considerable quantities during the winter months, packed in well-made boxes containing friable clay mould, sand, earth or rice husks, which must be thoroughly dry and devoid of moisture, otherwise germination and rotting is inevitable.

So much depends on the condition of the goods at time of shipment that identical stowage often fails to give the same results. Frequently the out-turn proves quite satisfactory; at others a considerable percentage turns out in a more or less deteriorated condition, such being due to one or other or the combination of the following: inherent vice, the presence of moisture in the packing material, defective ventilation, heat, sweat or moisture from goods stowed in the same compartment.

Being perishable goods, they are usually carried at shippers' risk and B/L should be claused accordingly. Nevertheless the greatest care must be exercised in selecting a suitable place for their stowage, as well as in the actual stowing, to enable the vessel to benefit to the full by such clause.

Dunnage well, stow in a dry, cool, well-ventilated space below the waterline, preferably No. 1 lower Hold, uppermost tiers, which is the place preferred by experienced shippers. Usually the cases are fitted with two thin battens on each side, which form air passages and facilitate circulation of air through the mass of the stowage; where that is not the case, interlay each tier with thin battens or laths. Avoid deep stowage or overstowing with bags or like "dense" cargo and stow entirely apart from all goods liable to give off heat and moisture.

Wet or damp packages should be rejected.

Meticulous attention to shippers instructions should be observed. Many claims are formulated where receiver claims bulbs have matured in transit. There is no damage but their shelf-life has been reduced and there is an alleged commercial loss in consequence.

Bulbs are also carried under refrigeration. European bulbs usually at a temperature of 1-4 degrees C, Japanese bulbs 4.5 degrees C and sometimes higher at shippers' request. Cool stowage with full air freshening.

Other carriage temperatures which may be acceptable:

Tulip 10 degrees C (50 degrees F)
Lily 0 degrees C (32 degrees F).

Commodity	*Characteristics*	*M3 per Tonne*	*Packaging*

LIME
CALCIUM OXIDE
UNHYDRATED LIME
WHITE ROCK

Quick lime, which is normally packaged in metal damp-proof drums, is prepared by heating limestone; is a very dangerous cargo, as it readily combines with water evolving great heat, which produces gases. Corrosive. May affect eyes and mucous membranes. Masks and goggles should be worn. See IMDG Code.

LIME (Unslaked)    Bulk

Must be loaded in dry conditions. If mixed with water produces hydrated lime which in turn creates intense heat and possible ignition if in contact with combustible materials.
(See IMO Code of Safe Practice for Solid Bulk Cargoes).

LIME, HYDRATED OR CALCIUM HYDRATE  1.11/1.17 Bags; 1.23/1.34 Casks

Slaked lime, this being the usual form in which lime is shipped — sometimes in bulk, mostly in drums or bags. Stow slaked lime in a dry compartment removed from moist goods.

LIME, BORATE OF  1.39/1.50 Bags

Shipped from W.C.S. America. Dry stowage.

LIME, CITRATE OF  2.23/2.37 Bags

Cool, dry stowage. Liable to attack tin plate and like products.

LIME, CHLORIDE OF  See Bleaching Powder

LIMES 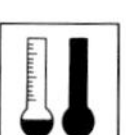

A citrus fruit, resembling a lemon but smaller, from which lime juice is obtained. Shipped in the green condition. See Citrus Fruit.

LIME JUICES 1.45/1.62 Cases; 2.09 Drums

In bottles, should be stowed in spirit room or special cargo locker, handled carefully and watched for broaching.

LIMESTONE 0.67/0.84 Bulk

A sedimentary rock containing calcium carbonate. Angle of repose 34-55 degrees.

Commodity	*Characteristics*	*M3 per Tonne*	*Packaging*

LINATEX

Is prepared from Rubber latex and resembles crepe rubber. See Rubber.

LINEN, FLAX		2.65/2.93	Bales
LINEN, TOW		2.65/2.93	Bales

See Bale Goods.

LINOLEUM		1.67/2.23	
		1.39	Rolls

Floor coverings usually made of cork refuse mixed with linseed or similar oils.

In cold weather is apt to crack; should not be overstowed except with the lightest of dry goods.

Shipped in rolls covered with stout paper or gunny and ends protected with soft material; occasionally it is crated. Stow rolls on end, protect from marking during slinging. Stowage requirements may be given on packaging. See IMDG Code.

LINSEED	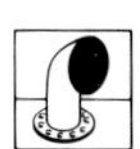  	1.59/1.70	Bags

Linseed or flax seed, the seed of the flax plant, from which linseed oil is obtained. (See Flax). It yields from 37 to 40 per cent of oil. It is one of the worst kind of seeds for shifting, its angle of repose being less than that of any class of seed shipped in bulk.

It is very liable to heat and should be well ventilated. See Grain and Seeds.

LINSEED CAKE	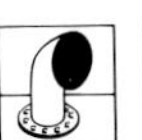  	1.39/1.53	Bag
(See Seed Cake)		1.34	Bulk

Made from the refuse of linseed after the oil has been extracted and is used for feeding cattle. Packed in bags or made up into packages covered with gunny or otherwise shipped in bulk.

Stow clear of strong smelling goods such as turpentine, onions, fruit, etc., and, as this commodity is given to heating, it must be stowed in a well-ventilated space clear of articles which are liable to be affected by the heat so generated. See IMDG Code. (See IMO Code of Safe Practice for Solid Bulk Cargoes).

LINSEED OIL		1.81/1.95	Drums
		1.07	Bulk

Linseed oil is obtained from flaxseed or linseed and although it is rich in some fatty acids that are important in the human diet the oil is always used for industrial purposes. The oil as produced by crushing and extraction has a dark amber colour and has a characteristic odour. The composition of the oil makes it suitable for applications in the paint and varnish industries, but the oil is also used in a number of other industrial applications. The ability of the oil to oxidise rapidly is a potential disadvantage for handlers of the oil, as oxidisation leads to the formation of a strong film which, if allowed to harden, is difficult to remove from surfaces. It is therefore generally advisable to clean tanks that have carried linseed oil as soon as possible after discharge of the oil.

The major producer and exporter of linseed oil is Argentina, but the oil is also produced on a significant scale in the European Union and in the U.S.A.

The density of linseed oil is 930 kg/m^3 at 15 degrees C. Crude linseed oil contains considerable quantities of mucilaginous material, which may precipitate on stowage.

See Part 2 Oils and Fats. See Appendix Table No. 11.

Commodity	*Characteristics*	*M3 per Tonne*	*Packaging*
LIQUEURS	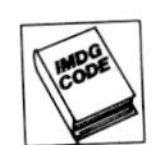	1.81/2.09	Bottled in Cases

Alcoholic liquors, flavoured, perfumed or sweetened in various ways. Stow in spirit or special cargo locker, careful watching being necessary to avoid pilferage.

LIQUORICE		See Licorice.	
LIVESTOCK		See Part 2, Livestock	
LOCUST BEANS	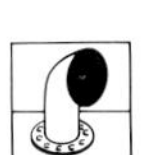	2.34/2.50	Bulk

Shipped in considerable quantities from Syria, Cyprus, Spain, etc. The flour obtained from these is made up into various classes of foodstuff for both human and animal consumption. Requires good ventilation. See Beans.

Locust Bean Oil, in bulk stows at 1.95 M3/tonne.

LOCUST MEAL	 	2.37	Bags

The flour of the bean of same name. Dry stowage. Keep apart from odorous or moist goods.

LOGS

See Timber. The average stowage factor of logs, varying in lengths and denuded of the bark, is 150 cu. ft per 1000 board feet and for logs of uniform length, denuded of the bark, is in the region of 135 cu. ft per 1000 board feet.

From W. Africa Abura stows at 75, Wara and Obeche about 130. Majority of W. Africa species av. 100 including Utele, Guaria, Agba, Iroko, Mahogany, Sápale, Okume, Afra.

In N.Z. — Japan log trade, Haakon Dahl scale gives estimated super feet by squaring one quarter centre girth and multiplying by length.

See Part 2, Techniques.

LOGWOOD		2.37/2.65	Bundles

Wood of a deep red colour, grown in Central America and W. Indies, used as a dye.

LOGSWOOD EXTRACT	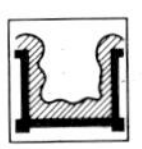		Drums

Dye from Central America. Liable to leak and stain. Cover container floor before loading.

LONGCLOTH			

A cotton fibre used for shirting; differs from calico inasmuch as the warp and woof are alike. See Part 2, Bale Goods.

Commodity	*Characteristics*	*M3 per Tonne*	*Packaging*
LUBRICATING OIL		1.25/1.34 1.48/1.62	Cases Drums

A lubricant obtained from petroleum. See Part 2 Petroleum.

Drums should not be stowed on their bilges as fillers or flatters in holds; neither is it prudent to stow two heights of drums on end over drums stowed on their bilges in 'tween decks, as the superimposed weight is too heavy for the bottom tiers. See IMDG Code.

LUCERNE		2.93/3.07	Bags

A forage crop. See Hay. (Alfalfa).

LUPIN SEED		1.81	Sacks

See Grass Seeds.

LUMBER See Timber. Also Part 2, Techniques, Neobulk.

MACARONI		1.20	Cases

A paste prepared from the glutinous granular flour of hard wheat, pressed through a perforated vessel into long tubes and then dried.

Stow as fine delicate cargo, well removed from garlic, onions, lemons or other green fruit, and from all cargo liable to damage edible goods either by tainting or by moisture thrown of.

MACASSAR OIL See Vegetable Fat

MACE		2.37/2.65 1.95/2.09	Cases Bags

A spice. The outside covering of the nutmeg, shipped from Singapore and adjacent islands. Treat and stow as a fine dry cargo liable to be damaged if stowed near moist or greasy goods.

MACE, OIL OF Or Nutmeg Oil. See Vegetable Fats.

MADDER **MADDER ROOT**		1.81/2.23	Bales

The dried root of a plant which grows in India, Turkey, Greece, etc., from which a red dye called garance or garancine is extracted.

Must be stowed in a dry place to avoid deterioration. It has a great affinity for water and attracts moisture.

MAGNESIA			Bottles in Cases

A white powder used medicinally and in certain manufacturing processes.

Commodity	*Characteristics*	*M3 per Tonne*	*Packaging*
MAGNESITE, DEADBURNED		0.70	Bulk

Used in the manufacture of refractory bricks. Very dusty but not odorous. Damaged if wetted. Liable to shift if wet; according to quantity, shifting boards may be necessary.

MAGNESIUM NITRATE			Bulk

Mixed with combustible material, will ignite easily and burn fiercely. (See IMO Code of Safe Practice for Solid Bulk Cargoes).

MAGNETITE		0.42/0.47	Bulk

An iron ore, containing about 52 per cent of metal, shipped principally from Scandinavia, Russia and Eastern U.S.A. See Part 2, Ores. See Appendix 13.

MAGUEY			

Local name (Philippines) for Hemp — which see.

MAHOGANY		0.75/0.84	Sq. Logs
ACAJON		1.11/1.25	Boards

The wood of one of the loftiest and largest of tropical trees, shipped from Cuba, Haiti, Jamaica, Honduras, West Africa, etc.

Avoid stowing near anything that will stain the timber. Chain slings, if used for lifting heavy sawn logs, should be well protected. Avoid rough handling of boards to prevent splitting. See Timber, also Jarrah Wood.

See Part 2, Techniques, Neobulk.

MAHWA BUTTER		See Vegetable Fats	

MAIL	  	2.79/4.18	
		3.34	Parcels

It is of the utmost importance that the receiving, stowing and redelivery of this cargo should be under the constant supervision of a responsible officer with independent ship tallyman.

The mail room should be well ventilated and rat proof. The mail rooms should be fitted with two locks with different keys.

Failing the lockers being equipped with gratings, it is necessary that the mailbags be well dunnaged.

Tally and check in the mail room before stowing, particular attention being paid to check for broken seals.

Delivery should never be made by the ships' officers except by order of the Master. Any dispute should be settled before delivery.

The keys of the mail room should be kept in the safe in the Master's and Chief Officer's or Purser's Rooms and should never be allowed in the hands of anyone except Senior Officers. Cases have occurred where duplicate keys were made through impressions of the keys proper taken while out of the custody of the responsible officer.

Commodity	*Characteristics*	*M3 per Tonne*	*Packaging*

All Way Bills should be carefully checked on receipt of the mail and should be forwarded to the appropriate Authorities, on delivery of the mail.

Where it is necessary to carry mails in open stow, reliable watchmen must be in constant attention.

ISO containers — Mail should only be carried in closed box containers. Doors must be sealed and/or locked on completion of stuffing. Ships' staff must inspect these seals when loaded on board and prior to discharge.

MAIZE **See Hominy**	 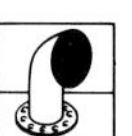	1.25/1.41 1.39/1.53	Bulk Bags

Indian corn, known as corn in America, is grown and exported in large quantities from Argentina, S. Africa, etc., as well as from the U.S.A., where it constitutes the greatest single crop.

Maize is the largest of the cereals and, besides its value as food for man and animal, is much in demand for the glucose and starch which it yields. It also yields 5 to 10 per cent of oil.

The grain is very liable to heat and sweat, resulting in sourness, and is a cargo prolific in damage claims.

In a great number of cases where maize fails to carry in good condition, the explanation is to be sought for in the pre-shipment history.

If not infrequently happens that maize is stacked in the open — either at the railside upcountry or on the barrancas off loading berths. Should it, at that time, be inefficiently dunnaged or inadequately protected from rain, deterioration will already have set in before shipment.

The greatest care should be exercised in protecting the ship against claims for damage or deterioration from causes beyond the control of those concerned in its sea transport. The only safe way of doing that is by suitably qualifying bills of lading. The furthest a Master should go in the matter of a clean bill of lading is to accept the words "Shipped in apparent good condition" or "weight, quality, quantity, and condition unknown".

Stow away from boilers and any similar source of heat. Keep well apart from any cargo such as flour, bran, oats, canned, dry and delicate goods, also wool, which are liable to be damaged by moisture — none of which should be stowed in compartments above maize.

Provided maize is suitably dried prior to shipment, it should carry well.

ISO containers — not normally suitable for carriage in closed box containers through marked changes in ambient temperatures due to the inherent tendency to sweat. Open-sided containers may be utilised for voyages of limited length, with tilts rolled up (when external conditions permit) to allow ventilation. Proprietary containers with double skins designed to protect the cargo from condensation damage have been used successfully on occasion.

See Part 2, Grain.

MAIZE CAKE	 	1.40/2.10	Bulk

See Hominy Chop. See Seed Cake.

(See IMO Code of Safe Practice for Solid Bulk Cargoes).

MALT		2.51/2.65 1.81/1.95	Bags Tanks Bulk

Barley or other grain immersed in water until it germinates, afterwards dried in an oven or kiln — a preliminary treatment of the grain to that of brewing and distilling.

Sometimes shipped in airtight tanks, at others in bags or bulk.

Stow away from dry, delicate and perishable goods.

Commodity *Characteristics* *M3 per Tonne* *Packaging*

ISO containers — suitable for dry bulk containers or as bulk (or bags) in general containers. Payload for 20 ft closed box container may be 16.5-17.5 tonnes. Dry bulk containers usually achieve less due to trimming.

MANDARINES See Citrus

MANDIOCA Barrels. Bags.

Brazilian arrowroot. Keep dry and away from taint. See Arrowroot.

MANGANESE FERRO 0.42/0.47 Bags

An alloy of Manganese and iron.

MANGANESE ORE   0.47/0.50 Bulk
0.61/0.70 Bags

The ore from which is obtained a metal not unlike and largely employed as an alloy of iron in the manufacture of steel. Two kinds of this ore exist, i.e. the black oxide and the red oxide — the former kind is also used for colouring pottery and glass.

When Chrome Ore and Manganese Ore are carried, they should be stowed in such a manner as to preclude any possibility of the smallest quantity of Chrome getting mixed with the Manganese. Admixture of even a small proportion of Chrome with Manganese renders the latter useless for important purposes in connection with the steel trade. Frequently very damp and gives off a lot of moisture during passage. Separate well from other cargoes. Dirty. May liquify. Check physical properties before shipping. See Part 2, Ores.

MANGROVE BARK 2.09/2.23 Bags
1.67/2.51 Bales

The bark of the mangrove tree. See Barks. Bales of 80 to 100 kg from E. Africa and Madagascar.

MANGROVE EXTRACT  See Cutch

MANI

Local name (Philippines) for Peanuts. See Groundnuts.

MANIOC OR MANDIOCA MEAL   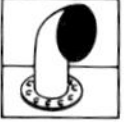  2.51/2.65 Cases. Bags.
1.67/1.81 Bulk/Pellets

An edible meal obtained from the manioca plant, from the roots of which are obtained tapioca and cassava.

Dry stowage away from odorous and moist goods.

Sometimes likely to be moist on shipment and can be liable to spontaneous combustion. Good ventilation is essential.

Shipments are frequently infested with the kaprha beetle, charterers usually being called upon to pay for the necessary fumigation after discharge.

See IMDG Code.

Commodity	*Characteristics*	*M3 per Tonne*	*Packaging*
MANJEET		1.81/2.23	Bales

A plant from which a red dye is obtained. Dry stowage.

MANNITOL	 		Bottles, Drums

A sugar product used for medicines, etc. Do not stow with poisons or goods liable to give off moisture or taint.

MANOLA		2.15/2.23	Bags

Dry stowage, removed from odorous and moist goods.

MANURES (Fertilisers)	  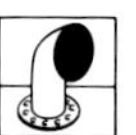	1.11/1.67	Bags
		0.90/1.40	Bulk

See Fertilisers.

MARBLE		0.42/0.47	Blocks
		0.50/0.56	Slabs
		0.56/0.61	Crates
		0.89/1.00	Marble Dust in Barrels
		0.70/0.72	Marble Clippings in Bags

Is a crystallised form of limestone which takes a high polish, valued for ornamental work in buildings, etc.

Being very heavy and usually shipped in large blocks, should be stowed on the ground floor. Care should be taken to see that marble slabs are compactly stowed and avoid overstowing with heavy goods.

Slabs of moderate size should be stowed on edge like glass, also slabs in crates.

Chain slings should not be used for slinging marble unless for heavy blocks, in which case they should be well protected with leather, etc., to avoid damaging the blocks.

Do not overstow marble with greasy oil or acid bearing cargo, as the marble will absorb any leakage from such and claims will quickly follow.

MARJORAM	 	5.57/6.69	Bales

An aromatic plant valued for its tonic and seasoning properties, also used as a stimulant. Dry stowage well clear of odours.

MASTIC

A gum resin obtained from the lentisk tree — a native of the Levant — by making incisions in the branches and stem. See Gums for Stowage, etc.

MATCHES		2.79/3.62	Cases

Packed in strongly made wooden case or cartons especially approved. See IMDG Code.

Stow in a dry place readily accessible.

Commodity	*Characteristics*	*M3 per Tonne*	*Packaging*
MATE			

Dried leaves, used as tea. See Tea.

MATHIC SEED METHEY	 	2.51/2.79	Bags

A very light aromatic seed, the stowage factor of which varies considerably. Valued for its medicinal properties. Usually shipped in small quantities. Dry stowage away from foodstuffs.

MATS MATTING		4.66/5.57	Rolls
		3.90/5.57	Bales
		2.79	Press Packed
		3.62/4.18	Tea Mats
		5.02/5.57	Ordinary Cargo Mats

Mats are woven from various fibrous material such as bamboo, grasses, leaves, seeds, sedge, straw, etc.

Some matting, like Japanese and certain Chinese matting, is valuable and should be given great care in stowing to avoid crushing the rolls, straining, etc.

The better class matting is made up into rolls and covered — they should never be overstowed with other cargo.

Hooks should not be used when handling mats.

The coarser matting like cargo mats, etc., are made up into bales or bundles.

MEALS, SEED	 	Variable	Bags

Seed meals. Residue after oil extraction from coconut, cottonseed, groundnut, linseed, maize, niger seed, palm kernal, rape seed, ricebran, soya bean, sunflower. Liable to heat. Moisture and oil content important. See IMDG Code. See Seed Cake. (See IMO Code of Safe Practice for Solid Bulk Cargoes).

MEAT MEAL	   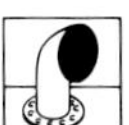	2.23/2.37	Bags

Usually packed in bags or multiple paper bags. Sometimes dusty and can be very odorous, stow well away from cargoes liable to taint. Keep dry and well ventilated. Certain grades with high oil content are liable to spontaneous combustion. See IMDG Code.

MEAT, PRESERVED		1.78	Cartons

Stow as for cartoned goods.

MEAT, SUNDRIES		1.81/2.23	Cases
		2.37/2.65	Bags
		1.81/2.23	Cartons

Offal includes oxtail, hearts, casings or intestine, livers, kidneys, etc., and are packed either in bags, cases or in cartons. Casings are often in kegs and offal is sometimes shipped in pails. They are all carried in the hard frozen condition. Carrying temperature −12/−9 degrees C. Stowage of cased offal, etc., is the same as other hard frozen goods in cases. Cartons may be stowed on the flat or edges.

Stow in 50 × 50 mm dunnage floored with 150 × 25 mm to ensure adequate support to packages. Cases are stowed on their flat; cartons hitherto stowed on the flat are now being stowed on their edge. In deep stowages it is advisable to break the stow at least at half height with suitable dunnage. Vessels equipped only with brine grid system should break the stow at least every four tiers high with 50 × 50 mm dunnage.

In the case of boneless meat in bags, adequate dunnage should be used to avoid the interlocking of the cargo, thus impeding the airflow. 50 × 50 mm battens should be laid in line with the airflow at least every four tiers high.

Bagged offal and meat should carefully be examined for mould, packages showing blood stains should be suspect, opened for inspection, and all mouldy packages rejected as well as any which are soft. This class of cargo is not suitable for over stowage by any other. Dirty, wet or cut cartons should be rejected.

Stow in 'tween decks if possible, as it does not carry well in deep stowage. Ordinary 50 × 50 mm floor dunnage is not sufficient with this and other class of bagged goods, which are likely to sag down to the deck and so impede circulation. One or other of the two following methods is adopted when stowing bagged offal to eliminate that risk:

1. Floor off with quarters of beef or cased offal, dunnaged as necessary; follow with a tier of quarters at the sides and bulkheads extending right up, such goods to be separated from the offal with suitable dunnage.

2. Lay flat dunnage boards fore and aft over the floor battens, spaced so as to allow air circulation, but not far enough apart to permit bags to sag on the deck.

See Part 2, Techniques, Refrigerated Cargoes.

MELONS 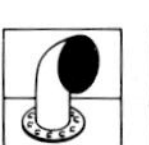  2.79/3.34 Crates
Cartons

Adequate ventilation, frequently carried on deck when they should be given suitable protection from excessive heat and bad weather. May also be held for limited periods under temperature control. Suggested carriage temperature +4.5 degrees C to +13 degrees C (40-55 degrees F). However shippers instructions should be obtained and maintained. Air freshening required. See also Fruit, Green and Refrigerated Cargo.

MENHADEN Bulk

Herring oil shipped in bulk, very odorous. Carried at 24-29 degrees C. Pumped out at 35-40 degrees C. Temperature must not be allowed to exceed 40 degrees C. See IMDG Code.

MERCURY   Cylinders
Flasks

A very heavy fluid metal (s.g. 13-59) shipped in cylinders or flasks.

Stow in a cool place, chock off well and do not overstow with heavy packages.

Very valuable, care against pillage.

METAL BORINGS AND CUTTINGS   0.70/0.84 Drums

See also Scrap Iron.

Metal borings and cuttings are frequently coated with oil. They will heat due to oxidisation. They should not be stowed in the same compartment, as that which contains any commodity which is known to heat in stowage, or in a compartment abutting or abreast of the engineroom, etc.

Commodity	*Characteristics*	*M3 per Tonne*	*Packaging*

Not infrequently this material contains woollen or cotton rags, paper, sawdust and other combustible materials which promote combustion.

Fires in this cargo must be dealt with by flooding the relevant holds.

The National Cargo Bureau Inc., New York, recommended that, before this commodity is accepted for shipment, it be carefully inspected by one of their inspectors to ascertain if materials of a combustible nature are mixed with the borings; to take the temperature of the metal prior to loading, and if it is found to be 43 degrees C or above that, the loading be not proceeded with until the temperature shows positive signs of decreasing. See IMDG Code.

METAL POLISHES (Liquid)  

Flammable liquids. See IMDG Code.

METAL SULPHIDE CONCENTRATES    0.31/0.56 Bulk

Sulphide concentrates of iron, lead, nickel, copper, zinc. Some may oxidise and self heat and produce toxic fumes. May cause corrosion.

Details should be furnished by shipper, prior to loading of any hazards or precautions to be taken during loading, carriage and discharges.

(See IMO Code of Safe Practice for Solid Bulk Cargoes).

METHANOL (Methyl Alcohol) 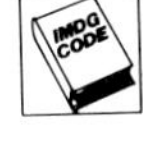 See Part 2, Chemical Cargoes.

METHYLATED SPIRITS  

	M3 per Tonne	Packaging
	2.0/2.06	Barrels
	1.95/2.01	Drums

Is a mixture of 9 parts alcohol and 1 part of wood spirit. Inflammable Substance. See IMDG Code.

METHYL TERTIARY- BUTYL ETHER (MTBE) See Part 2, Chemical Cargoes.

MICA TALC 

	M3 per Tonne	Packaging
	1.11	Cases
	1.25/1.53	Bags

A mineral glittering, easily split into thin plates.

White mica, also known as Muscovite or Siberian glass, is the form most generally in use as a substitute for glass. Shipped from India, Far East, etc. Dry cargo.

Commodity	M3 per Tonne	Packaging
MIDDLINGS SEMETIN	1.95/2.23	Bags
POLLARDS	1.48/1.56	Bulk, Pellets

A by-product of flour milling used for cattle food, dog biscuits, etc. Shipped in bags of about 40 kg. This requires to be well ventilated to prevent heating and deterioration especially by caking internally as distinct from external caking due to sweat and moisture. Dry stowage.

MILK, CONDENSED See Condensed Milk.

Commodity	*Characteristics*	*M3 per Tonne*	*Packaging*
MILK, DRY OR POWDERED		1.67/1.81	Cartons
		1.95/2.51	Cases
		1.67/2.27	Bags
		2.83/3.96	Pallets

Much of this cargo, Evaporated, Skim, Whole Milk, Buttermilk, etc., is shipped from ports where vessels normally load refrigerated cargoes in addition. Avoid wherever possible stowage over or adjacent to a frozen compartment. If this is not possible, at least 125 mm of dunnage should be laid on the deck and/or the adjacent bulkhead. It is the considered opinion of many shipping lines in the trade, that in order to avoid condensation damage, ventilation should be restricted throughout the passage.

Liable to taint damage. Liable to attract weevils, stow away from Barley, Flour and other cargoes likely to be weevil infested.

Milk powder from New Zealand usually shipped in 25 kg multiwall bags.

ISO containers — typical payload of 20 ft container:

bags 15.5 tonnes; consumer pack tins 14.0 tonnes (New Zealand)
bags 16.3 tonnes; consumer pack tins 11.0 tonnes (Australia)

MILK, FRESH 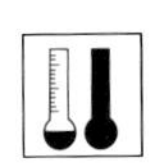

In cans or bottles, carries best at a temperature of + 1.5 degrees C. Homogenised may be hard frozen.

MILK, MALTED   2.51/2.65 Bottles in Cases

Fine cargo, very liable to pilferage. Must be stowed in cool place, well removed from moist goods.

MILLBOARDS  0.84/1.11 Crates

Asbestos, etc., Sheets, packed in crates which should be stowed on edge on a solid platform. See also Asbestos.

MILLET 1.37 Bulk
1.39/1.67 Bags

A species of small seed used as a substitute for rice, for poultry feeding, etc., of which there is a large variety.

MILORGANITE  1.53 Bulk

Heat-dried activated sludge. Very fine granular product. Angle of repose 40 to 45 degrees.

MINERAL OILS See Part 2, Petroleum

MIQUI

Chinese noodles made from Wheat.

Commodity	*Characteristics*	*M3 per Tonne*	*Packaging*
MIRABOLANS		1.95/2.01 1.95/2.06	Pockets Bags

The berry-like fruit of a tree allied to the maple, valuable for tanning and dyeing processes.

MIRABOLANS EXTRACT	 	1.25/1.39 0.98/1.03	Bags Cases

Ordinary dry stowage.

Extract and crushed Mirabolans should be given dirty stowage. Wet stowage, well dunnaged and matted. The commodity runs and melts in hot weather.

MOHAIR		6.97 2.79/4.18	Bags Bales

The hair of the Angora goat used for weaving certain kinds of cloth. Shipped in bags, also in bales of varying densities. Precautions against Anthrax.

Cashmere — may need inspection and disinfection. Adequate packaging essential, e.g. bales covered in double woven polypropylene. If packaging remains intact, there is no danger of Anthrax. If cashmere is visible through the covering, the adjacent cargo may need disinfection.

Goathair — tops and yarns (slivers and other rovings) most dangerous regarding risk of anthrax.

Cashmere yarns, bleached and dyed, mohair yarns, may need a certificate of approval from certain countries.

See Wool.

MOLASSES		1.39/1.67 0.74	Casks/Drums Bulk

The syrup obtained in the process of manufacturing brown sugar. When shipped in casks, hogsheads or puncheons, considerable leakage may occur; heavy rolling or pitching of vessel accentuating this loss.

Wet Cargo, avoid contact with dry and fine goods. Do not permit Molasses to be stowed over even the coarsest sugar.

MOLASSES, IN BULK		0.74	Bulk

This product is a by-product of sugar production and is used principally as a component in animal feeds. It is invariably shipped as a bulk liquid.

The viscosity of molasses is very dependent upon its temperature. It is common for shippers to stipulate a carrying temperature — normally a little above 30 degrees C and to require that the product is heated to not more than 40 degrees C to reduce the viscosity to facilitate discharge — normally by pumping through the ship's lines.

Molasses as shipped is a comparatively stable product. If it is diluted with water, microbiological fermentation can occur. This is associated with massive frothing. Some cargoes also froth due to a chemical process sometimes called inversion. This process is temperature sensitive, increasing rapidly as the temperature increases.

There are also instances, in land tanks, where there has been catastrophic decomposition of molasses from certain sources due to another chemical reaction. This is associated with the production of sufficient heat to carbonise the molasses affected. This reaction reputedly will not take place if the temperature of the molasses does not exceed 40 degrees C.

Commodity	*Characteristics*	*M3 per Tonne*	*Packaging*

It will be evident that the most important factor when carrying molasses is accurate temperature control. Preferably tanks used for carrying molasses should be fitted with temperature recording sensors at at least three different levels. If these are not available, the temperature of molasses in each tank should be measured at three levels (towards the top, in the middle and towards the bottom) at least three times per day at the outset of a voyage, daily once the temperature has stabilised and each watch during a final heating period before discharge, unless shippers/charterers provide different instructions.

Because molasses become very viscous when cool, there is a serious risk of localised overheating in the region of heating coils at the outset of cargo heating, prior to discharge. Thus it is normal for shippers/charterers to give instructions for a maximum rate of temperature increase when cargoes are being heated. It is essential that such instructions are carefully followed.

MOLYBDENUM ORE — 0.56 — Bags

See Ores.

MONAZITE  — 1.39 — Bags, Bulk

A sand obtained from Australia containing thorea from which thorium nitrate is obtained. For certain grades, holds must be exceptionally clean.

MONGOS

Vegetable Seeds, small globular. From Philippines. See Seeds.

MONKEY NUTS
ARACHIDES 

See Ground Nuts, also Nuts.

MONOAMMONIUM
PHOSPHATE   — 1.21 — Bulk

Can be highly corrosive in presence of moisture. Ammonium phosphates with pH greater than 4.5 are essentially non-corrosive. Continuous carriage may have detrimental structural effects over a long period of time. Angle of repose 35 degrees.

MONO ETHYLENE GLYCOL — See Part 2, Chemical Cargoes.

MOONSTONE

Or Selenite is a transparent form of gypsum, which see.

MOROCCO LEATHER 

M3 per Tonne	Packaging
5.57 Approx.	Rolls
2.79 Approx.	Bales
3.07 Approx.	Cases

Goat or sheep skins tanned with sumach and dyed. A valuable commodity. Shipped in bales, rolls and cases.

Special stowage in a dry place free from rats and well removed from oils, grease, greasy articles and acids. Avoid the use of cargo hooks.

Commodity	*Characteristics*	*M3 per Tonne*	*Packaging*
MOSS		8.36/16.72	Bales

Various kinds of mosses are shipped. Used for surgical dressings (sphaghnum), litter, etc. Dry stowage.

MOTHER OF PEARL		1.25/1.31	Cases

Iridescent sheets obtained in Australian waters, Arabian Sea, etc. (by diving); used in ornamental work.

Specially selected shells are of considerable value and should be cared for accordingly. Clean cargo. Special stowage to avoid broaching.

MOTOR CARS			
MOTOR VEHICLES		4.18/8.36	Unpacked
C.K.D.		2.79/3.07	Crates
M/CYCLES		3.34	Crates

These are shipped in cases or crates and in the unboxed condition.

In cases or crates, the cars may be shipped intact with certain parts dismantled to enable better use to be made of the stowage space. Cars Knocked Down, abbreviated to C.K.D., or Part Knocked Down — P.K.D.

Stow cases and crates on perfectly level and solid surfaces, never on their sides. Chock off and fill broken stowage with strongly built cases (not bales, which will be liable to chafe or light goods liable to crush).

If overstowing with these packages, the lower block of cargo must be closely overlaid with stout boards so as to provide a level and stable platform, each tier of cases or crates to be similarly overlaid in order that the superimposed weight is distributed as evenly as possible and so to avoid distortion and straining of the packages. If overstowing with other cargo, select light goods for that purpose.

Crates of C.K.D. are often stored in the pen prior to shipment, so that while the contents may themselves be protected by the packaging, coatings of grease, etc., the timber may have a very high moisture content especially if unseasoned or green timber is used in their construction. Ice or snow maybe adhering to the crates and, in some instances, pools of water may be lying on the waterproof covers. In these circumstances other cargo stowed in the same compartment or container may be affected by this source of moisture. When carried from temperate to hot weather, the protective oil coatings on some of the contents may liquify and leak out of the crates, to the possible detriment of cargo in the same compartment or container.

C.K.D. in *ISO containers* will achieve poor space utilisation unless the crates and packaging have been designed specifically to fit inside an ISO container. General Purpose containers and Insulated containers have different internal dimensions and the module developed for one will not necessarily successfully fit the other. Consideration must be given at time of stuffing for the problems of removing C.K.D. from the container at destination. A tight-fitting module that has to be pushed to the rear of the container with mechanical handling equipment, may be impossible to extract without damaging the crate. Where bearers exist, they should be in line with the direction of approach of any mechanical handling equipment and assistance can be given by leaving snotters or slings suitably positioned to help drag the cargo out. The concern over moisture migration from crates with high moisture content is even more important in ISO containers, which will develop their own micro-climate in varying ambient conditions. Crates should be absolutely dry at time of stuffing with a moisture content of less than 14 per cent.

Unpacked vehicles are normally left at the quayside, prior to loading, in a drivable condition with the windows closed, the doors and boot lids locked and all electrical systems switched off. Aerials should be in the collapsed position.

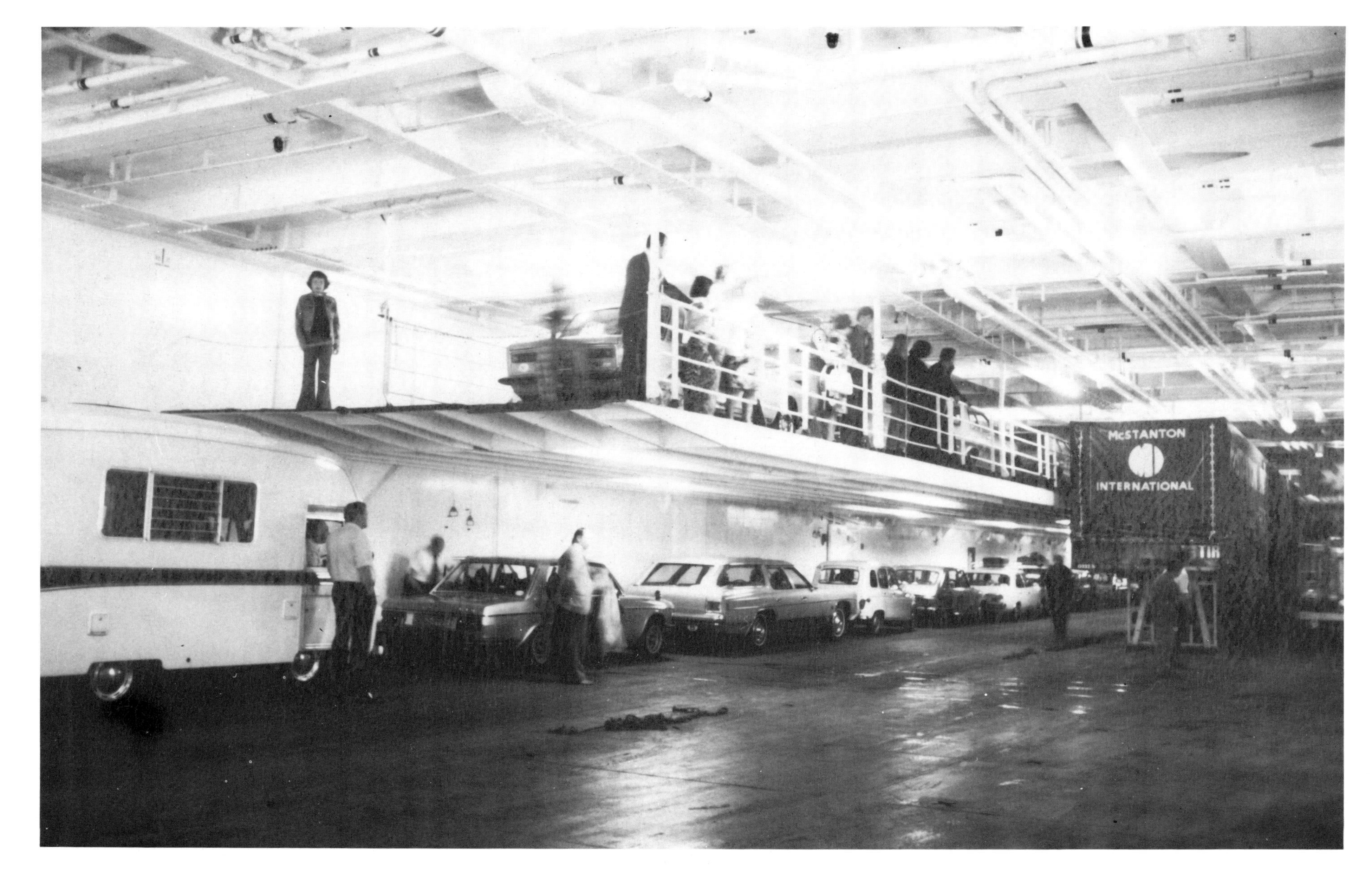

Upper and lower stern ramps provide access to two levels. (*Photo: MacGregor*)

Vehicles should only be left with their keys inside the car, if they are in a fenced area with adequate security.

Hand brakes should be fully on.

In the case of sports cars and convertibles, they may require the protection of a shed during their term on the quayside.

Identification markings must not be made on the hoods or windows.

Vehicles should never be moved by means of the starter motor.

Engines should never be started by using reverse or low gear or when moving vehicles down the ramps or pushing or towing them with other vehicles.

Vehicles should not be moved under their own power unless adequate ventilation is provided to dispel the fumes.

Where slings are used, spreader bars should be of sufficient length to prevent any contact of the lifting gear with the vehicle itself. When lifting nets are used, it must be ensured that the body work is not touched by any part of the net.

Steel work should be padded where necessary to prevent risk of damage to painted surfaces.

Drag ropes should be fitted where necessary to the platforms, front and/or rear of vehicles, to improve control and manoeuvering.

Scissor gear should never be used.

Vehicles must not be shipped as deck cargo without the shippers' knowledge and/or consent.

Engines should be switched off, together with all electrical systems. Hand brakes should be applied firmly. Vehicles stowed athwartships should be left in gear or in the parked mode for automatic vehicles. N.B. these latter requirements should not normally be carried out for vehicles parked fore and aft.

Check that the bonnet, boot lid or tail gate and windows are properly closed and that the keys are not left in the boot lid or door locks.

IMDG Code requirements should be followed regarding the disconnection of batteries, the taping of battery terminals and the requirements for petrol in the tank.

Vehicles should not normally be stowed closer than 230 mm (9″) to each other or any stanchion, ladder, bulkhead, etc. Some manufacturers require a greater clearance between each vehicle and the nearest obstruction. If a row of vehicles are being stowed one alongside each other, then the clearance for the end car (to allow the driver an exit) should be not less than 400 mm (16″). N.B. Australian labour require 450 mm clear between vehicles.

It is the Master's responsibility to ensure that all vehicles are correctly secured in the stow.

Each vehicle should be secured by at least 4 lashings, more if stowed athwartships. Approved lashing points may be provided by the shipper, however, normally the requirement is to secure by a point that will not be affected by the suspension; e.g. towing brackets; leaf springs (near shackle); axle beam; special brackets. However, heavy vehicles may require one or more lashing to be taken from the chassis.

Normally lashed vehicles should not be left in gear, except when the vehicle is parked athwartship.

Lashings should be taken down and away from the vehicle, so as not to foul any part of the body work, while providing both downward and horizontal restraint. On no account should vehicles be secured to each other.

In the process of stowing and securing vehicles, labour should not be permitted to climb over the vehicles, crawl across them or stand upon them.

Personnel employed to drive and secure vehicles should hold valid driver's licences. They must be issued with clean overalls which should not have metal buckles nor fastenings which could damage upholstery or paintwork.

Special attention should be taken when removing lashing equipment, so as not to damage the vehicles.

No attempt should be made to move or lift a vehicle while any lashings are still in place.

Commodity	*Characteristics*	*M3 per Tonne*	*Packaging*

Unpacked vehicles in *ISO containers* — it may not always be possible to drive vehicles into a closed box container, since the restricted internal width of the container may prevent the vehicle doors being opened. Securing is best achieved using both lashings and timber shoring. Lashings should be taken to the appropriate lashing points in the container floor and timber shoring nailed alongside the wheels, in front of the front wheels and behind the rear wheels. Timber across the front and rear should be chamfered to reduce wear and rub in the tyres. Other cargo should not normally be stowed in the same container, though, if space permits, suitable bagged stuff may be stuffed to form a level support onto which the vehicle may be rolled. Access to the lashing points must still be possible with this type of stow.

MOTOR SPIRIT  See Gasoline and Case Oil

MOULDING POWDER  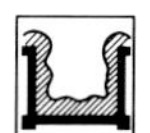 1.53 Bags

Any spillage will leave a strong residual phenolic taint. Great care must be taken to guard against spillage and clean up any residue.

MOWA, MOWRAH 1.76/1.84 Bags

The seeds of the mowrah plant. See Grain.

MOWA CAKE
MOWRAH CAKE  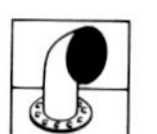   1.64/1.67 Bags

An oil cake which is very liable to heat and to spontaneous combustion, on which account it should be given 'tween deck stowage remote from source of heat and inflammable cargo, well ventilated and not overstowed with other cargo. See IMDG Code.

See Oil Cake.

MUD COAL  See Slurry

MUDSLURRY  See Slurry

MULES See Part 2, Livestock

MURIATE OF POTASH
(Potassium Chloride)  0.81/1.12 Bulk

White crystals, in granulated and powder form. Used for production of chemical fertiliser. Angle of repose 30 to 47 degrees. See Fertilisers.

Commodity	*Characteristics*	*M3 per Tonne*	*Packaging*
MUSK			Packed in Tin Lined Cases

A highly scented substance obtained from near the navel and other parts of the musk deer found in Central Asia, Indonesia, Ceylon, etc. The name is also given to plants with a similar scent.

A valuable commodity, especially the Tibet and Tonquin musk.

Must be stowed well clear of tea, edible and other goods liable to be damaged by the strong odour always present where musk is. Treat as Essential Oils, which see.

MUSK FUR MUSQUASH		2.79/4.18	Bales and Cases

The fur of the musk rat or musquash, a beaver-like animal but much smaller. See Furs.

MUSTARD SEED		1.81/1.95 1.56/1.62	Bags New Crop Bags Old Crop

An oil bearing seed, which yields about 25 per cent of oil. See Grain.

MUSTARD SEED OIL		1.37/1.42 1.48/1.53	Tins in Cases Drums

See Vegetable Oils.

MUTTON	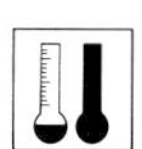	3.40 Mixed 3.15	Carcases Mutton/Lamb

See Lamb.

MYRABOLANS		See Mirabolans	

MYRRH			In Cases, Sometimes Tin Lined

A transparent, bitter gum resin which exudes from a small tree found in Nubia and Arabia. Valuable. See Gums.

NAILS		0.84 0.75 0.84/0.98	Cases Bags Kegs

Ordinary dry stowage. Bags are best stowed in a block to avoid chafe.

NAPHTHA	 		

See Part 2, Petroleum Cargoes.

Newsprint loaded at forward end of hold. Note extensive bridging down the wings.

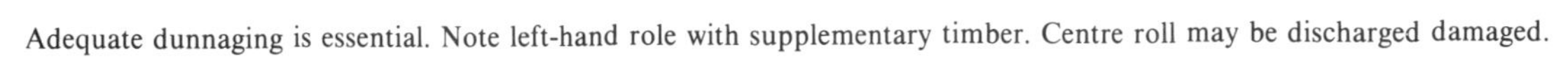

Adequate dunnaging is essential. Note left-hand role with supplementary timber. Centre roll may be discharged damaged.

Commodity	*Characteristics*	*M3 per Tonne*	*Packaging*
NEATSFOOT OIL		1.95/2.09	Drums Bulk

Shipped from Argentina, U.S.A., etc.

Avoid stowing over Tallow, Stearine, etc., often shipped at the same time. See Vegetable and Animal Oil constants, Appendix 9.

NEWSPRINT  See Paper. Also Part 2, Neobulk.

NICKEL BRIQUETTES			Pallet Loads

Dust given off is slightly toxic. *ISO container* — 18 tonnes payload 20 ft container.

NICKEL ORE		0.67/0.70 0.56 0.28/0.36	Barrels Bags Ingots

A mineral mostly obtained in Canada, France, Germany and Australia; alloyed with copper it forms nickel plate; with copper and zinc it forms German silver. Special armour steel (nickel steel) is made by adding about 4 per cent nickel to the steel. Nickel ore is normally shipped in bulk as a concentrate.

It may shift if the moisture content is above the acceptable "transportable moisture content".

(See IMO Code of Safe Transport for Solid Bulk Cargoes).

NIGER SEED		1.76/1.78	Bags

An Indian oil bearing seed. The seeds are black and shiny and yield about 40 to 45 per cent of oil.

NIGER SEED CAKE		1.50/1.56	Bags

An oil cake, the residue of niger seed after the oil is extracted. See Oil Cake. (See Seed Cake).

NIGER SEED OIL		1.67/1.78	Barrels/Drums

This pale yellow oil — not unlike rape seed oil in appearance — is obtained from the niger seed (which see), has a sweet taste but very little smell.

See Vegetable Oil Constants, also Barrels, Stowage of. Refer FOSFA International Procedures and Acceptable Previous Cargoes.

NITRATE OF SODA **(See Fertilisers)**		0.84/0.86 0.98/1.00	Bulk Bags

A natural deposit of sodium nitrate found encrusting the soil in certain regions of Chile and Peru — used as a fertiliser and in the chemical industry but usually prepared synthetically.

In the case of fire, all nitrates augment the rapidity and violence of the conflagration, so that this cargo should be stowed well removed from goods liable to spontaneous combustion.

Any fire in a compartment involving nitrate can best be extinguished by flooding with water. Selected hatch covers should be opened to prevent an explosion due to pressure.

Commodity	*Characteristics*	*M3 per Tonne*	*Packaging*

Clean nitrate of soda is an anhydrous salt, which, in its granular form, is almost completely dust-free and it is normally shipped in the dry granular form.

It is shipped from Tocapilla, Chile, normally in bulk (usually as full cargoes) by means of a fast mechanical loader. Some small amounts are, however, still shipped in jute or polythene-lined jute bags.

Moisture content of nitrate by analysis immediately prior to loading in Chile is usually 0.1 to 0.3%, after arrival at destination does not normally vary from this figure, although if discharged in particularly humid conditions, the hygroscopic nature of the material can result in a slight increase of moisture content of about 0.05%.

Masters of vessels carrying nitrate are recommended to keep all hatches and ventilators closed during the voyage to avoid any possibility of moisture pick up which would be prejudicial to the normal free running characteristic of this commodity.

It is shipped in a dry granulated form.

See IMDG Code.

NITRATES (See Fertilisers)	 	1.11	Bags, Bulk

In the case of fire all nitrates augment the rapidity and violence of the conflagration, and so should be stowed well apart from all cargoes liable to spontaneous combustion, and free from contact with sulphur, charcoal or acids. For various classes of nitrates, see IMDG Code.

(See IMO Code of Safe Practice for Solid Bulk Cargoes).

NITRE		See Nitrate of Soda	

NITRE, SWEET SPIRITS OF	 		

Its carriage is governed by the same Regulations as Petroleum Spirits. Stow in a cool place.

NITRIC ACID	 		

A corrosive liquid. Dangerous Goods. On deck stowage, clear of acids and infammable materials. See IMDG Code.

NITRO CHALK	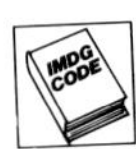	See Fertilisers	
NITROLINE		See Cyanide	
NITRO PHOSKA	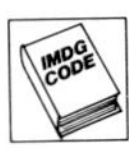	See Fertilisers	
NUTMEGS	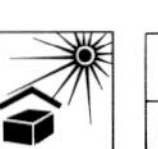 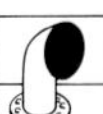	1.67/1.81 1.81/1.95	Large Cases Small Cases

The aromatic kernal of the fruit of an East Indian tree, valued for its spicy properties. An oil (nutmeg butter) is also obtained from same.

Commodity	*Characteristics*	*M3 per Tonne*	*Packaging*
NUTMEG OIL	   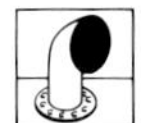		

Or Oil of Mace. See Vegetable Fats.

NUTS			Bags or Bulk

Many kinds of nuts valued for their edible or oil-bearing properties are dealt with herein. Nuts are very liable to heat and deteriorate. To guard against this they should be stowed in a cool, dry and well-ventilated space away from engine room and boilers, well dunnaged and not overstowed. They should also be kept apart from dry and delicate goods. Nuts which have been wetted by rain, etc., should not be shipped until they are thoroughly dry, and the bleeding of bags should on no account be permitted. See Part 2, Techniques, Bag Cargo, also Grain.

NUX VOMICA		1.67/1.81	Bags

The nut-like seeds of a tree growing on the Coromandel Coast, etc., from which strychinine is prepared — valuable for medicinal purposes, etc. Stow well away from tea. See Nuts for stowage, etc.

NYLON POLYMER		2.23/2.79	Paper Sacks
(Terylene Polymer)			Bags
(See Plastic Resins)			IBCs

Extruded into minute filaments. Granules made from "chopped strand". Clean and odour-free. Must not be contaminated by any other cargo — addition of foreign matter also spoils the flow characteristics. Different types of polymer must be kept well separated from each other.

May sift and contaminate adjacent bagged foodstuffs (e.g. Rice) if not adequately separated, hence rebagged sweepings can only be disposed of by salvage sale.

ISO containers — suitable for Close Box containers and insulated containers. May be carried in bulk; containers must be lined and/or properly cleaned. Payload varies from 15.5 tonnes (Insulated containers) to 17.8 tonnes in General Purpose containers. IBCs in containers may only achieve 14 tonnes.

OAK		1.11/1.17	Sawn (English)
		1.20/1.31	Sawn
		1.39/1.53	Staves

The stowage factor of oak, like that of all timber, varies considerably.

Avoid stowing green oak with steel or galvanised iron products as such gives off acetic acid gas which causes steel, iron, etc., to rust.

See Timber.

OAKUM	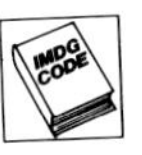	2.51/2.79	Ordinary Bales
		1.95/2.09	Pressed Bales

Tarred hemp in the "unlaid" condition sometimes obtained by unravelling and teasing old rope.

Oakum should be treated as wet cargo and stowed accordingly as the tar content is liable to run in warm weather. Though very light cargo, it should never be used for beam fillings or for overstowing fine cargo. See IMDG Code.

Commodity	*Characteristics*	*M3 per Tonne*	*Packaging*
OATMEAL		1.81/1.95	Bags

Dry stowage. Treat as for Flour.

OATS		1.81/2.06	Bags
		1.67/1.94	Bulk

The hardiest of all cereals, will grow in practically all countries from the Arctic Circle to the Tropics.

9-10% moisture content a normal figure at time of loading.

Like all grain, but more so than most, it is liable to heat and turn sour. Many heavy claims have resulted from the stowage of oats in compartments above that in which maize was stowed, damage by heating and sweat resulting therefrom, and the Courts have held that such is "Improper Stowage".

ISO containers — suitable for 20 ft bulk containers, achieves maximum weight payload. If loaded as bulk in closed box containers may give high side wall deflections. For this reason some plywood panelled containers may not be suitable for such cargo — too high a side wall deflection may cause the container to foul the cell guides or other containers.

See Grain.

OBNOXIOUS CARGOES See Part 2, Techniques

OCHRE	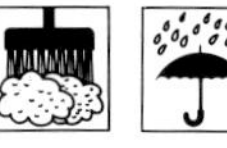 	1.39/1.67	Drums and Barrels
PUREE		2.23	Bags
			Polythene Lined Paper Bags

A clay-like substance strongly impregnated with oxide of iron, chiefly red and yellow. Is used as a pigment in colour making. Liable to stain other cargoes.

ISO containers — provided bags are sound the container does not require lining against contamination. Payload for 20 ft closed box containers may be 17.5 tonnes.

OFFAL — See MEAT, SUNDRIES

OILS AND FATS

The oils of commerce may conveniently be divided into the following principal classes:—

Essential oils, which normally move in small quantities. Many of these are in the nature of valuable cargo. Nearly all have a very strong odour hence leakage can taint other sensitive cargoes.

Animal and Fish oils. These are carried in plastic containers or drums and some varieties in bulk.

Vegetable Oils. These are carried in barrels or drums and some varieties in bulk.

Refer FOSFA Procedures and Previous Cargoes.

Mineral Oils. See Petroleum and Petroleum Products. Part 2.

A large number of oils and fats are referred to herein under their specific names. See also Vegetable Fats, Vegetable Oils, Vegetable and Animal Oil Constants, Essential Oils. See IMDG Code. See Part 2, Oils and Fats.

Commodity	*Characteristics*	*M3 per Tonne*	*Packaging*
OIL CAKES (Also See Seed Cake)	   	1.53/1.95	Bags. Varies considerably. Bulk, as Meal or Pellets.

Is the residue of various kinds of seeds, beans and nuts, from which the oil has been extracted, the "cake" being pressed into shape in hydraulic, etc., presses. It forms a very valuable cattle food.

Many varieties are marketed and shipped, the principal kinds being referred to herein.

Oil cakes are shipped both in bulk and in bags.

Permeated as it is with vegetable oil, oil cake is very liable to heat in some cases, to spontaneous combustion.

Stow well apart from odorous goods, such as hides, turpentine, etc., to avoid tainting of the cake, and as the cake throws off an oily, sickly smell which has a tendency to cling to substances, it is necessary that all foodstuffs, including green fruit, be kept well apart from oil cake. Prone to beetle infestation.

Oil cake should also be kept apart from moist goods, as moisture tends to soften it and produce a condition of mouldiness, furthermore the product may heat.

Good dunnaging is necessary throughout, to ensure as good circulation of air as possible, but avoid stowing on green or newly sawn wood, slate, marble or on damp ceiling. See IMDG Code.

Ventilation should be carefully attended to and temperatures recorded.

(See IMO Code of Safe Practice for Bulk Cargoes).

OIL CLOTH		1.39	Rolls

Certain kinds are classed as Dangerous Goods — see IMDG Code. See also Linoleum and Congoleum.

OITICICA OIL		0.92	Drums

Oil obtained from the oiticica nut, used in paint making. Shipped from Brazil, usually in drums. Wet stowage. See IMDG Code.

OLEO STEARINE	 	1.81/1.95 2.23	Bags Casks

An extract from lard. Cool stowage away from odorous cargoes.

See also Tallow and Stearine.

OLIBANUM		A gum — see Gums	

OLIVES		1.90/1.95	Kegs. Drums.

A fruit from which is obtained olive oil, the pulp yielding from 40 to 55 per cent and the kernels from 13 to 15 per cent, the latter being of inferior grade.

In the pickled form, olives are valued as a delicacy. Wet stowage.

Commodity	*Characteristics*	*M3 per Tonne*	*Packaging*
OLIVE OIL		1.67/1.73	Drums Bulk

See Part 2 Oils and Fats.

An edible oil obtained from the pulp and kernel of the olive (which see) shipped in considerable quantities from various Mediterranean, Spanish and Portuguese ports, etc., the coarser qualities being used for soap making, etc.

Colour varies according to quality, the best grades being of a yellowish tint, whilst the coarser quality is of a brownish colour and sometimes has a greenish tinge. See Vegetable Oil Constants, Appendix 12.

ONIONS	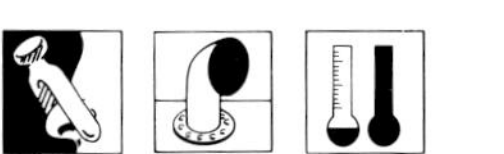  	2.23/2.29 2.6/3.8	Cases and Crates Bags, 40 = 1000 kilos

Are exported in large quantities from Spain, Portugal, Italy, Egypt, etc., and are packed in open cases or crates, or in bags of large open mesh.

Bags of onions should not come into contact with iron, so that dunnaging of decks, stringers, etc., should be very carefully effected on this account, to avoid bags coming into contact with the water which collects on such parts as a result of the large amount of moisture thrown off by onions. This also makes it necessary for the scupper pipes to be overhauled and cleaned before commencing to load onions as the bags are often rotted by the moisture thrown off in the course of short voyages.

Good ventilation is absolutely necessary, should receive constant attention and, for this purpose, the hatches should be kept open whenever possible.

Onions should be loosely stowed — and only goods which are known to be immune from damage by moisture or by tainting, stowed in the same compartment or in one connecting with that in which onions are carried. Best stowage factor obtained in open top containers. Onions have also been shipped successfully in standard containers on certain voyages with one door open to outside air.

Can also be carried under refrigeration — suggested temperature 0 degrees C to 1 degree C. See Fruits, Green, also Ventilation. Also Vegetables.

OPIUM			Small Cakes in Cases

The inspissated juice of the white or sleepy poppy — variously prepared for use. This highly narcotic drug is poisonous and is in great demand by the illegal drug market.

It is a highly valuable commodity and should be stowed in the strong room and carefully guarded against pilferage. Packages should be carefully examined before shipment and any showing signs of having been tampered with rejected.

The importation of opium and like narcotic drugs is only permitted to most countries under very strict regulations.

Opium smuggling by crews is to be most carefully guarded against. Many world syndicates have found this illicit traffic very remunerative and resort to all sorts of devices to attain their ends. Without the connivance of someone on board the ship, their operations are hampered.

OPOSSUM SKINS		6.25	Bales

See Furs.

Commodity	*Characteristics*	*M3 per Tonne*	*Packaging*
ORANGES	  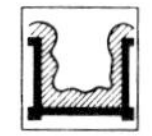	2.37/2.51 1.67/1.81	Cases Cartons

Are exported in large quantities from Spain, Portugal, Italy, Brazil, Azores, W. Indies, Florida, California, Israel, S. Africa, etc.

Jaffa oranges average 26/29 kg per case. Avoid stowing edible goods near or into the same compartment as Oranges. The odour which oranges leave behind after discharge has frequently been responsible for heavy claims for taint damage to such commodities as meat, eggs, flour, butter, etc., subsequently carried in the same compartments. Clean stowage. Usually stowed refrigerated.

Orange juice is also shipped hard frozen in plastic drums. Careful stowage and securing required. On passage these drums may chafe and leak. Check rim seals prior to shipment. Stow to avoid possible contamination or taint due to leakage.

See Citrus. See Fruits, Green, also Refrigerated Cargoes, also Tainting Damage.

ORANGE OIL	 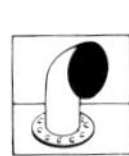 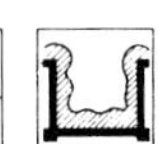	1.39 1.95	Barrels Cases

Shipped from E. Africa, etc. Pungent smell, stow well away from goods liable to taint. Good ventilation.

ORCHELLA		2.51/2.79	Bales

A lichen obtained from rocks on the coasts of S. Africa, S. America, Madagascar, etc., from which the blue dye known as archel or orchil is obtained. Must be stowed in a dry place and away from moisture or wet goods.

ORES	 		Bags/Bulk

Must be kept separated from other cargoes — particularly foods and fine goods — to avoid damage from sifting, moisture or contamination. Different types of ores should be kept separate from one another to avoid cross contamination.

By their very nature ores are heavy, and care must be taken to ensure appropriate distribution of weight — particularly in *ISO containers* and 'tween decks.

An increase in moisture content increases the chance of bulk ore shifting. See also Part 2, Techniques.

(See IMO Code of Safe Practice for Bulk Cargoes).

ORRIS ROOT			Bags or Bales

The dried, highly scented root of various kinds of the Iris plant, valued for its medicinal properties and in the perfumery trade, which is exported from some Mediterranean ports.

Dry stowage, remote from edible goods, and from oily materials.

OTTO OF ROSES		See Attar of Roses	
OX GALL			Drums

Odorous. Do not stow with foodstuffs or goods susceptible to damage by taint.

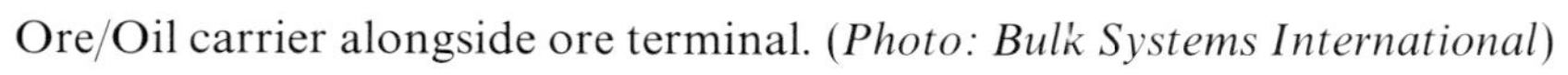

Ore/Oil carrier alongside ore terminal. (*Photo: Bulk Systems International*)

Commodity	*Characteristics*	*M3 per Tonne*	*Packaging*
PADDY		1.81/1.95	Bags

Is the name applied to rice before the husks are removed. See Rice.

PAINT	   	0.50/0.56	Drums

Ordinary wet stowage suitable to its weightiness. Oil based paints are inflammable and may be classed as Dangerous Goods. Water paints and some zinc silicate paints do not have dangerous properties but are damaged if frozen. There is a toxic danger from some solvents. See Dangerous Goods. See IMDG Code.

PALAY		See Rice	

PALM KERNELS		1.39/1.67 1.67/1.95	Bulk Bags

Are obtainable from the fruit of a certain palm tree grown in large numbers in W. Africa, etc.

The kernels yield an oil (about 45-50 per cent) which is much in demand for the manufacture of cattle food, soaps, etc. Like nuts, they are liable to heat and sweat; for stowage see Nuts.

PALM ACID OIL 

Palm acid oil is the by-product of the alkali refining of palm oil. In view of the fact that physical refining has largely displaced alkali refining in South-East Asia very little palm acid oil is now shipped. Its free fatty acid content is considerably lower than that specified for palm fatty acid distillate, the requirement being that it should contain at least 50% of free fatty acids. Palm acid oil is likely to be very dark in colour, and may contain traces of mineral acid.

Palm acid oil is likely to be used in the production of low-grade soap.

PALM FATTY ACID DISTILLATE (PFAD) 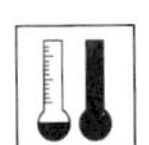

Palm fatty acid distillate is the by-product of the physical refining process used in South-East Asia to refine palm oil. It normally has a free fatty acid content of at least 70% and, due to its high content of palmitic acid, is relatively high-melting. It therefore needs to be maintained at a temperature above ambient when being shipped, but the high free fatty acid content makes it important to avoid overheating. The recommended temperatures for storage and shipping are 42-50 degrees C, and 67-72 degrees C during loading and discharge. The density of Palm Fatty Acid Distillate is approximately 840 kg/m^3 at 80 degrees C. It is normally bright orange in appearance, but if it has been mixed with palm acid oil the colour is likely to be dark.

Palm fatty acid distillate is used as a raw material for the production of fatty acids or in the production of soap.

Commodity	*Characteristics*	*M3 per Tonne*	*Packaging*

PALM KERNEL OIL 

Palm kernel oil is the oil obtained by crushing and extracting the kernels found in the palm fruitlet. It is therefore produced in those countries specialising in palm oil production. In contrast to palm oil, it is a relatively light-coloured fat which only melts completely when its temperature is raised to above 35 degrees C. The density of palm kernel oil is 890 kg/m³ at 60 degrees C. Palm kernel oil is exported principally from Malaysia, the destinations being Europe, North America and Japan.

It is possible that small quantities of palm kernel oil fractions, e.g. palm kernel oleine and palm kernel stearine, will also be exported. These fractions are normally produced as refined oils. Palm kernel oleine is partly liquid at ambient temperature, whereas palm kernel stearine is a hard fat requiring elevated storage temperatures (32-38 degrees C) to maintain it in a pumpable condition. The free fatty acid content of the refined oils is normally below 0.1%.

Palm kernel oil and its fractions are used widely in the production of edible fats but are finding increasing application in the oleochemicals industry for the production of fatty acids, fatty alcohols and their derivatives.

Commodity	Characteristics	M3 per Tonne	Packaging
PALM OIL	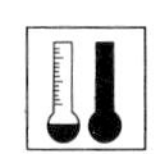	1.62/1.67	Drums
		1.05	Bulk

Palm oil is obtained from the fruitlets found in large bunches on the oil palm, grown extensively in Malaysia, Indonesia and other countries having a tropical climate. In order to avoid rapid deterioration of the oil after recovery from the fruit the complete bunch must be sterlised as soon as possible after harvesting. Crude palm oil is normally sold against a specification of "not more than 5% free fatty acid". It is a dark-red oil which will contain a considerable amount of solid (crystallised fat) at ambient temperature in regions of temperate climate, but the oil remains sufficiently liquid for pumping purposes at the temperatures prevailing in the regions where the oil is produced. The recommended temperature range for storage and bulk shipments of the oil is 32-40 degrees C, and for loading and discharge is 50-55 degrees C. Palm oil is exported on a large scale as the crude oil to various Asian countries but Malaysian palm oil is generally refined and processed for export to Europe as refined, bleached and deoderised (RBD) palm oleine and RBD palm stearine. RBD Palm stearine is a hard fat which should be maintained at a temperature of at least 40 degrees C, but not more than 45 degrees C, if handling problems are to be avoided.

At 50 degrees C the specific gravity of palm oil is approximately 890 kg/m³.

Palm oil is used extensively as an edible oil (after refining) but can also be used for the production of fatty acids and other oleochemicals.

See Part 2, Oils and Fats. See Appendix Table No. 11.

Commodity	Characteristics	M3 per Tonne	Packaging
PANOCHA			Bags

Philippine local produce. Solidified sugar syrup dehydrated into hemipherical shape in coconut shells. In bags. Susceptible to high temperature damage.

Commodity	Characteristics	M3 per Tonne	Packaging
PAPER		1.2/2.65	Reels
		1.3/1.8	Bales
	Kraft	1.67	Rolls
	Newsprint	1.81	Rolls
		1.39	Half Cases

Ordinary paper in rolls, stows, on average at about 2.51 M3/tonne. With a good selection of rolls and a suitable ship 2.37 M3/tonne will suffice, while with contrary conditions it may run to 2.65 M3/tonne.

Commodity	*Characteristics*	*M3 per Tonne*	*Packaging*

Paper is made from vegetable matter reduced to pulp, such as wood, esparto grass, flax refuse, straw, jute, also rags. Spruce, Balsam, Hemlock, cottonwood and other timber are used in great and increasing quantities for paper making, especially in Scandinavia, Newfoundland, Eastern Canada, United States and British Colombia.

Very vulnerable to mechanical damage, also to dammage by wetting. Distorted, torn or cut reels of paper may lose part or all of their value to the consignees.

Paper is usually shipped in rolls the ends of which are, in some cases protected by circular discs of wood; in other cases the rolls are simply wrapped with thick paper with extra layers over the ends.

Rolls of paper vary from 600 mm to 2,030 mm in length, i.e. width of paper while the diameters vary considerably, averaging in a mixed shipment about 900 mm.

Typical dimensions and weights might be:—

Diameter	*Height*	*Weight*
1535	835	
1125	1625	Upto 1200 kg each
1420	1220	
1050	1200	810 kg
800	1240	
960	960	
1010	960	769 kg
1010	900	756 kg
1010	650	716 kg

Rolls of paper should be stowed solid and well chocked off to avoid movement when vessel is at sea.

It is essential that the ground tier be stowed on a firm level floor, otherwise the bottom rolls will get badly distorted and claims for damage due to improper stowage will likely ensue. When mixed cargoes are carried, the floor can be levelled off up to turn of bilge by other suitable cargo — clean, dry lumber being particularly useful for this purpose. Clean, dry bagwork is very suitable.

In end holds, the greatest care should be exercised to ensure that the platform on which the ground tier of rolls is stowed is both level and firm. When a full cargo of paper is carried, the most satisfactory way of doing that is by building a platform consisting of a series of steps — width to suit the diameter of the larger rolls — the platform resting on substantial bearers.

All stanchions, ladders, etc., should be well covered with protective material to avoid chafage of rolls; and dunnaging should be thorough throughout.

Soft rope slings should be used for slinging paper rolls — alternatively rubber tubing is fitted over the rope sling to avoid tearing the paper. Loading or discharging of paper rolls by swinging derricks should not be resorted to owing to the difficulty of keeping rolls from banging against hatch coamings, ship's side, etc., which tends to destroy their shape, to avoid which is most essential. Large shipments of reels are commonly handled by fork lift trucks equipped with clamps.

The dragging of rolls from wings or ends of compartments to square of the hatch should be prohibited.

The use of cargo hooks, crow or pinch bars should never be permitted when handling paper.

The amount of broken stowage with a cargo of paper is very considerable, the smaller rolls may, with reasonable care, safely be utilised for "fillings" in the wings, etc. However much paper moves in full shipments. See Part 2, Techniques, Neobulk.

Reels of paper are not suitable modules for *ISO containers*, giving a very poor payload and causing handling problems at time of stuffing and stripping. Further, they may suffer moisture migration and sweat in a closed box container. Typical payload of a 20 ft container might be 16.8 M3, net weight 14.5 tonnes.

Chinese paper — is made from bamboo bark of fine quality, brownish in colour and used principally for lithographic work. Shipped in strong cases from China. Special stowage in a dry place.

Commodity	*Characteristics*	*M3 per Tonne*	*Packaging*

Japanese paper — is made from the bark of a certain mulberry tree and used principally for lithographic work. Shipped in cases. Special stowage in a dry place.

Recycled paper may be compressed into high density bales. These bales may themselves be grouped into units, usually about 12 bales to the unit. Stowage factor approximately 3.9 M3/tonne. Guard against risk of fire. See IMDG Code.

Rice paper in cases, stows at 2.23/2.37 M3/tonne. Prepared from the pith of a tree grown in Formosa.

Coated paper — one side gummed. Very susceptible to moisture damage. See also Neobulk Cargoes.

Carbon Paper — see IMDG Code.

Paper board on reels average 4 to 7 M3/tonne, and when packed in bales 1.5/2.3 M3/tonne.

Paper is frequently shipped from countries which have very cold winter climates. Large claims have resulted from such cargo being ventilated. The temperature of the reels should be checked at the time of loading and ventilation regime should be based on such temperature measurements.

Commodity	Characteristics	M3 per Tonne	Packaging
PAPER OR WOOD PULP	 	1.25/1.39	Loose Bales
		1.25/1.39	Unitised Bales
		1.45/1.56	Wet Pulp

Made from various kinds of timber and shipped in large quantities from Scandinavia, Canada, U.S.A., etc. Various processes are used in the manufacture of pulp according to the class of timber used; the principal processes being the sulphate and soda, ground wood or mechanical, and the sulphite processes.

Wood pulp is shipped in compressed bales both as "Dry", and occasionally as "Wet" pulp.

Wet pulp should not be stowed over any cargo liable to damage from drainage, not in the same compartment as dry or other goods, such as flour, seeds, canned goods (which see), etc., liable to be damaged by moisture.

Dry pulp should for the same reason be stowed apart from such goods.

Wood pulp is very liable to damage from any fibrous material. During handling it must be kept clear of any contact from ropes, etc., and should be loaded and discharged with wire or chain slings. Unless especially designed, the wire binding is not suitable as a lifting point. Holds should be clear of any matter that could become embedded in the bales.

Similarly especial care must be taken when cleaning up after a grain shipment. Large claims have had to be met in the past through contamination of wood pulp by grain.

In considering the stowage of this cargo, suitable stowage must be found away from any cargo containing or covered by fibrous materials, such as hessian on bales of wool, Sisal, Burlap, etc.

Bales of pulp may, in some ports, be strapped together into units. A typical configuration might be units made up of 12 bales each, with 12 units (weighing approximately 20 tonnes) forming a single lift.

Commodity	Characteristics	M3 per Tonne	Packaging
PARAFFIN WAX	 	1.39/1.53	Bags
		1.95/2.09	Barrels
		1.95/2.01	Cases
			Bulk

A white transparent substance obtained by distillation from petroleum, shale oil, coal, etc., and used for the production of petroleum jelly, candles and paper waxes.

Paraffin Wax is derived from the heavy end of the crude oil distillation process. Slack Wax (a dirty form of Paraffin Wax which still contains gasoil) is shipped to refineries, where it is processed to produce refined Paraffin Wax. The carriage of Paraffin Wax is generally undertaken by chemical tankers and carried in liquid form. Careful attention should be paid to shippers' instructions regarding carrying temperatures, as various grades of refined Paraffin Wax have a melting point between 50 degrees C and 71 degrees C. The discharging temperature is normally about 76 degrees C.

Commodity	*Characteristics*	*M3 per Tonne*	*Packaging*

FOSFA describes Paraffin Wax as a Fatty Alcohol along with Latex, Tallow, etc.

The earlier practice of carrying Paraffin Wax in dry cargo ships' deep tanks and dedicated oil tanks and allowing the product to solidify has now ceased. Frequent problems were encountered during the re-heating process, including uneven temperature distribution producing lumpiness and overheating resulting in the product being off-specification in colour.
See IMDG Code. See Part 2, Petroleum Cargoes.

PATCHOULI		4.46	Bales
PATCH		11.15	Loose

The dried branches of a tree grown in India and the E. Indies from which a very pungent perfume is obtained. Stow clear of foodstuffs.

See Essential Oils.

PATCHUK		See Putchok	

PEACAKE		1.39/1.45	Bale
			Bulk

See Seed Cake. Shipped from F. East.

PEACHES		3.78	Cartons
		See Part 2, Refrigerated Cargoes.	

PEANUTS		See Ground Nuts	

PEAPULP		1.81/1.95	Bag

Moist goods; shipped from F. east.

PEARLS		See Specials	

PEARS	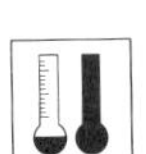	2.26/2.29	Cases
		2.05/2.96	Cartons

Pears are packed in boxes or cartons each fruit being wrapped in paper.

Stowage and carrying temperature same as apples — fluctuations of temperature being more injurious to pears than to apples.

Owing to the difficulty of maintaining correct temperature no other cargo should be worked through pear decks.

It is not advisable because of pressure damage, to give pears deep stowage. Very liable to bruising. Careful attention to handling and provision of suitable protective walking boards, etc., is essential.

Careful examination of cargo is very necessary before shipment to ensure that pears are in sound condition. Fans should be run to ensure maximum rate of air change during passage and CO2 content kept to below 2%.

Pears will taint other cargoes and whenever possible should not be stowed between the same bulkheads as meat, butter or cheese, flour, etc.

Other than the W.B.C. (William Bon Chretin) variety, apples and pears may be carried together in the same deck. All shippers must be agreeable to such conditions and also agree on the maximum permissible CO2 content. Carrying temperature + ½ degree C.

See Part 2, Refrigerated Cargoes.

Commodity	*Characteristics*	*M3 per Tonne*	*Packaging*
PEAS		1.45/1.75	Bags
		1.28/1.39	Bulk
	Quick Frozen	2.03	Cartons

Peas of various kinds are shipped in bags, but on occasions in bulk. May be liable to infestation.

Grain Regulations, which see, apply to their carriage. See remarks, on Beans, which generally apply, especially to peas in bulk. Also see Vegetables.

Quick frozen. Carriage temperature −15/−18 degrees C.

See Part 2, Refrigerated Cargoes.

PEBBLES (sea)		0.59	Bulk

Round pebbles, roll very easily. Should be overstowed with a layer of sacks. Angle of repose 30 degrees.

PELLETS (concentrates)		0.47	Bulk

Angle of repose 44 to 47 degrees.

See Part 2, Ores.

PELTS

See Furs, Hides, Skins. For Pickled Pelts see Wet Hides.

PEPPER

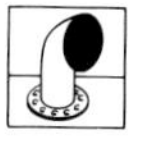

2.06/2.51 Bags

The principal kinds of pepper, all valued for their spicy properties, are:—

Black pepper, the dried fruit of a small E. Indian tree.
White pepper, the same fruit with the outer covering removed.
Cayenne Pepper, Chillies, which see.

Other varieties are long pepper, shipped from India, bell pepper, bird pepper, etc.

Black and white pepper are shipped from the principal ports of the Far East, notably Singapore and Penang. The two kinds should be stowed well apart, preferably in separate compartments.

Pepper is very liable to heat and sweat (though white pepper produces less sweat) and for that reason it should not be stowed with fine and delicate goods such as tea, tapioca, etc., and it should, also, be stowed apart from moist and other goods liable to heat such as copra, jelatong, gambier, and from cubebs, onions, etc.

It should be well dunnaged and thoroughly protected from contact with beams, beam knees and other iron — avoid stowing too many heights — also, if possible, arrange to overstow suitable cargo with pepper so as to lessen the weight on lower tiers of pepper. If rattan is carried, they should be stowed in conjunction with pepper, i.e. rattan on the floor, up the sides and across bulkheads, three or four tiers of pepper, then a tier of rattan and finish with pepper. This gives ideal conditions for the proper carriage of the latter. See Dunnage.

Ventilation should be given constant attention.

Commodity	*Characteristics*	*M3 per Tonne*	*Packaging*

ISO container — 2.22 M3/tonne approximately may be achieved. Open-sided containers, with tilts rolled up during transport and when weather protected. Dunnage (50 × 50 mm, 2″ × 2″) laid across the floor, and similarly half-way up the stow if a full container load. Pallets may be used instead of dunnage. Either must be suitably dry and clean. For short voyages may carry in closed box containers, with the doors hooked back while on board. The stowage position on board must allow adequate ventilation to flow around (and through) the containers, while maintaining protection from rain and spray. N.B. a refrigerated container with fans operating and without temperature control may also be suitable.

This is a valuable cargo and should receive suitable stowage and careful supervision during loading and discharge.

PEPPERMINT 

A pungent aromatic liquid, extracted by distilation from the mint plant. See Essential Oils.

PERFUMERY 

The raw materials from which perfumery is prepared are referred to under Essential Oils, which see.

The finished article is very valuable and is generally put up in sealed bottles packed in cases.

Special stowage in strong room.

See Part 2, Techniques, "Specials".

PERLITE ROCK  0.98/1.06 Bulk

Light grey, odourless, clay-like and dusty. Angle of repose 45 degrees.

PERSIAN GUM 

A gum exported from Middle East. See Gums.

PERSONAL EFFECTS AND ANTIQUES  2.83 Crates

Valuable. May be fragile. Advice should be sought before carrying out in container fumigation, as Methyl Bromide may linger in articles containing horsehair and other types of stuffing material. Great care should be taken when stuffing a container. Stow should be tight and well secured without damaging the goods. See also Furniture.

PERUVIAN BARK  See Cinchona

Commodity	*Characteristics*	*M3 per Tonne*	*Packaging*
PETROLEUM COKE	 	1.25/1.68	Bulk

Calcined — Transported hot. Essential that temperature is known prior to loading. Material in excess of 107 degrees C should not be loaded. Regulations exist for loading over fuel tank containing fuel with flashpoint under 90 degrees C. Essential reference to IMDG Code and IMO Code of Safe Practice for Solid Bulk Cargoes.

Uncalcined — Liable to heat and ignite spontaneously. Not to be loaded when temperature of material exceeds 55 degrees C. If temperature exceeds 44 degrees C, should be monitored and recorded on passage.

PETROL, PETROLEUM   See Part 2, Petroleum Bulk

Petroleum, under the Systems and Technique section, should be studied in conjunction with the International Convention for Safety of Life at Sea (SOLAS Convention), the IMDG Code, the International Safety Guide for Oil Tankers and Terminals (ISGOTT) also the International Convention for the Prevention of Pollution from Ships, the MARPOL Convention and its Protocol together with the Clean Seas Guide for Oil Tankers.

PETROLEUM PRODUCTS  

Cases
Drums
Deeptanks

See Part 2, Petroleum — See sub-heading "Carriage of Petroleum Products in General Cargo Vessels" (Drums/Cases etc.) also "Carriage in Deeptanks".

PETROLEUM JELLY  

See Vaseline
See Paraffin Wax

PHORMIUM  2.51/2.79 Bales

Also known as New Zealand hemp, is a strong leaf fibre. See Hemp.

PHOSPHATE defluorinated  1.12 Bulk

Granular, similar to fine sand. Angle of repose 30 to 35 degrees.

PHOSPHATE ROCK   0.79/1.20 Bulk

A mineral fertiliser which absorbs moisture very readily and so should not be exposed to rain or moisture.

Very dusty cargo.

PHOSPHORIC ACID
(Orthophosphoric Acid) 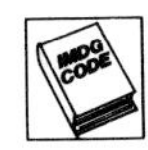 See Chemical Cargoes, Part 2.

Commodity	*Characteristics*	*M3 per Tonne*	*Packaging*
PHOTOGRAPHIC MATERIALS	  		Cartons Fibre Board Drums

Do not stow developed and sensitive paper in the same ISO container or near one another. Susceptible to temperature damage — protect from direct sunlight.

Some materials may be Dangerous Goods. See IMDG Code.

PIANOS		5.57	Cases

Stow right way up or as marked on cases. 'Tween deck stowage or top stowage Lower Hold. Away from hot bulkheads or moisture inherent cargo.

PIASSABA		2.79/3.07	Bales

A stiff course fibre in considerable demand for brush making — grown in Brazil and other tropical S. American countries, also in Africa. Care should be taken with tallies as there may be problems with short shipment. May be damaged by moisture. Keep dry.

PICKETS		See Timber	
PIGS		See Pork	
PIG IRON		0.30	Bulk

High-carbon iron. Angle of repose 36 degrees.

PIMENTO		3.34/3.62	Bags

Also called Jamaica pepper and allspice, the unripe berries of a West Indian tree. Dry stowage, but away from delicate goods.

PINEAPPLES FRESH		1.53/1.73 1.95/2.09	Tins in Cases In Crate/Carton

See canned goods. Should be carefully watched to avoid broaching.

May be carried chilled. Strictly limited life both during and after temperature control.

PINEY TALLOW		See Vegetable Fats See Part 2, Oils and Fats.	
PIPES Iron and Steel Tubing		1.67	Bundles

See Part 2, Iron and Steel Products.

Untreated steel pipes in stow.

Note shipside/coaming tomming
and hoop iron lashings.

Commodity	*Characteristics*	*M3 per Tonne*	*Packaging*
PISTACHIO NUTS		1.95/2.12 1.67/1.81	Cases Drums

The fruit of the pistachio tree has a very delicate flavour and is greenish in colour. See Nuts.

PITCH	 	1.25/1.39 0.84/0.98	Drum Bulk

Is a tar or bitumen product — partakes of the same characteristics as Asphalt, which see.

It melts with heat — stow clear of fine and dry goods and away from boilers, etc. Pencil Prill or Prill Coal Tar when shipped in bulk care should be taken to minimise dust. Cargo should be softlanded etc. Turpentine residue can cause intense irritation to skin and eyes. Protective clothing should be worn. See IMDG Code. (See IMO Code of Safe Practice for Bulk Cargoes).

PITPROPS		6.41/7.25	Bundles

Pitprops are short, straight lengths of timber, mostly fir, denuded of the bark, and are exported in large quantities from Scandinavia and Russia to many coal mining countries.

The stowage factor varies according to lengths and moisture content. The unit of measurement in the Pitprops trade to the U.K. is the English (cubic or pile) "Fathom", i.e. 6 ft × 6 ft × 6 ft = 216 cu. ft, freight being payable on that basis.

The Props vary from 75 mm to 300 mm in diameter, and are usually shipped in standard lengths, viz. 3.5 ft, 4 ft, 4.5 ft, 5 ft, etc., up to 11 ft. Crooked or split Props should be rejected.

A fathom of pitprops varies considerably in weight, viz. from 2.3 tonnes or less to, on occasions, as high as 4 tonnes, such depending partly on the class of timber but more particularly on the moisture content of same. When receiving from wharves or railway trucks, the timber, being relatively dry, approximates the lesser weight per fathom, but, when the props are brought alongside in leaky barges which, not infrequently, are partially waterlogged, or are drafted alongside, the weight per fathom is consequently high. In general, however, it may be stated that, with the modern type of ship, the limit of the deck load she will carry is determined by consideration of stability rather than by that of deadweight.

See Part 2, Techniques — Timber.

PLANTAINS		3.62	Bunches

A banana-like fruit shipped from the W. Indies. See Bananas.

PLASTER		0.98	Bags

Keep dry. Dusty.

PLASTER OF PARIS		See Gypsum	

Commodity	*Characteristics*	*M3 per Tonne*	*Packaging*
PLASTIC RESINS		Various	Bags Drums Containers Bulk

These may be in the form of liquids. They may be dissolved in solvents, their carriage usually being covered in the IMDG Code as Class 3 flammable liquids. They may also be in the form of chemical intermediates — typical examples being isocyanates which are covered in the IMDG Code under their specific names.

Apart from hazards referred to in the IMDG Code, partially polymerised materials frequently continue to polymerise when exposed to the atmosphere to produce solid sticky materials which are often difficult to remove from ship's structures. Hence any leaking products should be absorbed on sand or sawdust as soon as possible after a leak is detected. However all members of a crew carrying out this work should take necessary precautions such as wearing respirators and protective clothing, as indicated by the specific IMDG Code entry.

Needless to say no leaking containers should be accepted at the time of loading.

Plastics resins may also be solids in the form of powders or beads. Plastics materials shipped in this form are polyethylene (polythene), polypropylene, polyvinyl chloride, nylon, polyester (terylene), polyacrylates, polycarbonate etc. These substances are normally packaged in paper or plastics sacks.

They must be kept free from any contamination and hence particular care should be taken during loading and discharge to prevent broaching of packages. Stows should be secured to minimise risk of damaging packages. Any material which may leak from packages should be collected and rebagged making every effort to avoid contamination. Plastics resins of the same chemical type are produced in a number of grades e.g. there are separate grades of polyethylene for the manufacture of films, plastics bottles, rigid mouldings etc. Care must be taken to keep each grade of resin in a separate stow. Material rebagged must be properly labelled.

Some resins, particularly nylons, are moisture sensitive.

PLUMBAGO	 	See Graphite. See IMDG Code.	
PLUMS		2.34/2.41	Cartons

May require change of temperature settings during transit. See Part 2, Techniques, Refrigeration.

PLYWOOD	 	2.09/2.23 1.95/2.12	Crates Best Bundles

Plywood consists of three or more layers of thin wood laid crosswise and joined together by glue or cement under pressure.

Must be treated as fine cargo which is very liable to deteriorate if allowed to get damp. Stow apart from moist goods, including sawn lumber, or those which give off moisture when heated such as seeds, nuts, maize, etc. Protect from dust of all kinds and other staining mediums. Wet or stained bundles should be rejected or mates receipts/bills of lading claused accordingly. Stow away from logs or sawn timber. Secure well without damaging edges as cargo slides easily.

Sheets usually measure 8 × 4 ft or 6 × 3 ft. When banded together into bundles or units, waste timber is used to protect the edges, thus making the package size slightly in excess of 8 × 4 ft or 6 × 3 ft.

ISO containers — the 8 × 4 ft sized sheet is not suitable for ISO closed box containers — the size makes it awkward to handle and impossible to achieve a worthwhile payload. Open sided containers and flatracks may be used, but very careful attention must be paid to secure the packs in such a way that they cannot move in a sideways direction. 8 × 4 ft packs will almost certainly make the container

Commodity	*Characteristics*	*M3 per Tonne*	*Packaging*

"overwidth" and it must be ascertained that such containers can be stowed on cellular vessels without fouling the cell guides or adjacent containers. Weather protection for the plywood must be provided — especially for flat rack stowage, a thin plastic sheet is insufficient against driving rain. If the container is stowed on deck it must not be on the outboard slots.

PODOPHYLLUM ERNODI  2.51/2.79 Bags

A root possessing medicinal properties from which is also obtained the dye, Quercetin. Dry stowage.

POLLARDS  See Middlings

POLYTHENE GRANULES 

See Plastics Resins.

POONAC  See Brunack, also Oil Cake.

POPPY SEED  1.95/2.01 Bags

The very small seed of the poppy plant, its oil yield being about 45 to 50 per cent. Two kinds of this seed are shipped, i.e. the white and the blue, the latter being bluish white in colour. Shipped in bags which, owing to the smallness of the seed, should be made of close woven material to avoid loss of contents. The bleeding of bags should not be permitted, and care should be taken to avoid mixture with other seeds. See Bag Cargo.

POPPY SEED CAKE   1.53/1.62

See Seed Cake. IMDG Code. (See IMO Code of Safe Practice for Solid Bulk Cargoes).

PORK (Hard Frozen)  1.67/1.73 Tierces; 2.51//2.65 Sides

Fresh Pork. Pickled, salt; wet cargo. Stow well apart from dry and odorous goods like turpentine, etc., and in a cool place.

Pig carcases, or frozen pork, are shipped in bulk in the hard frozen condition, the carcases being wrapped in gauzy cotton material. Inspection, stowage, etc., as for Lamb, which see.

POTABLE SPIRITS 

See IMDG Code.

Commodity	*Characteristics*	*M3 per Tonne*	*Packaging*
POTASH		0.98/1.06	Bags
Muriate of Potash		0.94/1.04	Bulk
Potassium chloride		1.10	Bags (Preslung)
Dead Sea Potash			

This is obtained from the water of the Dead Sea and elsewhere, and is exported in fairly large quantity from the Middle East.

It should be shipped thoroughly dried (water content averaging 0.25 per cent), it is not hygroscopic, does not give off any fumes when heated or under pressure. Under ordinary conditions, is not liable to taint edible goods.

It does not affect bare steel or iron to any greater extent than does common salt.

May be shipped in bulk as Standard, Fine, Flotation, Granular. Stowage factors for each may vary:

Standard and Fine 0.94 M3/t; Flotation 1.04 M3/t; Granular 0.95 M3/t.

See Fertilisers.

POTASH, CARBONITE (Pearl and Pot) 

An alkali obtained from vegetable matter by burning, from which is obtained caustic potash, and largely used in soap making.

Dry stowage remote from oils and greases.

POTASH, CAUSTIC		0.75/0.84	Drums

Wet stowage, remote from textile, leather and like goods.

POTASSIUM CHLORATE OF		1.53/1.59	Kegs

See IMDG Code.

POTASSIUM NITRATE (Saltpetre)	 	0.88	Bulk

May be readily ignited if mixed with combustible materials. See IMDG Code. (See IMO Code of Safe Practice for Solid Bulk Cargoes).

POTASSIUM SULPHATE		0.90	Bulk

Hard crystals or powder. Used in fertilisers and manufacture of glass, etc. Angle of repose 31 degrees.

See Fertilisers.

Bags of potatoes in reefer stow.

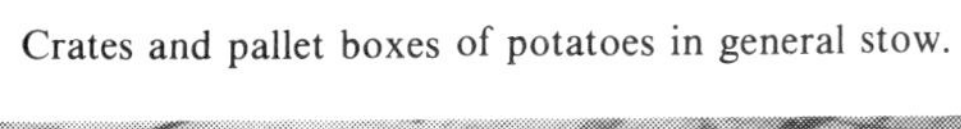

Crates and pallet boxes of potatoes in general stow.

Commodity	*Characteristics*	*M3 per Tonne*	*Packaging*
POTATOES		1.53/1.81	Bags
		1.53/1.67	Bulk
		1.62/1.90	Cartons

Potatoes are generally shipped in bags if the passage is of considerable length. For coastal passages by small vessels they are often shipped in bulk, in which case the holds should be thoroughly clean and well sweetened. A "live" cargo may give off heat and moisture. Liable to mould growth and rot if not shipped in a dry condition.

On a long voyage through the tropics packages are best stowed fore and aft on athwartship laid dunnage. Air spaces of about 305 mm in width under the upcast ventilators, and extending right across the ship should be provided.

When fans are fitted air conduits, formed of boards, crates or canvas over wooden frames should be led to bottom of holds from the fans. The use of windsails should not be neglected whenever conditions are favourable. Care must be taken to avoid overheating, potatoes stowed under the weather deck should have ample clearance from the top of the stow to the underside of the deck. Suitable arrangements should be made to register temperatures and these should be recorded. Ideally the temperatures should be below 15.5 degrees C. Normally the temperature of potatoes should remain steady in relation to that of the surrounding air. If, however, the production of heat becomes excessive the temperature of the stow will soon rise above the ambient temperature and unless provision has been made for adequate airflow when necessary, it may become so hot that it can no longer be remedied by corrective cooling, with the result that the stow will overheat and rapidly deteriorate.

Immature (new) potatoes are frequently carried under controlled temperature, although considerable quantities are shipped for on deck or under deck normal stowage. At about 4.5 degrees C sprouting is retarded. Do not give a deep stowage. Must not be stowed with fruits, dairy products, eggs, etc.

Too much top weight should be guarded against and the ventilation should be ample. In frosty weather there is great danger of frostbite setting in — especially under downcast ventilators — this can be avoided by covering with mats or straw. Care should be exercised not to receive potatoes that are frostbitten. The loss of weight with this cargo is appreciable. Check and reject wet bags. Avoid loading during rain.

ISO containers — open-sided containers with tilts rolled up on passage to allow through ventilation. Stowage on board should be protected from rain, spray or direct sunlight. May on occasions (e.g. short voyages) be carried in closed box containers, with pallets laid on the floor and doors open to assist ventilation.

POULTRY		1.67/2.23	Crates. Cartons.

Should be shipped in the hard frozen condition or deep frozen. Crates should be examined carefully for soft carcases (all of which should be rejected) the presence of which is readily detected by prodding the exposed contents of crate with a clean rod or dowel. If the outside carcases are hard frozen it may reasonably be assumed that the inside ones are so. At least 50 mm dunnage laid in line with air flow to keep crates and cartons clear of the deck.

See Part 2, Techniques, Refrigeration.

PRAWNS		2.37	Bags
(Also see Fish)		2.79	Cartons

Odorous cargo stow away from cargo liable to taint. More often carried under Refrigeration (Deep Frozen Products). Careful and detailed inspection of cartons should be carried out on random basis prior to shipment. Frequently it has been found that there has been a delay between catching and freezing also that the contents of cartons are not at proper temperature when presented for shipment.

Multi-sling cradle of woodpulp.

Commodity	*Characteristics*	*M3 per Tonne*	*Packaging*

Receivers may complain of discoloration or the prawns being mushy. Careful log of temperatures prior to and during carriage should be maintained. Prawns are a frequent source of claims against the carrier.

See Part 2, Refrigerated Cargoes.

PRECIOUS STONES,
PEARLS, GEMS,
CURRENCY NOTES,
BONDS,
POSTAGE OR REVENUE STAMPS,
HIGH CLASS JEWELLERY,
ETC.

High value cargo. See Part 2, "Specials".

PRESERVED MEATS 

See Meat, Preserved.

PRESERVES See Chinese Groceries.

PRIME WOODS

See Timber, Terms, etc.
This class of timber being valuable, should be carefully protected from stains.

PRUNES	 	1.50/1.56	Casks
		1.39/1.45	Cases
			Bags

A dried fruit, very liable to broaching like most fancy edible goods. May be damaged by crushing (bags). May be infested.

PULP  See Paper Pulp

PULPWOOD		3.34/4.18	Var. Lengths

Pulpwood is cut in various lengths to the requirements of the mill to which it is consigned. Usually lengths vary between 1067 mm to 2134 mm or 1 to 2 metres; the diameter, like pitprops varying between 76 mm and 254 mm. Generally speaking any particular shipment is confined to one length. Pulpwood is a dry clean cargo, devoid of all traces of bark, and must be protected from oil, coal and other stains. Holds should be scrupulously clean. It is measured and stowed the same as Pitprops, which see.

Actual condition of cargo upon loading must be reflected on Mates' Receipts and Bills of Lading.

Commodity	*Characteristics*	*M3 per Tonne*	*Packaging*
PUMICE		2.93/3.21 1.90/3.25	Bags Bulk

A light and sponge-like stone of volcanic origin, used by polishers of metal stone, etc. Shipped from Lipari, Italy, Sicily, etc. C/P may stipulate bulk stowage factor to be not greater than 1.45 M3/tonne.

PUREE
YELLOW OCHRE 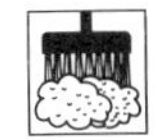

See Ochre Puree.

PUTCHOK
(Patchuk)  3.34/4.18 Bales

The leaves and root of an Indian plant, used for burning as incense. Odorous — dry stowage but apart from tea and all delicate goods.

PYRETHRUM
FLOWER SEEDS  1.81/2.09 Bales

Shipped from E. Africa. See Seeds.

PYRITES  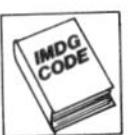 0.30/0.55 Bulk

Sulphide of iron (iron pyrites) or sulphide of copper (copper pyrites) used in the manufacture of sulphuric acid — they throw off sparks if struck with steel. May liquify and shift if shipped "wet". Check physical properties before loading. See Ores.

A very dusty cargo, which ruins paintwork. Engine room skylights and accommodation ventilators should be kept tightly closed when loading or discharging pyrites.

See IMDG Code.

PYROPHYLLITE  0.50 Bulk

Chalk-white natural hydrous aluminium silicate. Used in ceramics, slate, pencils, etc. Angle of repose 40 degrees.

QUARTZ 0.67 Bulk

QUARTZITE 0.64 Bulk

Angle of repose 35 to 40 degrees.

QUEBRACHO 1.67/2.23

A very heavy wood extensively shipped from the Argentine for use for tanning purposes, etc. Usually shipped in short lengths but owing to its crookedness the stowage factor is most variable.

Commodity	*Characteristics*	*M3 per Tonne*	*Packaging*
QUEBRACHO EXTRACT	 	1.11/1.23	Bags

In powder form. Dry stowage, but not over delicate goods.

QUERCITRON		2.23/2.51	Bales
		2.34/2.51	Bags

The inner bark of a tree grown in the States of Carolina, Georgia, etc., from which is obtained a yellow dye used for calico printing. See Barks.

QUICK-FROZEN			
	Asparagus	1.98	Cartons
	Beans	2.95	Cartons
	Peas	2.03	Cartons
	Sweetcorn	3.09	Cartons

Carriage temperature −15/−18 degrees C. Handling at time of stowage or stuffing should be as quick as possible to avoid absorbing heat into the packages and product. See Part 2, Refrigeration.

QUILLIA			Bales

The bark of a tree used in the soap making industry, washing silk and printed goods. See Bark.

QUININE BARK		3.62/4.04	Bales

See Chinchona, also Barks.

RABBITS	 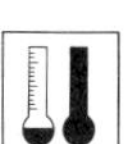	2.23/2.68	Crates Cartons

Hard frozen. Susceptible to damage by taint. May be stowed with butter, carcase meat, carton meat, etc., if temperature compatible −12/−15 degrees C. Rabbits are packed in open crates generally two dozen to the crate.

They should be scrutinised carefully for soft carcases, the presence of which is easily detected by prodding the exposed carcases with a clean rod or dowel.

See Part 2, Refrigeration.

RABBIT SKINS	 	4.46/5.02	Bale

Stow well clear of edible and delicate goods to avoid tainting by smells given off. See Skins.

RADIOACTIVE MATERIALS		Special Containers	

Shipped subject to complex international regulations. Specialist advice should be sought if the documentation supplied causes any concern. On deck stowage away from crew accommodation and susceptible cargoes. See IMDG Code. It is also necessary to consult Port and Canal Authorities as to their local or Government Regulations.

Commodity	*Characteristics*	*M3 per Tonne*	*Packaging*
RAFFIA GRASS		4.46/5.02 3.34/3.62	Bales Pressed Bales

A grass fibre. See Fibres.

RAGS	  	4.18/5.57 1.53/2.09	Bales Pressed Bales

Very objectionable cargo for amongst others, the following reasons:

(a) It is peculiarly liable to spontaneous combustion — probably due to carelessness in permitting oily material to be included in the bales, the detection of which by the ships' officers is practically impossible.

Many cases of fire in rag cargoes have occurred with, in some instances, disastrous results. The acrid smoke from rags often makes it impossible to fight and subdue the fire.

(b) The insanitary sometimes even verminous condition of rags where they have not been thoroughly disinfected. See Part 4, Smaller Vermin.

Most countries insist upon the production of a Sanitary Certificate before rags are permitted to be landed so that in such cases, if a certificate properly endorsed has not been obtained before shipment, the ship may be involved in much loss of time and money.

(c) The inadequacy of packing. Usually rags are packed in bales with the scantiest of coverings and poorly bound together. In such cases the bales will not stand ordinary handling with the result that difficulty is experienced with consignees who, holding Bs/L for a number of bales, very naturally refuse to accept delivery of a quantity of loose rags — the relief afforded by endorsing the bill of lading with remarks re condition of bales is, at best, but partial.

The bales should not be slung by the binding.

Stow well clear of grease, oils and oily goods, near the upper hatchways if at all possible and away from foodstuffs and delicate cargo. See IMDG Code.

RAILWAY IRON		0.36/0.42 1.11/1.39	Rails Ties

Stow fore and aft, as solid as possible, well chocked, tommed and wedged off. With full cargo of rails in a 'tween decker at least a fourth of the weight should be stowed in the 'tween deck (all over) well tommed down from the deck beams. The stowage and securing of this cargo calls for a considerable supply of timber.

In order to increase the rate of discharge, the stow should be "glutted" with strong dunnage laid 'thwartship. This practice considerably increases the stowage space required and is really only suitable for small shipments. Another practice is to leave wire loops on the ends in order to facilitate the quick slinging of the railway irons.

RAISINS	  	1.39/1.45	Cases

Dried grapes. Stow apart from wet, moist and odorous goods, also goods liable to heat. May be subject to infestation.

RAMIE		2.23/2.37 1.39/1.53	Ordinary Bales Hard Pressed

Or Rhea, the flax-like fibre of China grass, used in the manufacture of cordage and textiles, also of incandescent mantles, etc. See Fibres.

Commodity	*Characteristics*	*M3 per Tonne*	*Packaging*

RAMTIL — See Niger Seed

RAPE SEED   1.67/1.81 Bags

The seed of a plant grown in large quantities in India, Russia and throughout Middle Europe, from which rape seed or colza oil is obtained, the average oil yield being from 35 to 45 per cent.

This seed is very apt to heat — especially if new — so that good ventilation is essential. Stow clear of hemp, jute and like cargo, as in contact with these fibres the danger of spontaneous combustion exists.

See Grain. See IMDG Code.

RAPE SEED CAKE/PELLETS    1.45 Bulk; 1.39/1.50 Bags

The residue of rape seed after the oil has been extracted. Used as cattle food. Very strong smelling, liable to heat, and spontaneous combustion, soften and become mouldy. Cake should be weathered and aged according to oil content. Certificate of analysis should be issued by exporting body or country. Careful monitoring of temperatures essential prior to and during loading, also on passage. Ideally cargo temperature should not exceed 5 degrees F of ambient temperature. Cargo received around 50 degrees F should be watched carefully and any cargo heating in excess of 55 degrees F should be rejected and discharged. See Seed Cake. See IMDG Code. Also see (IMO Code of Safe Practice for Solid Bulk Cargoes).

RAPE SEED OIL 0.98 Bulk

The production of rapeseed oil has grown very strongly since the development of new forms of rapeseed in the 1970s made it possible to use the oil more extensively in food products. As a result of this development two types of rapeseed oil are now available — the new type, often referred to as canola oil, which is used as an edible oil, and the older variety, which may be traded as an oil to be used for industrial purposes. Although Europe is the largest producer of rapeseed (canola) oil now Canada remains the major exporter. India and China are large-scale producers of the older variety of rapeseed oil, but this is largely used for domestic food supplies and is therefore not exported.

Rapeseed (canola) oil has a density of approximately 920 kg/m^3. The crude oil can be expected to contain up to 1.0% free fatty acids and is also likely to contain a small quantity of gums, which may precipitate during shipment where the moisture content of the oil is close to or at its permitted maximum.

See Part 2, Oils and Fats.

RATTANS  3.34/4.46 Bundles

An E. Indian cane — the long stems of a plant which grows to a very great length. Rattan is in great demand for basket, mat and furniture making, etc. Shipped in bundles of 100 canes in distinct grades.

The coarser grade may safely be used for dunnage under fine cargo and, as rattan absorbs moisture, they are particularly useful for that purpose with pepper, sago, tapioca, sago flour and other commodities in bags which are liable to heat. See Part 2, Dunnage.

White rattan, however, should not be used as dunnage except with cargo, that will not sweat, and only then if B/L does not prohibit their use for this purpose.

Rattan, if wetted by rain, should be rejected; if stowed in that condition, they will become badly marked and depreciate in value: result — claims.

Commodity	*Characteristics*	*M3 per Tonne*	*Packaging*

The coarser grade rattan may be stowed almost with any cargo, even on top of jelatong (with good dunnage).

Both classes of rattan are very suitable for filling hatchways on completion of loading; loosely packed, they readily permit the heated air from the holds to pass through and so escape when hatches are open. With that object in view, a sufficient quantity of suitable length and grade of rattan should be reserved to complete the loading of each hatch. When shipped in containers, care should be taken that interior is clean and dry and rattan is free from moisture. Green mould grows readily on damp rattan.

RATTAN CORE   3.90/4.18 Bales

Usually packaged in strong waterproof paper, double at ends, an outer cover of hessian. Cool stowage, remote from boilers and other sources of heat, is indispensable to avoid the core drying, which, if permitted, produces a condition of brittleness and makes the core unsuitable for the purpose for which it is intended. Do not crush. Care during handling.

RATTAN (Split)
RATTAN (Peel) 3.90/4.46 Bales

REFRIGERATED CARGO See Part 2, Techniques

REPTILES See Part 2, Livestock

RESIN (Natural)
(See Rosin)    2.50/3.29 Bags

An exudation from plants and trees of which there is a great variety. Resins are insoluble in water, but generally dissolve in alcohol. All are of an inflammable nature and melt if heated.

Epoxy resin in plastic pails: 18 to 25 kg. Non-toxic. Adequate ventilation. See Gums. See IMDG Code.

RESINS (Synthetic)

Polymeric materials may be shipped in drums. The term sometimes refers to plastic pellets. See Plastics Resins.

RETURNED EMPTIES See Empties

REVERTEX Drums/Bulk

This commodity is a concentrated form of Latex (which see) and having a substantially higher rubber content.

The anti-coagulent or stabiliser used is an alkaline mixture of caustic potash, whereas an ammonia solution is used with Latex. Ordinarily, Revertex is composed of (about) 73 per cent rubber, 2 per cent caustic potash, balance water.

The viscosity is substantially greater than that of Latex, so that the pumping rate will be proportionately lower. Revertex partakes of the character of Latex as regards non-inflammability.

This commodity does not move in large volume and for the most part is shipped in steel drums, devoid of any protective internal coating, without any adverse effect on the metal containers.

Small bulk shipments are however made with satisfactory results. See Rubber Latex.

Discharge of bulk rice.

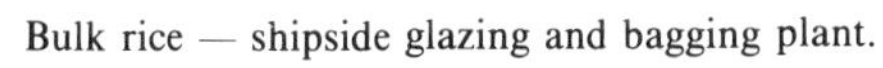

Bulk rice — shipside glazing and bagging plant.

Commodity	*Characteristics*	*M3 per Tonne*	*Packaging*
RHEA FIBRE		2.23/2.37	Bales
		1.39/1.53	Bales Hard Pressed

See Ramie.

RHUS		See Shumac	

RICE			Bags or Bulk
		1.81/1.95	Paddy
		1.34/1.39	Cargo Rice
		1.39/1.45	White Rice
		1.48/1.56	Broken Rice
		1.78/1.84	Rice Dust (Boussik)
		1.81/1.90	Rice Meal
		1.67/1.73	Rice Bran (Beaten)
		1.95/2.09	Rice Bran (Unbeaten)

Originally an E. Indian grain but now extensively grown in countries, between Latitudes 45 North and 36 South.

Probably the choicest grain is grown on the banks of the Tiber whilst the bulk of world supplies is obtained from India, Burma, U.S.A., Australia, S. America and Thailand.

Rice is shipped in bags in the forms mentioned above.

Paddy is the name by which the rice is known before the husk is removed. This is sometimes called rough rice.

Brown rice (also sometimes called rough rice) is the paddy after the husk is removed.

White or polished rice is the rice in its most refined form.

Rice is dried before shipment. Losses due to mould growth will occur if drying is insufficient.

Rice is susceptible to damage by strong odours, so that scented or smelly goods should not be stowed in the same compartment.

Particularly from Australia, it is possible that rice will have to be stowed in close vicinity to wool or sheepskins. It is not considered that polished rice will take taint from either of these two commodities.

Bagged rice must be properly ventilated to prevent condensation.

Rice that is damp or wetted in stowage quickly rots, generates great heat and emits an evil stench, all of which affect other rice in that vicinity.

Rice Bran is used for animal feed.

It is the experience of the Ministry of Agriculture that rice bran, even originating in the U.S.A., is often infested and is, in general, the most heavily infested single commodity handled in world trade.

Rice bran should therefore be stowed away from other goods liable to be infested by movement of insects during the voyage. It may be "clean" from the point of view of taint, but not so as far as insects are concerned.

Bagged rice cargoes should be dunnaged as normal bagged cargoes.

All iron surfaces, including shipside, frames, beams, bulkheads, stringers, etc., with which the rice bags are likely to come into contact, are to be well covered, but the use of mats at the side over spar ceiling is to be avoided, as they interfere with the air circulation. Bamboo suitably placed, is much better in this position as already referred to. (However certain qualities of rice have been and are being carried in bags, dump or random stowed, with only kraft paper protection from metal-work, with satisfactory out-turns resulting).

Shipped in bulk as brown rice, with the husk off, or white rice, with bran off. Can also be shipped as milled rice. Holds should be prepared as for grain cargo and loaded in compliance with the Grain Rules.

Commodity	*Characteristics*	*M3 per Tonne*	*Packaging*

Australian rice (1.51 M3/t for random stow; 1.32 M3/t for conventional stow of bags; 1.25 M3/t for bulk). Full cargoes may be shipped in bags usually using "random stow" method, without dunnage or spar ceiling and only kraft paper to protect where necessary from ships sides, frames, etc. Also shipped as full cargoes of bulk — see Grain. Parcels of milled rice in poly lined bags often in containers without ventilation.

ISO containers — Both white and brown rice may be shipped bagged or in bulk. Containers must be clean and free from residual taint. Rice may have tainted, and the taint not discovered, until after it has been cooked.

RICE PAPER 2.23/2.37 Cases

Is prepared from the pith of a tree grown in Formosa.

RICE SEED CAKE   1.39/2.09 Bulk

Rice Bran — See Seed Cake. (See IMO Code of Safe Practice for Solid Bulk Cargoes).

RIVETS 0.84 Bags and Kegs

Any suitable bottom stowage. Being heavy, cargo should be well distributed over *ISO container* floor.

ROCK ASPHALT See Asphalt

ROOFING FELT

Roofing sheathing. Shipped in rolls. Stow on end as pressure may cause bitumen content to adhere.

ROPES	**Coir**	2.51/2.79	Coils
	Manilla	2.23/2.37	Coils
	Wire	0.70/0.98	Coils
	Sisal	2.23/2.37	Coils

Stowage factors vary with size of coils and with the quality of the rope.

The oil used in the manufacture of ropes and the tar, with which the burlap wrapping usually is impregnated, gives off strong smells, rendering this class of cargo quite unsuitable for stowage with fine and edible goods.

Dry stowage firmly chocked to prevent chafage. See IMDG Code.

ROPE SEED  1.67/1.81 Bags

Cool, dry stowage.

ROSEMARY LEAVES  Bags

Leaves of the rosemary shrub shipped in bags from Mediterranean ports. Has a very pungent smell and should be stowed well away from foodstuffs and other cargoes likely to taint.

Commodity	*Characteristics*	*M3 per Tonne*	*Packaging*
ROSEMARY OIL			Drums

See Essential Oils.

Commodity	*Characteristics*	*M3 per Tonne*	*Packaging*
ROSEWOOD OIL			Drums

See Essential Oils.

Commodity	*Characteristics*	*M3 per Tonne*	*Packaging*
ROSIN	  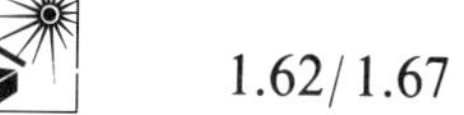	1.62/1.67	Barrels

An inflammable substance obtained by distillation from certain kinds of trees such as the pine, larch, etc., packed in barrels for shipment.

It is only in special cases that the loading Regulations of U.S.A. ports permit rosin to be stowed in the same holds as cotton, and even then such permission is not given unless airtight floor of 50 mm boards is laid so as completely to cover the rosin before cotton is stowed over same. At times rosin is carried on deck, in which case it should be carefully borne in mind that it is inflammable. See IMDG Code.

Commodity	*Characteristics*	*M3 per Tonne*	*Packaging*
RUBBER	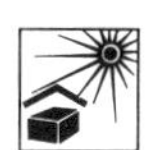	1.90/1.95	Cases
		1.81/1.87	Bales
		1.81/1.87	Bags
		1.67	Sheet
		3.34	Crepe

Also known as caoutchouc and india rubber, is the coagulated sap or milk obtained from a variety of tropical trees and climbing plants, the bulk of commercial rubber being obtained from the true Para rubber tree of Brazil, now extensively cultivated throughout Malaysia, Indonesia, India, Sri Lanka, etc., whence it is exported to Europe and the U.S.A. in increasingly large quantities.

Crude rubber is shipped either in the sheet, block or crepe form and is packed in cases, bales or bags — such packages from the Straits ports measuring 0.146 M3 weighing about 105 kg equalling 7 cases per freight tonne of 1 M3.

Rubber (dry) is a clean cargo and suitable for stowage with ordinary dry goods, but not with tea or other delicate cargo; smoked rubber is liable to taint fine cargoes such as coffee, etc.

This commodity is readily damaged by heat or by contact with oils, greases or acids. Will deteriorate and/or get mouldy if stowed near or with wet or moist goods of those which are liable to heat and sweat such as jelatong supar, copra, etc.

Some natural rubber liable to "coldflow" and blocking. Blanket crepe has a strong odour and sticks easily. SMPT rubber may have a coloured line on the bale which must be stowed vertically. Polythene separation is usually better than talc; the latter may cause labour problems at port of discharge.

Stained packages: a very careful watch should be kept for wet stained packages, all of which should be rejected if claims at destination are to be avoided.

Full particulars of all rejected packages returned for shipment after reconditioning should be endorsed on mates' receipts.

Mould damage: many claims are advanced for deterioration due to the formation of mould on rubber, such condition most likely arising, either because the contents were not thoroughly dry when packed or because the packages were wetted by rain or spray before shipment, and as the outside of packages so wetted quickly dries under the influence of a tropical sun, it is exceedingly difficult, often quite impossible, for their true condition to be noted at time of shipment, more especially as regards bale and bag rubber.

In such cases it is generally found, on opening the packages at destination, that the wrapper sheets and edges of inside sheets are covered with mildew (mould).

Commodity *Characteristics* *M3 per Tonne* *Packaging*

Mould growth may also result, should heavy condensation occur during the voyage, and should be safeguarded against by careful dunnaging and matting and by constant attention to ventilation.

Further to obviate claims, it is also necessary to avoid the staining of wrapper from whatever source, especially by coal dust, grease, flour, etc.

Crushing damage: another prolific source of claims and disputation is due to the frailty of the cases in which rubber customarily is packed. These cases, ordinarily, are made of thin three ply (Venesta cases) which, under very ordinary pressure, are apt to collapse, with the result that splinters penetrate the contents or get embedded in the wrapper sheets, and the sheets becoming massed or matted under pressure necessitating the use of knives to separate them.

Dunnaging: correct dunnaging, with proper material, will go a very long way towards ensuring the correct delivery of a rubber cargo.

Neither case nor bale rubber should be stowed on rough uneven surfaces such as are afforded by ingots of tin, copper, etc., bag cargo, rattans or unevenly disposed case goods, unless such surfaces are boarded over so as to produce a firm and level floor for the rubber to rest upon.

Cordwood, bamboos — split or whole — sticks, rattan bundles or like irregularly formed dunnage is most unsuitable for case or baled and bag rubber as their use will inevitably result in deformation of the underside of lower tier packages, etc., causing splinters from the cases to become embedded in, and the massing of, the rubber contents by the superimposed weight.

For filling broken stowage, cordwood and like dunnage is correct and proper, provided it is firmly packed and well boarded over and none of the rubber packages come into contact with same.

Nothing but fairly wide boards should be used, for dunnaging or for flooring over other cargo, with rubber.

Such should be closely spaced, so that each package is well supported along most of its underside and not suspended by the ends or sides between two boards or poised on one board by its middle.

Great care should be exercised to avoid using dunnage that is not thoroughly dry. Wet or even damp dunnage in contact with packages of rubber may result in mould growth on their contents.

Cases should be stowed square and upright in all position, never canted, and as solid as possible; ample dunnage should be used at the turn of bilge, in way of the crown of tunnel casing, and of stringers and brackets. See Part 2, Techniques, Cases.

If at all possible, avoid stowing rubber in the fine ends of forward and of aft holds, neither should case rubber be stowed on top of bale or bag rubber as the latter has too much "life" to make a stable platform and is likely to settle under pressure.

If overstowing with other cargo, light goods should be selected for that purpose.

Separations — for discharge to B/L or mark is best effected by laying double layers of matting. Talc should only sparingly be used.

Separations are sometimes made by overlaying the cargo with a plastic film or woven plastic sheeting. Rubber which has come into contact with any oil is useless. Adequate separation must be provided from oily bag cargoes.

Certain shipments of rubber have shippers' own markings on the bales indicating special stowage such as to ensure that the sheets within the bales lie flat. Advice of these special stowage marks should be sought from the Agent/Stevedore.

Bale (covered with talc powder, matting or gunny) and bag rubber: this should be stowed in a cool space removed from boiler and engine spaces. Stowage should be as compact and firm as possible. It should be well dunnaged and protected from contact with ladders, pillars, stiffeners, brackets, etc., as well as from sweat, by dunnage and matting.

In stowing in aft holds, the side and crown of tunnel should be overlaid with crossboards to ensure an air course between cargo and the tunnel plating which, at times, is apt to be heated.

If stowing in the fine ends of forward or aft holds — which should be avoided if at all possible owing to the danger of crushing, massing and consequent overheating and deterioration of rubber — the spar ceiling should be overlaid with boards, hung vertically, to avoid bales part hanging on edge of

Commodity *Characteristics* *M3 per Tonne* *Packaging*

ceiling boards and leaving the part not so supported to sag under the superimposed weight, resulting in "massing" of contents and the compressing of wrappers into the rubber, both of which will likely result in claims.

It should be borne well in mind that deep stowage of bale or bag rubber or overstowing with weighty goods tends to "mass" the contents of bale and bag rubber. Too much pressure on bottom tiers should be avoided.

'Tween deck and upper hatches should always be placed in position to avoid adding unnecessarily to the weight on lower tiers of cargo.

Crepe rubber is frequently shipped in a tacky condition and, to avoid blocking, should be given top of 'tween deck stowage. Stow away from warm bulkheads and, to avoid exposure from the sun, do not stow in square of hatch. Mats should not be used between tiers of crepe; use as little dunnage as possible. Mats should be used to prevent sticking to other cargoes. Use talc in the stow.

Discharging: care must be exercised during discharge to ensure that slings are made up to include all the bales; accidents have occurred where bales not included in the sling nevertheless stick to others and are lifted by the gear. They are then liable to fall off as the load is swung ashore, with possible danger to nearby personnel.

Commodity	Characteristics	M3 per Tonne	Packaging
RUBBER LATEX	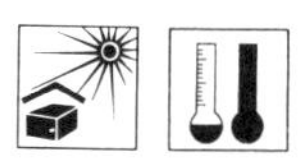	1.03 1.38/1.53 0.94/0.98 S.G.	Bulk Drums

Rubber latex is the sap of the rubber or caoutchouc tree which formerly was treated on the rubber estates immediately after collection, the crepe or sheet rubber so obtained being packed in cases, bales or bags for shipment.

It quickly coagulates if exposed to the air, so that the addition of some preservative possessing alkaline properties is necessary if to be desired to keep it in a fluid state.

To enable it to be shipped and transported in bulk, latex is preserved in the fluid condition by the addition of certain chemicals.

Ammonia is considered to be the best preservative, especially for large scale use and is now mostly in use for this purpose, about 0.7% of ammonia on weight of 60% natural latex. See Revertex. Latex, when preserved with ammonia, is not affected by temperatures as low as 10 degrees C or 15.5 degrees C below freezing point. In some instances creosote is also used for this purpose.

Drums: internal surfaces of drums may be coated to help preserve the quality of the latex. It is most important that these drums are handled with care to avoid damaging the contents.

Bulk: in view of the fact that bulk shipments will be preserved with ammonia, the following remarks on the conditions and considerations to be observed for the carriage of latex in bulk are confined to ammonia preserved latex. See Revertex.

Deep tanks which have been designed and constructed for the carriage of oil are the most suitable for its carriage.

The danger of the latex coagulating — partly or wholly — through loss of ammonia, consequent upon the straining of ship developing leaks in tanks, cannot be ignored. If this happens, the recovery of the coagulated rubber would be impossible by ordinary means — in fact nothing short of removing shell plating would make its recovery possible.

Peak tanks: latex has been successfully carried in these tanks.

Cutch: avoid stowing cutch in close proximity to latex, especially on top of a tank containing latex.

Cases have been known of liquid cutch having seeped into the tank through manhole packing and seriously contaminating the latex.

Recommended precautions to be taken when carrying ammonia preserved latex in bulk:—

(1) Tank must be thoroughly clean and free from traces of rust, oil or acid. See Vegetable Oil in Bulk, cleaning of tanks, etc.

Commodity	*Characteristics*	*M3 per Tonne*	*Packaging*

(2) It must not contain any brass, copper, yellow metal or galvanised material in any form. (Ullage caps and spring loaded valves to be of Stainless Steel).

(3) It must be proven absolutely air tight with all suctions, air and sound pipes and ventilating shafts effectively blanked or plugged.

The ventilating air shaft — one each side — should be blanked at the highest point compatible with air tightness of tube and must be free of rust on the inside. Heating coils, if any, are to be removed.

(4) Each tank is to be fitted with a spring-loaded relief valve — one each side of tank is better — capable of being set at .35 to .49 kgf/cm2 pressure, to protect the tank from injury through excessive pressure due to expansion of ammonia or of gaseous compounds evolved, if, through loss of ammonia, putrefaction of latex ensues. The valves, which sometimes are fitted to the blank flanges on ventilator shafts (as above) to have on-deck discharges designed to exclude water.

(5) The tank, after it is passed under pressure and for cleanliness, etc., is to be coated all over with Paraffin Wax having a melting point of about 63 degrees C.

It takes about 0.8 kg of wax to cover 1 square metre of surface and it is best applied by fine bristle brushes whilst boiling hot.

(6) In the case of Peak Tanks the manholes should be as large as possible to ensure easy access and egress for the men. A small square steel hatch with steel air tight cover is a much better arrangement. A shipside 100 mm connections for discharge will generally be necessary, such to be fitted as near level of tanktop as is practicable but above Tropical Load Line.

(7) The loading of latex, which is done with the lids screwed down, should be as expeditious as possible to avoid loss of ammonia. Shippers should make up any such loss of ammonia occurring during transit and loading. Men handling the hoses, etc., should be supplied with gas masks.

Receiving: latex is pumped on board by shippers' plant from tank barges or railway tank cars and, in some of the large rubber exporting ports of the Far East, from large storage tanks.

Discharging: it is discharged to consignee's tank by pump or compressed air, usually at consignee's expense.

After discharging latex: the tank should be swilled down with water as soon as possible after it is empty. Water has a great affinity for ammonia and thus frees the tank of ammonia fumes for entry by cleaners.

A thin film of latex will be found to adhere to the wax coating. On plain surfaces it is often possible to tear the latex off in sizeable sheets; elsewhere, scrapers, etc., have to be used to remove latex and wax. If the tank is not immediately required for liquid cargo or fuel oil, no purpose is served by removing the wax from which the latex film has been removed, as the wax acts as a good preservative for the steel work.

Synthetic Latex, with a specific gravity of 0.98, is easily washed from tanks using detergent and fresh water.

Both natural and synthetic Latex are suitable for carriage in Bulk bags in *ISO closed box containers*. See Part 2, Techniques, Containers.

RUBBER MERCHANDISE  Cartons. Cases.

May be adversely affected by heat, oil moisture and pressure. Stow accodingly.

RUM 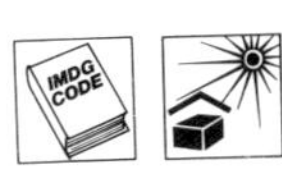

1.84/1.95 Hogsheads and Casks, Drums, Cases, Tank Containers

A spirit distilled from the juice of the sugar cane, principally shipped from the W. Indies and South America.

As this cargo is frequently shipped along with sugar, it should be borne well in mind that, though there is no objection to the stowage of both in the same compartment, provided the dunnaging is proper, rum should not, on any account, be stowed over sugar. Considerable claims arising from that class of stowage have had to be met in the past. Rum is about the only spirit that takes no harm from heat. This, like all wines and spirit cargo, is very liable to broaching, especially when in bottles. Cases of rum should be stowed in special cargo locker. See Barrels, Stowage of, also Wines. See IMDG Code.

Commodity | *Characteristics* | *M3 per Tonne* | *Packaging*

RUSH HATS  5.02/8.36 Bales

Shipped from F. East ports — a very light cargo with a stowage factor varying from 5.02/8.36. Should not be overstowed with other goods.

RUTILE SAND  0.70/0.84 Bags
0.56 Bulk

A mineral obtained from certain sands in Australia and Brazil. Used as a source of Titanium Dioxide in steel production.

Bags should be given dry stowage; care should be taken to avoid loss through dragging out and hook holes. Sand is very fine and will seep. Very clean stowage space required. Preferably protected by polythene sheeting.

RUTIN, GROUND Fibre Drums

Extract from buck wheat. Used for medicines. Do not stow with poisons or toxic substances.

RYE 1.25/1.35 Bulk
1.53 Bags

A cereal not unlike barley in appearance which is capable of enduring greater cold and thrives in much less fertile land than wheat. It constitutes, perhaps, the most important cereal crop of N. and C. Europe. It has the steepest angle of repose of all the food cereals, and as it does not run freely it requires much more trimming than other grain and more attention on the part of the officers to ensure that the cargo is properly trimmed.

See Grain.

RYEGRASS SEED 3.62 Sacks

See Grass Seeds.

SABICI Logs

A very hard wood grown in Central America. Shipped in square logs.

SABLES 

One of the most valuable of all furs. Usually shipped in well made tin lined cases. See Furs.

SAFFLOWER  1.81/1.95 Bales

The dried flower of an Eastern plant allied to the thistle, from which is obtained a red dye. Ordinary dry stowage. The seeds are sometimes called carthamus seeds.

SAFFLOWER OIL 1.73/1.78 Barrels. Drums.

A vegetable oil obtained from the seed of the safflower plant, the production of which is rapidly increasing in India, etc. See Part 2, Oils and Fats.

SAFFLOWER SEEDS 1.81/1.95 Bags
1.89 Bulk (Australia)

The seed referred to in the preceeding paragraph.

Commodity	*Characteristics*	*M3 per Tonne*	*Packaging*
SAFFRON		1.95/2.06	Cases

The bud of a bulbous plant of the crocus order from which is obtained a deep yellow dye. Treat as special cargo — dry stowage.

SAGE OIL		See Essential Oils	

SAGO **SAGO FLOUR**	  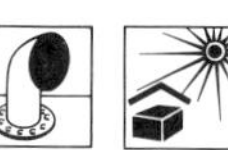	1.48/1.56 1.39/1.45	Bags Bags

Is obtained from the pith forming the inside of the sago palm which grows in the E. Indies and F. East principally. Shipped in considerable quantities from Singapore.

Sago is liable to heat, is readily damaged by moisture or strong odours. Stow apart from jelatong, gambier and other cargoes, which throw off moisture, and dunnage well on the deck and up the sides. Do not stow more than 8 heights without laying a tier of ratten over to prevent heating; better results will be assured by interleaving with ratten at shorter intervals.

Good ventilation is essential if the effects of moisture migration are to be minimised. Smaller shipments should have the bags stowed to form channels in stow. Keep away from ship's side and bulkheads.

It stows well with tapioca and is useful to stow with tea and other fine goods.

ISO containers — may be carried in closed box containers, but should be stowed away from heat or direct sunlight. 20 ft containers can achieve a payload of 16.5 tonnes.

SALMON	 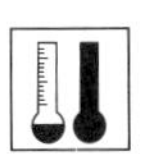	1.39/1.45	Tins in Cases

Salmon in tins packed in cases. See Canned Goods.

Frozen salmon usually is shipped in boxes or cartons but on occasions has been carried in bulk, being dipped in freezing fresh water after hard freezing, in order to glaze and give them a glistening appearance.

In bulk, stow fore and aft and on their sides rather than "on edge". Avoid deep stowing. See Fish. See also Part 2, Techniques, Refrigerated Cargoes.

SALT	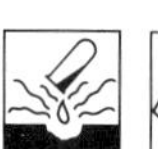 	0.98/1.11 1.06/1.11	Bulk Drums Bags

Common salt or sodium chloride is obtained from sea water, which contains about 3 per cent of it, by a process of evaporation; or from large deposits to be found in many countries.

In large quantities it is carried in bulk, small quantities being usually packed in bags.

With bulk cargoes it is usual to limewash the holds prior to loading. On long voyages the loss of weight is 5 per cent or more, this being due to evaporation, from which it follows that dry goods liable to take harm from moisture should not be stowed with salt, neither should salt be stowed near to wet or moist goods. Salt used for salting fish must not be contaminated with metallic compounds particularly copper. Do not stow in insulated compartments or refrigerated (or insulated) containers.

ISO containers — suitable for closed box containers. Poly lined fibre bags 50 kg may obtain a payload of 15 tonnes in 20 ft containers. Any spillage should be removed immediately on discharge due to corrosive properties of the salt.

SALT CAKE		0.89/0.95	Bulk

Impure sodium sulphate. Used in ceramic glaze. Granular. Angle of repose 30 degrees.

Commodity	*Characteristics*	*M3 per Tonne*	*Packaging*
SALTPETRE		See Nitrate of Soda	

Commodity	*Characteristics*	*M3 per Tonne*	*Packaging*
SALT ROCK		0.98/1.06	Bulk

Small white granules. Angle of repose 30 degrees.

Commodity	*Characteristics*	*M3 per Tonne*	*Packaging*
SAMP	  	1.39/1.50	Bags

Crushed corn (maize) and, like Maize (which see), it is very liable to heat and sweat; do not stow near flour or other fine goods nor with odorous goods such as turpentine, etc.

Commodity	*Characteristics*	*M3 per Tonne*	*Packaging*
SAND	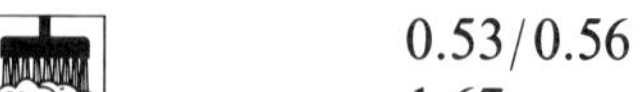	0.53/0.56	Bulk
See "Ilmenite Sand"		1.67	Cases
"Rutile Sand			
"Zircon"			
"Baryte"			
"Spodumene"			

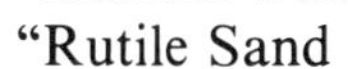

May be used for glass and steel making processes. It should be borne well in mind that any admixture of foreign matter such as China Clay, lime, coal, coke, seeds, etc., may result in claims. Very clean holds are normally required by shippers of bulk cargo.

Bilges and wells should be carefully protected against sand.

Industrial sand may be coated with resin and will cake or clog if exposed to heat (55–60 degrees C). Might be similarly affected if stowed under extreme top pressure.

Rutile sand stows at 0.42/0.56 M3/tonne.

Foundry, Quartz, Silica, Potassium, Felspar, Soda Felspar; in bulk stow at 0.50/0.98 M3/tonnes, angle of repose 30 to 52 degrees.

Commodity	*Characteristics*	*M3 per Tonne*	*Packaging*
SANDAL WOOD		3.21/3.34	Bales
SAPAN WOOD		2.79/3.07	Bulk

The valuable and very heavy wood of a tree grown in India and F. East used for ornamental purposes, also for perfumery. Packed in bales usually but sometimes shipped in bulk. It is very easily stained and so ruined for ornamental work, etc.

Stow well clear of oily goods, Chinese groceries, etc.

Commodity	*Characteristics*	*M3 per Tonne*	*Packaging*
SANDAL WOOD POWDER		2.51/2.79	Cases

Shipped from Far East.

Commodity	*Characteristics*	*M3 per Tonne*	*Packaging*
SARDINES (Canned)		1.23/1.34	Cases

Treat as Cartoned Goods, which see.

Commodity	*Characteristics*	*M3 per Tonne*	*Packaging*
SARSPARILLA ROOT		4·46/5.02	Bale

The root of a climbing plant growing in Mexico, Honduras, Jamaica, Brazil, Guianas, etc., from which a starch is obtained which is in demand for medicinal purposes. Dry stowage away from greases and oily goods.

Commodity	*Characteristics*	*M3 per Tonne*	*Packaging*
SAUCES	 	1.67	Cartons

Care to avoid crushing — cartons usually contain bottles and jars. Any broken or spilled contents may form a source of taint affecting adjacent cargo. Stow as for cartoned goods. See part 2, techniques

SAUSAGE SKINS		See Casings	

SAWDUST			Bulk

May cause oxygen depletion. Test before entering space/hold. Must be clean and dry or becomes liable to spontaneous combustion. See IMDG Code. (See IMO Code of Safe Practice for Solid Bulk Cargoes).

SCHELLITE ORE		0.70/0.75	Bags

Ore from which tungsten is obtained. Shipped in bags, drums and in bulk from E. Africa and Australia. See Ores.

SCRAP IRON		0.98/1.04	In Bulk, also Heavy Bales

See Metal Borings and Cuttings.

More especially, scrap iron moves in large quantities, mostly as full cargoes by chartered ships.

Charters are, usually, based on a stowage factor of 1.4 M3 bale per tonne; loading, stowing and discharging free of expense to vessel.

Owing to the very rough, uneven character of this cargo, it is entirely unsuitable for overstowing by other cargo and, of course equally unsuitable for stowing on top of other goods.

In most cases, the loading is done by magnetic cranes. The scrap is dropped in the hatchway and allowed to pile up in pyramidal form until it sides are steep enough for larger sections to roll down into wings and ends.

The presence of so large a volume of magnetic material on board should lead Masters to even great assiduity in checking Compass Errors.

Before commencing to load scrap iron the lower strakes of spar ceiling should be well protected from damage by covering with dunnage boards, that being preferable to unshipping the lower stakes as, if left in position, they protect the ship's side against severe battering by large sections of metal as it shoots down the sides of the steep pile.

Most scrap iron cargoes are composed of sections very diversified as to form, size and weight. The larger sections have a comparatively high stowage factor while the smaller sections have a very low factor, e.g. rivet punchings about 0.22 M3 tonne. These smaller sections, of low stowage factor, tend to gravitate to the bottom of the pile in the hatchway, leaving the large clumsy sections of high stowage factor to roll down sides and ends. In this manner the weight accumulates in the hatchway from which it follows that generous trimming to sides and ends is necessary to avoid subjecting the ship structure to undue strains and so should be insisted upon by the Master.

This condition is particularly serious as regards scrap iron received into the 'tween decks. In such a case the weight is localised — in the absence of proper trimming — on the 'tween deck hatches. Many 'tween decks have been set down as a result of that condition. It follows, therefore, that generous trimming to sides and ends be insisted upon in all compartments and that the Master satisfies himself that the 'tween deck hatches are not over-weighted. Occasionally difficulty is experienced in getting Charterers' Stevedore to do this. In such a case, the Master should, without loss of time, protest, by letter to the Charterers, holding them responsible for any damage to decks which may result. See Part 5, Procedures.

Commodity	*Characteristics*	*M3 per Tonne*	*Packaging*

A suitable proportion of scrap iron should be received into the 'tween decks and so raise the centre of gravity and reduce the G.M. thus avoiding the vessel being too stiff and uneasy in a seaway.

ISO containers — not suitable for containers on account of the damage caused to the container, the difficulty of securing the cargo, poor payload and (with the exception of Open-top containers) the difficulty in stuffing this type of cargo.

SEA WATER

See Fluids, Weights and Capacities, Appendix 7.

SEAL OIL		1.53/1.59	Casks

See Fish Oil, also Oils and Fats.

SEA WEED	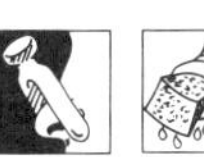 	6.69/7.25	Dried Bags or Bulk Bales

May taint and give off moisture. Dried and reduced to dust and small particles, then shipped in bags or bulk. Angle of repose about 35 degrees. Suitable for carriage in closed box *ISO containers* — see Part 2, Techniques.

SEED CAKE **Seed Expellers** **Oil Cake** **Meal**	 	1.40/2.10	Bulk

Derivative after oil has been extracted mechanically from oil-bearing seeds of cottonseed, coconut, groundnut, linseed, maize, palm kernel, rape seed, rice bran.

Shipped as flake, pellet, meal, may self heat slowly and if wetted may ignite spontaneously. Risk of oxygen depletion in carrying space, also presence of carbon dioxide.

These cargoes must be loaded under competent supervision. A certificate from authority should state oil content and moisture.

The risk factor of these cargoes is determined by the percentage of these measurements.

It is essential that the IMO Code of Safe Practice for Bulk Cargoes be consulted for further details.

SEED LAC		See LAC.	

SEED MEALS		1.53/1.81	Bags

See Meals.

Commodity	*Characteristics*	*M3 per Tonne*	*Packaging*

SEEDS Var

All seeds contain oil to a greater or lesser extent, some as much as 60 per cent.

In Bags

Seed	*M3/tonne*	*Seed*	*M3/tonne*	*Seed*	*M3/tonne*
Alfalfa	1.59	Croton	2.23	Mustard	1.67
Ajwan	2.23	Cummin	3.48/3.62	Onion	1.81
Alsike	1.28	Dari	1.48	Poppy	1.95/2.01
Alpia	1.67	Durra	1.48	Rape	1.62/1.73
Alpiste	1.67	Fennel	2.65/2.68	Sesame	1.67/1.95
Aniseed	3.34	Flax	1.59	Shursee	1.67
Bayari	1.56	Gingelly	1.67	Sorghum	1.39
Canary	1.39/1.67	Gram	1.39	Spinach	1.95
Caraway	1.67/1.78	Grasses	1.39/2.51	Sugarbeet	3.65/3.67
Cardamon	2.09	Hemp	1.89/1.95	Sunflower	2.09/3.07
Carthamus	2.51	Jowaree	1.34/1.59	Surson	1.67
Castor	1.95/2.23	Linseed	1.53/1.67	Taro	1.39
Cebadello	2.37	Lucerne	1.81	Tares/Vetch	1.37/1.79
Celery	2.12	Mafurra	2.51	Teel	1.67
Clover	1.34/1.67	Millet	1.39/1.53	Timothy	1.95
Common Sds	2.79/3.07	Mirabolans	1.95	Trefoil	1.67
Coriander	3.48/3.62	Mowrah	1.70	Tokmari	1.64
Cotton	2.09/2.51	Mustard	1.67	Turkish Millet	1.48

Bulk

Cotton	1.81	Flax	Linseed1.39	1.42
Millet	1.25/1.42			

The properties, stowage, etc., of various kinds of seeds are described herein, each under its own name, and they are further dealt with under Part 2, Techniques: Bulk Cargoes, also General Cargoes, bagged.

SEEDLINGS  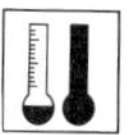 Cases. Crates

May give off moisture. Controlled temperature carriage + 5 degrees C, with air freshening. Delivery air must not fall below + 2 degrees C.

SELENITE 

A transparent form of Gypsum, which see.

SEMETIN  See Middlings

SEMOLINA   1.67/1.73 Bags

A wheat byproduct — a flour. Stow as for Flour, which see.

SENEGAL

An African Gum. See Gums.

Commodity	*Characteristics*	*M3 per Tonne*	*Packaging*
SENNA LEAVES		3.21/3.48 5.02/6/13 6.97/9.75	Pressed Bales Bags Unpressed

The dried leaves of a tree which grows in India, Arabia, Egypt, etc., valued for their medicinal properties. Dry stowage away from greasy, moist and odorous goods. These leaves quickly ferment if permitted to get damp or wet.

SESAME		See Gingelly	

Sesame seed paste may be carried in drums, deck tanks or tank containers.

SHEA BUTTER 

Is obtained from the nuts of a W. African tree. Not unlike Palm Oil, which see. Is used for similar purposes. See Vegetable Fats.

SHEA NUTS		1.95	Bags

See Nuts for Stowage.

SHEEP		See Part 2, Livestock	

SHEEPSKINS	  	3.62/4.45 2.23/2.51	Bales Bales dumped

Useful overstow for tallow when dumped. Adequate ventilation and sufficient ground dunnage. Liable to moisture and rust damage. Difficult to fumigate centres of the bales. When skins settle, some banding may become slack and interlock in the stow. Bales from New Zealand average 254 kg.

SHEEPWASH OR SHEEPDIP	 	1.25/1.53	Drums

A poisonous liquid, which should be stowed away from foodstuffs of all kinds and from other delicate goods. Shipped in steel or iron drums. See IMDG Code.

SHELAC		See Lac	

SHELLS	  	1.67/1.95 1.17/1.34	Bags Cases

Shells in bags and cases. Those being exported for collectors, e.g., from Philippines, extremely fragile. May be valuable. Do not overstow with heavy cargo. Shells in bags and bulk for fertilizers may be infested. Possible strong odour. Keep away from foodstuffs. See Mother of Pearl and also IMDG Code.

SHINGLES			Bundles

Slats of wood used for roofing — one end being thinner than the other. See Timber.
Shipped in bundles 10,000 slats = 1000 board ft and stow in about 2.83/2.97 M3.

Commodity	*Characteristics*	*M3 per Tonne*	*Packaging*
SHOOKS		2.65/2.79	Bundles

Bundles of staves or boards which are used for making casks and boxes respectively. See Timber.

SHOYU		See Soy	

SHROT		1.87/1.90	Bags

Sunflower seed meal. See Seed Cake.

SHUMAC		1.95/2.01	Bags

Ground leaves and buds of a Sicilian tree used for tanning, also for dyeing fabrics yellow. As this commodity readily absorbs moisture, it should be stowed apart from wet, moist goods and edible goods.

SHURSEE

Also known as Mustard Seed, which see.

SILICOMANGANESE	   	0.18/0.26	Bulk

Used as an additive in steel making process.

Contact with water produces alkalis or acids, which may evolve hydrogen, a flammable gas. Other highly toxic gases may be produced. A certificate is required, either stating that the cargo does not possess chemical hazards which could produce a dangerous situation on board, or alternatively that the cargo was stowed under cover but exposed to open air for not less than three days prior to shipment. Ventilation should not allow escaping gases to reach crew accommodation. See IMDG Code. (See IMO Code of Safe Practice for Solid Bulk Cargoes).

SILK	 	2.79/3.07 Raw	Bales
		3.34/5.57 Waste	Bales
		1.53/1.81 Waste	hyd. Presses
		3.07/3.62 Piece	Tin Lined Cases
		4.46/5.02 Punjam	Tin Lined Cases

A valuable commodity which should be stowed in special cargo locker well removed from moist or wet goods, carefully tallied and watched to avoid broaching. Shipped in bales and in cases. Easily damaged by damp or crushing. Use no hooks.

SILK COCOONS		5.85/6.13	Bales Unpressed
		1.67/2.23	Bales Pressed

SILK WASTE		3.07/3.34	Bundles

Dry Stowage. Valuable. See Part 2, Techniques, "Specials".

Commodity	*Characteristics*	*M3 per Tonne*	*Packaging*

SILVER LEAD ORE CONCENTRATE

See Appendix 13.

SINAMAY 

Abaca Fibre — see Hemp.

SISAL  See Hemp

SKINS (DRIED)  

A very large variety of animal skins are shipped. The most valuable skins are those of the fur bearing animals and are dealt with herein as Furs, which see. Skins which are shipped in the wet or pickled condition are also dealt with separately, as Hides, as also are tanned skins under Leather. The following remarks are in respect of dried skins only.

Cattle, horse, sheep and goat skins form the bulk of "skin" shipments, in addition to which are buffalo, deer, dog, cat, rabbit, etc.

Dried skins are shipped in ordinary bales, machine pressed bales, crates (the smaller kinds like the rabbit, etc.) or loose, in which condition they require very careful tallying.

Calf Skins	4.18	Bales	All tending to increase in size
Goat Skins	2.65	Bales	
Kangaroo Skins	2.79	Bales	
Sheep Skins	3.79	Bales	

All skins should be stowed apart from wet goods, as they readily absorb moisture and soften as the result; they are odorous and so not suitable for stowage with fine goods such as tea, flour, coffee, fruit and other edible goods.

Stow on the flat, particularly ground tier, but wing tier above ground tier on Edge and so confine any damage which may occur to the smallest numbers of skins possible. Dunnage well under skins that are stowed to the deck and protect by dunnage and matting from contact with iron or any part on which moisture is likely to condense or collect; rust marks on skins inevitably result in claims.

Skins from African countries are a frequent source of claims. Invariably these are of preshipment origin. Parasitic infection — packed when damp etc. Humidity inside containers exacerbates the problem.

Anthrax: men who are handling dry hides, hair or wool, run the risk of being infected by Anthrax.

Many Health Authorities insist that they be supplied, at stevedores' expense, with gloves and other protective clothing.

SLAG, BASIC 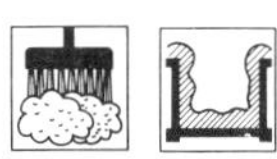 0.84/0.89 Bags. Bulk

The fused scum of, or waste, obtained during the smelting of iron, which on account of its high phosphoric acid content is largely used as a fertilizer.

Dry cargo free from objectionable properties generally associated with fertilizers, but is very dusty; stowage accordingly. See Manures.

Commodity	*Characteristics*	*M3 per Tonne*	*Packaging*
SLATES		0.50/0.56	Loose
		0.67/0.72	Cases

Roofing slates can be shipped loose and in wooden cases. They require careful handling in both forms to avoid breakage. Stow on edge with plentiful use of straw. Slate slabs should be treated as for marble, which see.

Commodity	*Characteristics*	*M3 per Tonne*	*Packaging*
SLEEPERS		1.00/1.06	Jarrah
(Creosoted)		1.11/1.25	Oak
		1.45/1.53	Pine
		0.81/1.03	Metal

In order to preserve wood railway sleepers against weather and the attacks of insects, they are usually impregnated with creosote — a coal tar product.

Creosote in any form is a very objectionable cargo owing to the very strong and pungent smells which it throws off. Fine or edible goods, dry or wet, should not be stowed in or near to a compartment carrying creosoted timber and the greatest possible care should be observed in removing all traces of creosote or its fumes (after the creosoted goods are landed) before loading grain or any other cargo liable to taint or damage. Cases have occurred, where vessels have been declared unseaworthy because of the presence of these fumes, and large claims for damage to subsequent cargoes have been paid. See Creosote. Sleepers are made from various classes of timber: pitchpine, oak and jarrah wood principally. See Timber.

Steel Iron and concrete sleepers are extensively used; usually nested for shipment.

SLUDGE ORE  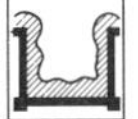

Quantities of which are shipped in bulk in Spanish ports. The Committee of Lloyds have drawn attention to the dangerous nature of sludge ore cargoes loaded in Spanish ports, when in a wet condition. The fine nature of this class of ore is such that, unless it is in a perfectly dry condition and precautions are taken by erecting longitudinal bulkheads, the cargo is liable to shift. Vessels have been lost and cases of serious listing have occurred, which have been attributed to the damp condition of the cargo. See Part 2, Concentrates, also Ballast. See IMDG Code.

Sludge (dried), containing Nickel and other valuable metals, may be shipped in *ISO containers* using disposable plastic liners. Payload approximately 18 tonnes per 20 ft container.

SLURRY  Mud coal or Duff

The moisture content of the slurry tends to rise to the surface and cause the upper portion of the cargo to become mobile and this tendency is assisted by the vibration and movement during passage. In Coasting ships the risk of a serious casualty is increased because of their more rapid motion and smaller freeboards.

The following recommendations, based on past experience and accidents, are made for the loading and carriage of slurry:—

(a) holds containing slurry should be properly trimmed in such a manner, approved by the Master, as will ensure the safety of the ship;

(b) where, for the present, slurry is loaded with a free moisture content of more than 20% in addition to taking even greater care in trimming the cargo, special steps must be taken to prevent the cargo from shifting, such as by

Commodity	*Characteristics*	*M3 per Tonne*	*Packaging*

(i) fitting a transverse bulkhead in the hold whereby the part of the hold containing the slurry cargo, because of its smaller space, can be completely filled into the hatchway which will serve as a feeder; or by

(ii) any other equally effective means;

(c) the effects of exposure to heavy rain during loading after the analysis for moisture content has been made or of leakage into the holds during the voyage will enhance the danger and must be taken into account;

(d) all openings through which water might gain access to the cargo spaces should be securely closed and made effectively weather-tight. Ventilating cowls should be removed and the coamings plugged and covered; and

(e) hold bilges should be sounded frequently, but if this should be impracticable for any reason (e.g. bad weather) the pumps should be tried on the bilges every few hours. See IMDG Code.

SOAP		1.95/2.23	In Cases, Fancy
		1.39/1.53	Ordinary
		1.11/1.25	Kegs

Many varieties of soap are shipped, the finer and more expensive article being peculiarly liable to broaching and pilferage.

Soap odours are penetrating and are present with most classes of soap if the voyage is through the tropics. Stowage accordingly.

Soft soap — wet stowage.

SODA		1.23/1.34	
SODA ASH		1.11/1.25	Bag
MAGADI			Bulk

SODIUM NITRATE	 	0.88	Bulk

Although non combustible, mixtures combined with combustible material are readily ignited and burn fiercely. See IMDG Code. (See IMO Code of Safe Practice for Solid Bulk Cargoes).

Or sodium carbonate, extensively used in the manufacture of glass and soap, also for bleaching and washing, is not suitable for stowage near dry goods. Requires clean holds.

SOLARINE			Cases. Cartons

Metal polish, paste in tins. See IMDG Code.

SOOJI		2.51/2.65	Bags
SUJI			

Is obtained from wheat and used in the confectionery trade.

Dry stowage remote from moist and odorous goods.

SORGHUM		1.24/1.37	Bulk

See Millet. Grain Regulations apply.

Commodity	*Characteristics*	*M3 per Tonne*	*Packaging*

SOTANJON

Chinese noodles made from rice.

SOY		1.25/1.39	Casks. Drums

Or soya sauce, a Chinese delicacy, prepared from the soya bean.

Wet cargo from which there is, generally, very considerable leakage — requires special stowage on account of the containers being usually of frail construction. See Chinese Groceries. See also Sauce.

SOYA BEANS	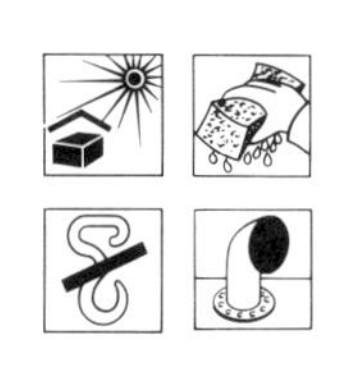	1.23/1.28	Bulk
		1.39/1.48	Bag, F. East
		1.81	Bags, W. Africa
		1.35/1.39	Bulk, U.S. Gulf
		1.59/1.62	Bags, U.S. Gulf

An oil bearing bean extensively grown in North and South America, etc., shipped thence in bulk and in bag. The beans are prepared for shipment by drying by kiln or by natural means. Soya bean meal is a valuable product used in production of animal feed.

This cargo is liable to cause a lot of trouble arising from deterioration, which inevitably sets in if the beans are shipped in a damp condition, when they heat and sweat profusely and then ferment. Claims in respect of this cargo have been many, large and vexatious. When damage to soya bean occurs, the principal effect on the meal is that it is a darker colour; this does not affect the value of sale as the crude protein content is not altered. Particular note should be taken of the climatic conditions under which they are shipped and, if any doubt exists as to their condition, B/L should be qualified. In any case it is desirable that B/L should have the written endorsement, "Shipped in apparent good condition", if nothing stronger is endorsed thereon. Care must be taken not to heat high quality soyabean used for sauce making as the meal very temperature sensitive. Stow away from contact with engine room bulkhead and heated fuel oil double bottom tanks.

The use of hooks in handling this cargo should not be permitted. The dunnaging should be most thorough and the ventilation constantly attended to.

SOYA BEAN CAKE (Meal/Pellets)	 	2.01/2.09	Bags
		1.81/1.95	Bulk

An oil cake made from the residue of the bean of that name after the oil is extracted — see Seed Cake. See IMDG Code. (See IMO Code of Safe Practice for Solid Bulk Cargoes).

SOYA (BEAN) OIL		1.39/1.53	Drums
		1.08	Bulk

Soya oil (soy oil, soyabean oil) is the most important vegetable oil, being produced on a very large scale in many parts of the world by extraction of the oil from the soyabean. At present South America (Brazil, Argentina) is the principal source of soya oil shipped for export, but the U.S.A. also exports substantial quantities of the oil. Export destinations for the oil cover all continents. The oil, which is light in colour, is always exported in crude form and can normally be expected to have a free fatty acid content of 1.0% or less. The high content of poly-unsaturated fatty acids in the oil makes it vulnerable to oxidation but also means that the oil remains liquid in all conditions normally experienced, and it therefore requires no heating during shipment or at the time of loading or discharge.

The oil is generally degummed before leaving the extraction plant, but may nevertheless form a deposit of insoluble material during storage. The crude oil has a density of 920 kg/m^3 at 25°C.

See Part 2, Oils and Fats. Appendix Table No. 11.

Commodity	*Characteristics*	*M3 per Tonne*	*Packaging*
SPAGHETTI	 	2.51/2.79	Cartons Cases

Foodstuff of the Macaroni order. See Macaroni.

SPECIAL CARGO — See Part 2, Techniques, "Specials".

SPECIE

Gold, silver and copper coins. See Part 2, Techniques, "Specials".

SPELTER		0.22/0.33	Ingots

Is the name by which the metal zinc is generally referred to.
Requires careful tallying.

SPERMACETI		1.81/1.95	Cases. Cartons.

White fatty substance obtained from the Sperm Whale. Stow and treat as for Wax.

SPERM OIL — See Fish Oil

SPICES   

This term covers most vegetable products which are fragrant and pungent — such as cinammon, cloves, mace, nutmegs and others, most of which are dealt with herein under their respective name.

SPIRIT OF WINE  — See Alcohol

SPODUMENE **(See Sand)**		1.64	Bulk

Dry, white to biege in appearance, avoid exposure to dust. Protect eyes. High cleanliness required.

SPONGES		3.90/5.02 3.34	Bales Bales, Pressed

Many varieties are shipped, usually in bales, which in some instances, are pressed to a moderate density.
If thoroughly prepared, this cargo is free from objectionable properties, otherwise it is apt to smell.

STAINLESS STEEL GRINDING DUST		0.42	Bulk

Angle of repose 45 degrees.

STAMPS

Revenue or Postage. See Part 2, Techniques, "Specials".

Commodity	*Characteristics*	*M3 per Tonne*	*Packaging*
STARCH		1.39/1.50	Cases Bags

Is a grain-like substance, obtained from various kinds of plants, such as corn (maize) potato, etc. Dry stowage, away from wet or moist goods.

STAVES		2.65/2.79	Bundles or Loose

Narrow pieces of timber prepared for assembling to form casks, etc.

Shipped in bundles — see Shooks — and loose, when they come in very useful to fill broken stowage.

When staves are used for broken stowage or dunnage, much time, money and trouble will be saved at discharge if, so far as possible, the mixing of different shipments and marks is avoided. An effort should be made to confine certain lots to its own compartment. See Part 2, Techniques, Bales and Bundles.

STEARINE		1.39/1.50	Bags Bulk

Obtained from natural fat.

Similar but superior to tallow. When carried in tanks the heating coils must be completely covered. Sudden changes in temperature will cause damage. Temperatures should not be raised more than 3 degrees C (5 degrees F) in any 24 hours. Daily temperatures may have to be taken and logged. Top and bottom stow temperatures should differ as little as possible. Pumping temperature 65.5-74 degrees C. Solidifies at c. 40 degrees C.

See Tallow. Also Oleo Stearine. See Oils and Fats Part 2.

STEEL-IRON		0.33/0.45	Bars
COILS		0.28/0.39	Billets
PILING		0.28/0.33	Plates
PIPES		0.28/0.33	Pig
PLATE			
SHEET			
STRUCTURAL			
TUBES			
WIRE-ROD			

See Part 2 — Iron and Steel Products.

STIBNITE

See Antimony and Ores.

STICK LAC		See Lac.	
STOCKFISH		See Codfish.	
STONE CHIPPINGS		0.71	Bulk

Angle of repose 55 degrees.

Disposable pallets for sheet steel in envelopes.

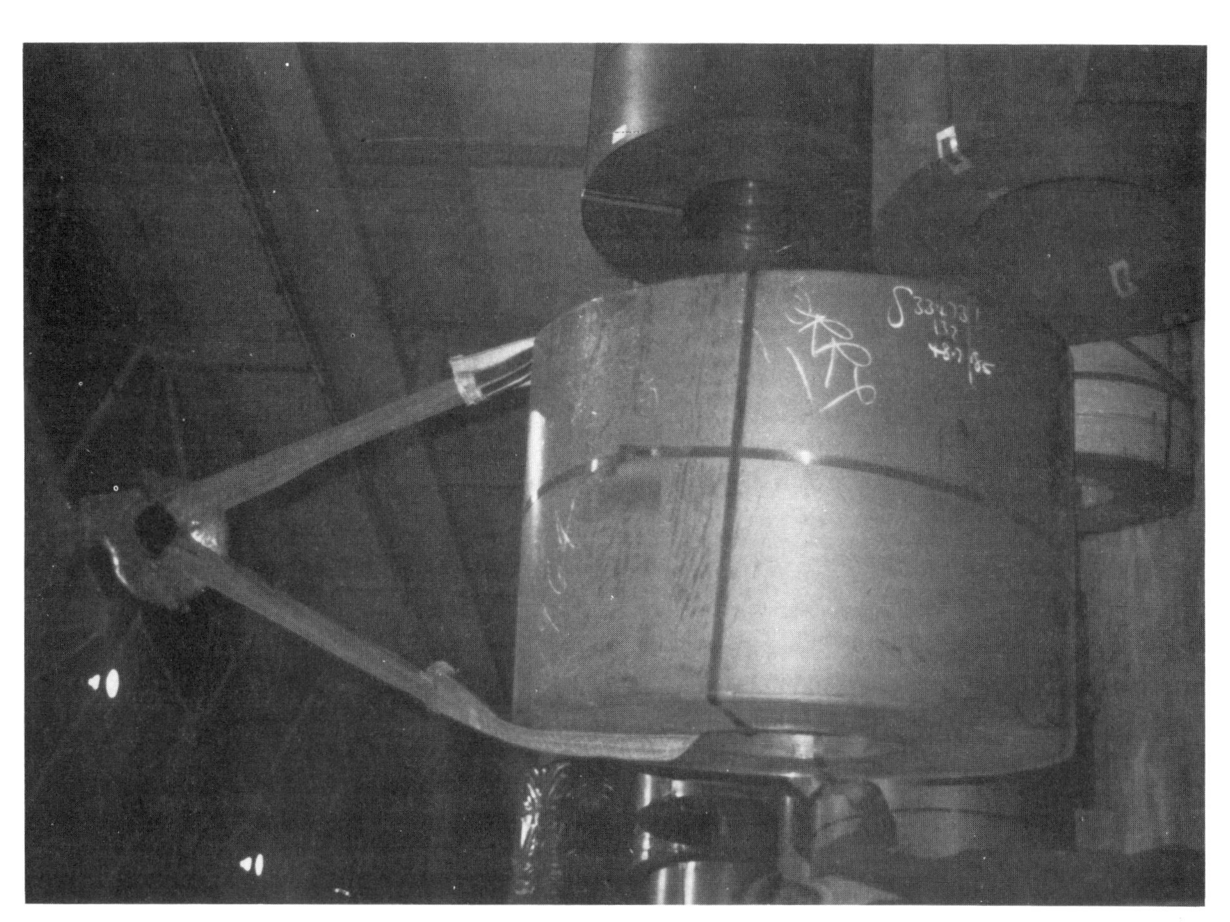

Nylon Webbed sling for coiled steel.

Grab discharge of bulk sugar.

Commodity	*Characteristics*	*M3 per Tonne*	*Packaging*
STONES		See Granite, Marble.	

STRAW	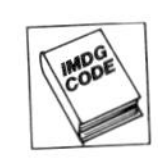		Bales

See IMDG Code.

STRAWBRAID		5.02/6.97	Bales
STRAWPLAIT		4.46/5.57	Cases

The Far East export a great deal of this commodity for hat making, etc., packed in lightly-made cases or bales. The packages are never pressed, to do so would injure the braid. Not suitable for over-stowing with any other cargo.

STYRENE MONOMER	   		Bulk

See IMDG Code. See Part 2, Chemical Cargoes.

SUGAR	 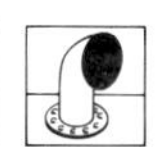	1.28/1.34 Dry	Bags
		1.11/1.25 Dry	Bulk

Angle of repose about 40 degrees.

Bagged Sugar, stowed in 'tween decks, may only achieve 1.67/1.81 M3/tonne.

Refined Sugar is normally shipped in plastic lined bags. The loading temperature must be considered when assessing the necessity for ventilation. Cargo may be loaded in Northern Europe in freezing conditions for shipment to Middle East, etc.

The various kinds of sugar are, for shipping purposes, divided into two general classes, viz:—

Dry Sugar, i.e. free of syrup, is devoid of objectionable properties and should not be stowed with or near to wet goods or those which are liable to heat and throw off moisture such as coconuts, etc. Some of this class of sugar is apt to lose weight.

Green Sugar (raw, wet, soft sugar) is a moist commodity from which a considerable amount of syrup drains resulting in loss of weight up to as high as 10 or 12 per cent — the worst form being Jaggery, which see. Green Sugar should not be stowed with, near or above dry sugar, as the latter is readily damaged by the syrup which drains from the former. Do not permit stevedores to stow rum over any class of sugar. See Rum.

When receiving sugar from lighter, a sharp lookout should be kept for packages wetted with salt water — all of which should be rejected; any doubt on the matter can easily be settled by tasting.

Holds and bilges should be thoroughly cleaned and all traces of oils, acid, or odours removed from both before receiving sugar. Loading sugar in wet weather should be avoided, as sugar readily takes damage from moisture.

For bag or mat sugar, all steel work requires to be covered.

When loading mat sugar, it is a good practice to buy spare mats — similar to the kind in use for the cargo — say 1000 mats per 1000 tons of sugar, as they will come in very useful to hold loose sugar, of which there is present a considerable quantity at discharge. See Part 2, Sweepings.

The fumes given off by sugar will taint delicate and edible goods, which, therefore, should not be stowed in the same compartment. Thorough ventilation of sugar cargoes, especially raw or partly refined sugar is very important.

Commodity	*Characteristics*	*M3 per Tonne*	*Packaging*

When receiving considerable quantities of sugar, it is often necessary to separate one parcel from another. Sometimes paint markers are used, which is a practice fraught with danger, several large claims having been paid because of contamination from this source. See part 4, Damage.

Sugar cargoes are not free from danger. A number of instances of sugar cargo being fired by ignition (not spontaneously) have been recorded. Such fires are very difficult to extinguish except by chemicals or by completely flooding the compartment.

After discharging green sugar, holds and bilges should be thoroughly washed with salt water (see Molasses) and, if possible, swilled down with fresh water to ensure proper drying.

Bulk Sugar: Unrefined sugar may be carried in bulk and discharged by grab cranes with satisfactory results. For this reason conventional self trimming bulk carriers are principally utilised in the raw sugar trades wherever possible. However, when 'tween deckers are used, it is recommended that if for stability reasons sugar requires to be shipped in the 'tween decks, although lower holds may not be completely filled, then 'tween deck hatch boards should be shipped to prevent 'tween deck sugar entering the lower hold.

Raw sugar is almost always shipped in bulk and loaded in temperate climates and this should be taken into consideration when ventilation is being contemplated. Normally this cargo should not be ventilated.

Carbon dioxide gas can be detected by the odour of alcoholic gas. At sea, vent covers, when in use, should be removed daily and, should there be any indication of this gas being present, the holds are to be given full ventilation and the use of naked lights in the vicinity prohibited. Any cargo space suspected of containing carbon dioxide gas should not be entered until after being certified gas-free by a qualified person.

Note — bulk sugar has a stowage factor between 1.17–1.25 cu. ft per ton depending on the moisture content, but, owing to the difficulties of trimming the sugar close up to the deck head in the wings and ends of holds, the stowage factor of 1.39 M3/Tonne should be allowed where a compartment is to be completely filled.

ISO containers: closed box containers suitable for refined or semi-refined bagged or bulk sugar. Payload for 20 ft containers (bags) c. 15 tonnes dry sugar. Container should be lined with plastic or similar extending well up the walls when green sugar is carried.

SULPHATES 

1.19/1.62 Bags, Drums

See Fertilisers, Sulphate of Ammonia, Sulphate of Copper, Sulphate of Soda and Sulphate of Potash — dry dusty cargo.

When exposed to damp, the dust liquefies and may become very corrosive. Do not stow over other cargo — particularly tin plate.

Sulphate of Ammonia stows at 0.70/1.11 M3/tonne.

SULPHUR 

   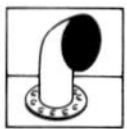

SULPHUR	0.84/0.89	Bulk
Brimstone	0.98/1.03	Bags

A mineral substance found free in volcanic countries, from which large quantities are shipped in bulk. It is yellow in colour, brittle, insoluble in water, but readily fusible by heat.

Sulphur readily ignites in contact with flame, even of a match or spark, and numerous cases of sulphur cargoes fired by these means have occurred with disastrous results; this being the reason why it is classed "Dangerous Cargo". See IMDG Code. Proper precautions to prevent the fine dust finding its way into the bilges should be observed as well as to protect other cargo from sulphur dust if loading part cargo of sulphur in bulk.

Should sulphur catch fire, it should be smothered with more sulphur or with a very fine spray of fresh water — not salt water. The danger of dust explosion exists with a cargo of powdered sulphur during loading and whilst the holds are being cleaned after discharging sulphur.

Commodity	*Characteristics*	*M3 per Tonne*	*Packaging*

The danger may be avoided by preventing the atmosphere becoming laden with fine sulphur particles by adequate ventilation or, when cleaning the holds after discharging sulphur, by hosing down instead of sweeping.

Holds should be well cleaned prior to loading. Wet sulphur is potentially corrosive, particularly if the condition of the holds is poor. A thorough coating of lime wash put on sufficiently thick and allowed to dry should constitute a barrier to minimise corrosion. No metal objects which could be picked up by grabs during discharge should be left in holds. Bilges should be caulked, any chinks being covered with wooden battens, burlap or paper must not be used for covering bilges, limbers, etc. On discharge, holds should be thoroughly swept and washed out before being used for other cargo; particular attention should be paid to sweeping ledges, box beams, etc., where the dust can lodge. Dry sulphur is harmless, however wet sulphur is extremely corrosive to bare metal; this may be aggravated by the chlorides present in salt water when washing down the holds with salt water etc. Sulphur, lump and coarse grained powder or fine grained powder, is classified under IMO as Class 4.1. However may be exempt if it is commercially formed sulphur in prills, granules, pellets, pastilles or flakes. See IMDG Code. (See IMO Code of Safe Practice for Solid Bulk Cargoes).

SULPHURIC ACID — See Part 2, Chemical Cargoes.

SULTANAS  1.28/1.39 Cases

A form of raisin, which see.

SUMAC  See Shumac.

SUNDRIES, MEAT  See Meat, Sundries.

SUNFLOWER OIL

1.73/1.78 Barrels
1.09 Bulk

A vegetable oil obtained from sunflower seeds, which is extensively produced in Russia. See Vegetable Oils. See Part 2, Oils and Fats.

SUNFLOWER SEED  

2.19/2.50 Bulk
2.79/3.07 Bags

One of the lightest of the oil bearing seeds, its oil yield being about 20 per cent. Liable to spontaneous combustion if shipped wet. See Grain and Seeds. See IMDG Code.

SUNFLOWER SEED CAKE (See Seed Cake)   1.62/1.67 Bulk

Liable to spontaneous combustion. See IMDG Code. (See IMO Code of Safe Practice for Solid Bulk Cargoes).

Commodity	*Characteristics*	*M3 per Tonne*	*Packaging*
SUNFLOWER SEED EXPELLERS (See Seed Cake)	  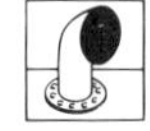	1.39/2.09	Bags Bulk

Risk of spontaneous combustion. Degree of danger is dependent on the oil and moisture content. Must not be shipped in bags and wet. Should not be exposed to rain or damp. May require ventilation. May be limits on quantities carried in one compartment. Can be successfully carried in ISO closed box containers. See IMDG Code. (See IMO Code of Safe Practice for Solid Bulk Cargoes).

SUPERPHOSPHATE		1.06/1.11 0.84/1.00	Bags Bulk

A fertiliser composed of phosphate treated with sulphuric acid. The gas given off by superphosphate attacks and damages textile goods, foodstuffs, galvanised iron, etc., and causes the jute bags in which it is packed to rot. Angle of repose 30 to 40 degrees. Stowage factor varies according to grade.

See Fertilisers.

SURSEE OR SURSON 

Native names for the Mustard Seed, which see.

SWARF	 		Drums

Metal drillings, usually of oily nature. May be liable to spontaneous combustion. See IMDG Code.

SWEET SPIRITS OF NITRE		See Nitre, Spirits of	
SWINE		See Part 2, Livestock	
SYRUP		1.39 1.23	Barrels Hogshead

A sweet variety of molasses. (Not carried in bulk). Often shipped in drums liable to leak. Ground stowage.

TACONITE PELLETS		1.53/1.67	Bulk

Round steel pellets. Angle of repose 30 degrees.

TAGUA		See Corozo Nuts	
TAILED PEPPER		See Cubebs	

Commodity	*Characteristics*	*M3 per Tonne*	*Packaging*
TALC		1.06/1.11 0.64/0.73	Bags Bulk

A natural hydrous magnesium silicate, used in ceramics, electrical insulation, etc. Angle of repose 20 to 45 degrees. Soiling dust. Cool, dry, stowage.

TALC MICA		5.57	Bags

A magnesian mineral. See Mica. Powdered talc.

TALLOW		1.95/2.23 1.06/1.11	Drums In Bulk

Tallow is the principal by-product of slaughter house rendering and is therefore produced in large quantities in the major meat-consuming countries. Beef and mutton tallow account for most of the fat output of the rendering industry. Only a small proportion of the tallow produced in the USA (the major producer) and Western Europe is suitable for edible purposes. Tallow is classified into a number of quality grades, which are distinguished by the titre of the fat (the titre measures the melting behaviour of a fat and is similar to the melting point used to characterise many edible fats), the maximum permitted free fatty acid content and the colour. The better qualities of tallow have a titre of 41-42 degrees C, whereas poorer qualities are more likely to have a titre of approximately 40 degrees C. Tallow as shipped is likely to be at least semi-solid. Whereas an edible grade of tallow is likely to have a free fatty acid content of less than 1.0 per cent lower quality material can have a free fatty acid content in excess of 20 per cent. The colour of the fat depends on the grade, edible grades being off-white to pale yellow. The presence of highly unsaturated fatty acids can lead to rapid deterioration of tallow unless it is handled properly but deterioration in quality can also occur as the result of failure to separate various micro-organisms from the fat during rendering.

Edible tallow is traded on a limited scale internationally but inedible tallow, which is an important raw material for the oleochemicals industry in North America and Western Europe and a relatively low-cost fat ingredient for animal feeds in many parts of the world, is traded extensively on international markets. Both fats are shipped in bulk.

The density of tallow of good quality is about 890-900 kg/m^3 at 40 degrees C.

See Part 2, Oils and Fats. See Appendix Table No. 11.

TAMARINA		1.11/1.25 1.06/1.11 1.39	Barrels Cases Bags

A tropical fruit from which a very refreshing drink is made. Wet stowage in a cool place. It is often carried on deck protected from the sun.

TAMARIND SEED		1.67/1.95	Bags

Seed of the tamarind tree used in the manufacture of condiments. Prone to beetle infestation.

TANGERINES		See Oranges	

Commodity	*Characteristics*	*M3 per Tonne*	*Packaging*
TANKAGE	  	1.67/1.81	Bags Bulk

The dried sweepings of animal matter (offal) from slaughter house floors. Very dusty and odorous, the odour becoming very strong if allowed to get damp.

Stow apart from goods liable to taint damage; also away from fine goods generally, to avoid claims for soiled packages due to the settling of dust from tankage, which is hard to remove.

In bulk, may be subject to spontaneous heating. See IMDG Code. (See IMO Code of Safe Practice for Solid Bulk Cargoes).

TANNING EXTRACT			Bag. Boxes.

Stow well clear of all commodities liable to be damaged by taint. Sawdust between each tier. Line ISO container floor with plastic or similar. See Gambier and Shumac.

TAPIOCA	  	1.50/1.56 1.90/1.95 1.73/1.78 1.73/1.78	Bags Flour Bags Flake Bags Pearl Bags Seed

A farinaceous starchy substance obtained from the roots of the Manioc (which see) or cassava plant.

It is easily damaged if stowed with or near to odorous goods or those which are liable to heat; it is also liable to heat.

It is used for edible purposes and in large quantities in the manufacture of glue, sizing, etc. As the edible and commercial grades usually are not designated, all shipments should be treated and stowed as the edible kind.

If rattan is available for stowage with tapioca, rattan should be laid to form the ground tier, up the sides (wing tier) also to form a vertical tier amidships. See Dunnage. When tapioca is stowed to about half depth of 'tween deck — another tier of rattan should be laid, completing with tapioca. With such safeguards for ventilation, it should carry without injury. Too much topweight should be avoided if stowing otherwise. When given deep stowage, rice type ventilation should be effected or bags should be laid so as to provide channel ventilation.

Stow away from timber and other cargoes liable to give off excessive moisture or be weevil infested.

ISO Containers: may be carried in closed box containers, but extremely liable to sweat and moisture migration.

TAR	  	Pine 1.53/1.59 Coal Tar 1.34/1.45 1.02/1.05 S.G.	Drums

The vegetable variety is impure turpentine obtained by distillation from the roots, etc., of various pine trees, this being the kind used for preserving ropes, etc.

The other variety is obtained from coal.

Both kinds are inflammable and, as the smell of these commodities is very injurious to all edible and to fine goods generally, tar must be stowed entirely apart from such. The leakage, when carried in poor quality receptacles, is usually very great, and to avoid choking bilge suctions the pumps should not be put on the bilges until after they are thoroughly cleaned, if that course is at all possible. See Bitumen and IMDG Code.

Commodity	*Characteristics*	*M3 per Tonne*	*Packaging*

TAR OILS 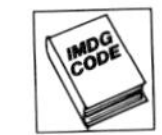

See Flammable Liquids in Dangerous Goods and Part 2, Petroleum. See IMDG Code.

TARA — Bags

Pods of a Peruvian tree shipped in bags. The pods are after crushed to make Tara powder which is used in tanning. Dry stowage.

TAROS  — Slatted Crates, Bags

Root crop exported from S. Pacific Island. Similar characteristics to potatoes. Must be clean and free from moisture at time of shipment. Short voyages: ventilation only required. Longer voyages: temperature control and air freshening. Carriage temperature may be + 10 degrees C (50 degrees F). Shelf life after refrigerated stowage approximately 10-14 days. Payload in refrigerated 20 ft *ISO containers* approximately 9 tonnes.

TEA   Black Tea — 2.79/3.07 — Sacks (Paper), Chests

The dried leaf of the tea shrub shipped in large quantities from Taiwan (Keelung), North and South China, Japan, Ceylon, India, etc.

Tea is the most delicate cargo of the East and tea claims, when they occur, are usually very heavy.

Tea must be stowed well apart from all kinds of odorous or moist cargoes; such commodities as Copra (which see), sugar, tumeric, cassia, oils of all kinds, especially essential oils, hides, etc., should never be stowed in the same or in any compartment connected by ventilators with that in which tea is stowed.

The greatest care is necessary in preparing holds for the reception of tea. All oil, etc., stains on tank top ceiling and elsewhere should be carefully removed by the use, if necessary, of a caustic solution, afterwards coating with limewash; the bilges thorougly cleaned, disinfected and deodorised by chloride of lime and coated with cement wash, and the compartment thoroughly ventilated and dried out before commencing to lay dunnage.

Dunnage well (using dunnage that is free from oil stains) especially on brackets, stringers, box beams, etc.; the spar ceiling, when widely spaced, to be faced over with crossed bamboo, which is also to be placed against bulkheads, tunnel sides, etc.; after which the whole is be well matted with clean dry mats. Intermediate dunnage should be used in deep stows, about every four tiers.

Palletised chests should be stacked no more than 3 tiers high or damage may occur. Palletised cartons are not normally sufficiently robust to be overstowed with other pallets. Final pallets of cargo stowed in the hatch square should be pre-slung to assist breaking out.

Tea is packed in chests, half-chests, cases, cartons and woven fibre plastic polythene bags and, in a few instances, in bales; the sizes and gross weights of these packages vary considerably, even for the same port, as the following will illustrate:—

Tea chests ex Taiwan:— Olong, Punchong, Gunpowder, Black Bup, 500 mm (20″) × 500 mm (20″) × 600 mm (24″) 38-45 kg. Lapsang, Stoughong, Boxer Leaf 480 mm (19″) × 480 mm (19″) × 600 mm (24″) 45-50 kg. Dust and Fanning 380 mm (15″) × 480 mm (19″) × 580 mm (23″) 45-50 kg.

Most teas now shipped in paper sacks and stowed in *ISO containers*.

TEA DUST  — 2.51 — Boxes

Usually packed in bales; of no great value. Ordinary dry stowage.

Commodity	*Characteristics*	*M3 per Tonne*	*Packaging*
TEA SHOOKS		1.67/1.78	Cases

The boards or staves from which tea chests are made. Dry stowage, remote from moist and odorous goods.

TEAK		2.09/2.23	Boards Squared
		2.23/2.37	Logs
		2.37/2.51	Logs, Deck Planks
		2.65/3.07	Assorted
		2.65/2.79	Scantlings

A very durable wood, shipped in large quantities from Burma ports, etc. 1 M3 of teak weighs from 737 to 833 kg.

TEEL SEED See Sesame

TEJPATA 6.97/8.36 Bales

The leaves of an Indian tree used for medicinal purposes, also for dyeing processes. Dry stowage away from moist goods.

TERRA JAPONICA 

Pale Cutch or Catechu.
See Cutch, also Gambier.

TERYLENE POLYMER  See Plastics Resins.

THORIUM  See Monazite Sand.

THUS 

A gum also called Frankincense. See Gums.

THYME   4.18/5.02 Bales

An aromatic plant from which is obtained an Essential Oil. Chiefly used in culinary processes. Dry stowage remote from foodstuffs.

THYMOL OIL 1.45/1.50 Cases

Is obtained from the Ajwan Seed. Wet stowage, guard against leakage.

Shortsea trade with timber deck cargo. (*Photo: Fotoflite*)

Packaged sawn timber in stow.

Sheets of plywood with damaged edges.

Commodity	*Characteristics*	*M3 per Tonne*	*Packaging*

TIES 

This is the name by which Sleepers are known in the United States. See Sleepers.

TIRES, TYRES		2.79/5.57	Loose
		3.62/4.18	Bales

Stow flat and away from foodstuffs.

TIL SEED (Teel Seed)

See Sesame, Gingelly.

TILES, FIRECLAY		0.98/1.39	Crates
ROOFING		2.23	Crates
		1.11	Bulk

Stow as for Slates.

TIKITIKI  Bags

Byproduct from milled rice and corn. May sift. Keep dry.

TIMBER

Softwood — Normally shipped in bundles or packages of sawn timber of various sizes. Normally shipped unprotected unless dried in which case may be wrapped in kraft paper. Susceptible to staining and fungal growth when not kiln dried. Claims may arise from wetness, staining, and bundling marks. Where these defects exist upon loading, bills of lading to be endorsed accordingly. Some shippers agree to standard phrases reflecting the above being incorporated on the bill of lading.

Hardwood — Normally shipped in bundled units of planks, strapped by metal bands.

Sawn timber should be loaded in clean holds and away from rusty surfaces. Careful ventilation to avoid ship sweat causing rust stains.

Logs — Do not normally produce claims.

Older timber measurements are the most complicated and laborious of all the various measurements in use for shipping purposes.

The unit of measurement in use in the U.K., Northern European countries, etc., is the Standard, of which there are many varieties, bearing no relation one to the other as shown in Appendix 10.

In the U.S.A. and Canada, the unit of measurement is the 1000 Board Feet; in France, Italy, Belgium, etc., the unit is the Stere, equivalent to the Cubic Metre of the Metric System; while the Petrograd Standard is almost exclusively used in the U.K. — wholesale transactions in battens, boards, deals, planks, etc., being on that basis.

See Part 2, Timber. See also Appendix 10.

See also IMO Code of Safe Practice for Ships Carrying Timber Deck Cargoes (1991).

TIMOTHY SEED  1.90/1.95 Bags

A grass seed, fine and small, usually shipped in small quantities, always bagged, which should be of close woven material and well made to avoid loss.

Commodity	*Characteristics*	*M3 per Tonne*	*Packaging*
TIN		0.22/0.28	Ingots

A highly malleable and valuable metal. Shipped in ingot form in large quantities from Malaysia, Australia, etc.

A Malaysian ingot weighs about 51.25 kg.

Ground space should be reserved for tin, which should be very carefully tallied (on board — not on the lighter) as cases are known of lightermen dropping ingots overboard in the hope of being able to recover same after the vessel leaves. It also requires careful watching in the hold until well covered up with other cargo. When discharging tin and ingots of other valuable metals on to wharves, etc., it is sound practice to arrange for them to be formed into piles of a certain fixed form and number of ingots. Stacked thus, they are easily counted and the removal of any ingots by unauthorised persons quickly detected; but ship tally should always be taken at the rail.

ISO containers — because of its high density, this cargo should be distributed evenly, making maximum use of that part of the container floor with the best load-bearing properties, i.e. next to walls and end.

TINCAL		1.50/1.56	Casks
		1.17/1.25	Bags

The crude borax, which see. Ordinary dry stowage.

TIN CLIPPINGS		1.06/1.11	Pressed Bales

The portions of tin plates that are sheared or clipped off during the process of manufacturing tin cans, etc.

TIN FOIL		0.61/0.70	Bundles
		0.84/0.89	Cases

From China.

TIN MUD		0.50/0.61	Bulk

May become contaminated. Not suitable in bulk in *ISO containers* (unless specially designed) as it removes paintwork and leaves a deposit difficult to remove and clean.

TIN ORE		0.50/0.61	Bags

Known also as Cassiterite. Valuable cargo. See Tin. See Part 2, Ores.

TIN PLATE		0.28/0.39	Steel Bulk Packs
		0.42/0.50	Boxes

Tinplates are thin sheets of steel coated with tin, the demand for which is continually increasing as additions are made to the already long list of edible goods, etc., which are packed in cans.

A box of tinplate is the standard unit of the tinplate trade and is referred to as a basis box, weight approximately 49 kg.

The majority of tinplates are now exported from the U.K. in steel bulk packs or stillages, each containing an equivalent of 10 units or say 1120 sheets of tinplate.

Commodity	*Characteristics*	*M3 per Tonne*	*Packaging*

Such is packed in a complete metal envelope, which affords good protection to its contents in as much as there is no visible sign of the tinplates themselves once the steel cover is in position.

The steel cover is strapped by wire strapping with platform underneath which is placed on wooden bearers to facilitate lifting by fork lift trucks.

Tinplates packed in wooden boxes or crates are covered with oilpaper and require very careful handling to avoid crushing plate edges and rendering them unsuitable for the can-making industry.

The use of chain slings in loading or discharging tinplates should not be permitted and trucking over plank ways (if working off the ceiling) should be insisted upon.

Tinplate cargoes are often productive of claims either in respect of improper handling or of rust.

To obviate the former, the plates should be loaded and discharged by means of trays of suitable dimensions and the slings fitted with suitable spreaders to avoid crushing the topmost cases.

The individual package being too heavy to be carried or lifted in the hold, it is the practice, with some stevedores, to land the draught on a small raised platform composed of the cases, situated in the square of the hatchway, and depend on the winch to disengage the sling or tray — the cases then being worked on their corners to their required positions, for which rough usage they are but ill-fitted.

Too deep stowage is also responsible for crushed cases (though not so much for damaged plates) this involving heavy cooperage expenses at destination, etc. Loading stevedores should be held responsible for cooperage of all cases damaged during loading.

In view of the weightiness of this cargo and possibly that of the overstow, it is essential that the cases be stowed perfectly level and firm, more particularly at the turn of the bilge where plenty of dunnage is necessary.

Rust damage — from time to time claims of this order are encountered, especially in respect of plates packed in wood boxes or crates. A rusty condition of the plates may result from one or other of the following conditions which are to be guarded against:—

(a) the cases being made of green, wet, or imperfectly seasoned timber which condition reveals itself, occasionally, by the formation of mildew on the box;

(b) stowing the plates on wet or damp ceiling;

(c) handling plates in the open during wet weather;

(d) heavy condensation due to unavoidable conditions for which the ship cannot be blamed if the ventilation is sufficient; or to cargo having the tendency to heat and throw off moisture having been stowed in the same compartment.

See Part 2, Steel and Iron — for stow and securing advice.

TITANIUM OXIDE  Paper Bags

Paint pigment. Dry top stowage. Dirty. Liable to sift.

TITANIUM SPONGE 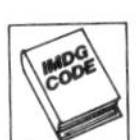

Covered with an inert gas in hermetically sealed drums. Valuable. Great care in handling and stowage since any entry of air might make the cargo valueless. See IMDG Code.

TOBACCO 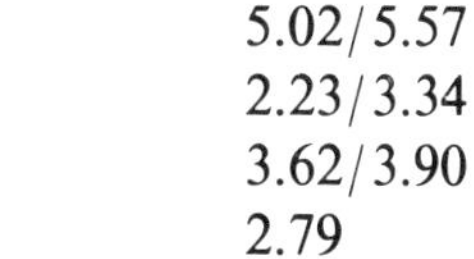  

		5.02/5.57	Hogshead
		2.23/3.34	Cases
		3.62/3.90	Ctn. Cs.
		2.79	Small Fl. Cases

In kegs, cases, cartons and bales. Susceptible to damage from taint, moisture, staining and, in some cases, crushing. Compartments and *ISO containers* must be thoroughly clean and dry before loading. Deli Tobacco: used for cigar wrappers. Value depends on leaves remaining unbroken. Breakage may occur through crushing or when tobacco becomes very dry. Must not come into contact with steel or timber. May be carried with non-odorous cargo, i.e. crepe rubber in cases, trocas shells, hard paraffin wax. Do not stow with coffee or canned foodstuffs.

North American raw or leaf tobacco is usually packed in cartons and tierces, the weight of which varies between 590 kg and 816 kg for Kentucky leaf and 272 kg to 544 kg for Virginia leaf.

South American leaf is, usually, packed in bales and rolls.

Leaf tobacco is liable to heat and sweat and become mouldy. Tobacco both in the leaf and manufactured form is readily damaged by moisture or strong odours so that it must be stowed outside the influence of wet goods or strong smelling commodities. Good ventilation may also be necessary to avoid mould growth.

Java Tobacco: three categories — Estate, Semi-Estate and Native. Hang Krossok: covers Estate and Semi-Estate. Shipped in "hands" and should be stowed on the upper tiers where possible. Campong Krossok: also shipped in "hands" but not hung for drying. It is a produce of the smaller farmers and is liable to be inferior to Hang Krossok.

Scrap (Gruis): is derived from Hang Krossok or Campong Krossok. Estate Tobacco, exclusively shipped by Perusahaan Perkabunan Negara (PPN) is assumed to be free from infestation (Lasioderma) in view of all the precautions on the Estates to prevent the presence of these tobacco beetles in the bales. Semi-Estate tobacco and Native grown tobacco should be disinfected prior to shipment, but some consignments may still be infested or develop infestation through the hatching of beetle eggs during the voyage.

Estate tobacco must be in a different container to other categories of Java tobacco. Semi-Estate and Native tobacco must not be stowed in the same container. Tobacco must not come into contact with steel or timber. Protect against sweat with mats.

Tobacco should be stowed on the flat and bales must not be stowed on their sides. Only clean non-tainting cargoes may be stowed in the same container, i.e. rattans, fibre, kapok, teak and sheet rubber. Keep well away from copra or copra products.

Deli and Java tobacco bales may be marked with coloured ribbons:

Red ribbon bales may be stowed top tiers.
White ribbon bales must not be top or bottom tiers.
Blue/Green ribbon may be used for bottom tiers.

Tobacco is subject to high import duties in most countries and hence, even if slightly damaged, it may be uneconomic for the receivers to import the cargo.

TOKMARI SEED  1.53/1.64 Bags

An edible seed, which, if permitted to get damp becomes very messy.
Dry, cool stowage remote from moist goods is indispensable.

TOLUENE (See Benzene) see Part 2 Chemical Cargoes

TOMATOES 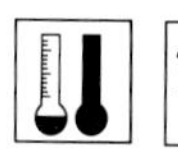  1.95/2.09 Cartons. Boxes. Trays

Shipped from the Canary Islands and near Mediterranean ports, Guernsey and Holland etc.
Some as deck cargo.

When showing signs of ripening, or for long voyages, it is best carried in refrigerated compartments at a temperature of 10/12.5 degrees C.

Keep covered from wet weather or spray.

May also be carried under temperature control conditions. Suggested carriage temperatures:—
Coloured + 7 degrees C. Unripe + 13 degrees C.
Delivery air temperature must not fall below + 1 degree C at any stage.

Commodity	*Characteristics*	*M3 per Tonne*	*Packaging*
TONKA BEANS		2.51 1.95	Barrels Boxes

Seeds with a vanilla-like odour, used for flavouring tobacco and in the manufacture of artificial vanilla extract.

Shipped from the W. Indies and S. America. Dry stowage away from commodities likely to take taint.

TORTOISE SHELL — 3.76/4.18 — Cases

The tortoise shell of commerce is the horny shell of the hawksbill turtle and of the carett tortoise. The best quality is shipped from Singapore, but the shell is also obtained from the W. Indies, Mauritius, Ascension, C. Verde Islands, etc. Clean cargo — treat as valuable goods.

TOW — NATURAL 

2.65/2.79 — Bales

Hemp refuse. Stowage as for hemp, which see. Also see Acetate Tow. See IMDG Code.

TOYO 

Black liquid sauce from Soya beans. In bottles. May leak.

TRAGACANTH 

A gum shipped from Asia Minor in flake form. See Gums.

TREACLE — See Molasses

TREFOIL SEEDS

The seeds of a clover-like herbaceous plant. See Grass Seeds.

TREPANG 
See Beche-de-Mer

TRIPOLITE — 2.09 — Bulk

A Siliciferous and porous rock. Shipped from Australia. Stow away from oils and greases.

TUNA — See Fish. 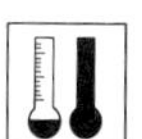

TUNG OIL — Bulk
(See China Wood Oil)

Tanks must be washed clean and dried. Tung Oil in contact with water causes isomerisation. The oil becomes cloudy and thickens. Avoid exposure to direct sunlight. Ship should take and retain samples during loading.

Commodity	*Characteristics*	*M3 per Tonne*	*Packaging*
TUNGSTEN ORE		0.45/0.50	Bags

A brownish or greyish black ore from which the metal tungsten is obtained. Known also as wolfram. See Part 2, Ores.

TURKISH MILLET		See Durra	

TURMERIC	 	2.23/2.37	Bags

The root of an E. Indian plant from which is obtained a yellow powder used as a dye and for testing for alkalies.

Throws off a strong pungent smell and is dusty, both of which conditions should be guarded against.

TURPENTINE	  	1.53/1.56 1.53/1.56	Tins in Cases Drums

A transparent spirit obtained by distillation of the resinous substance obtained from a variety of trees, principally the pine and larch.

Turpentine and its vapour is highly inflammable, so the greatest care should be exercised to prevent sparks or lights coming into contact with same and for this reason it is not suitable cargo to handle by night.

Turpentine smells are highly pungent and penetrating and many serious claims for damage, through tainting by turpentine fumes, have had to be paid in the past. So real is this danger, that it is always recommended that turpentine should be separated by the engine and boiler room space from flour, tea and similar fine and edible goods.

Special regulations have to be observed in most ports with regard to the stowage of turpentine on a vessel which also carries cotton (owing to the liability of the latter under certain conditions to spontaneous combustion). Such regulations prohibit turpentine from being carried in the same compartment as cotton and demand, where practicable, that these goods must be separated by two steel bulkheads — never less than one.

The same principle should be respected when arranging the stowage of turpentine along with other goods, i.e. stow as far away as possible from all goods liable to spontaneous combustion. See Part 2, Petroleum and IMDG Code.

TURTLES			

Shipped in hard frozen condition. Usually stowed on their backs.

TUTH-MALANGA		See Tokmari Seed	

TWINE (Binder)		1.67/1.95 2.39/2.50	Bales Cartons

The twine used in harvesting machines for binding the sheafs of corn — made of sisal hemp and similar fibres — more or less oily; do not stow over fine goods. Flammable.

Commodity	*Characteristics*	*M3 per Tonne*	*Packaging*
UMBER		1.11/1.17	Bulk

A dark brown ore used as a pigment, etc. See Ores.

UNITISED CARGOES See Part 2, Techniques, Unitised Cargoes.

URANIUM ORE		0.47/0.50	Bags

The ore from which the rare metal uranium — a nickel white metal — is obtained. See Ores.

UREA		1.17/1.56	Bulk

Granules, beads and prills. Dusty. Angle of repose 28 to 45 degrees. See also Fertilizers.

VAILA		See Galls	
VALONIA	 	2.79/2.93 2.56/2.62	Bags Bulk

The large acorn cups of a dwarf oak, exported from the Levant, etc., and used in the tanning industry. Very liable to heat and sweat. Stow clear of fruit and other delicate goods liable to be affected by heat or moisture.

VALUABLE CARGO See Part 2, Techniques, "Specials".

VANADIUM ORE	 	0.64/0.70 0.56	Bags Bulk

Shipped from E. Africa and S. America. See Ores. See IMDG Code. (See IMO Code of Safe Practice for Solid Bulk Cargoes).

VANILLA	 	1.67/1.81 1.39/1.53	Cases Bags

The fruit of a parasitic orchid, which flourishes in the hot and moist forests of the Orinoco and Amazon basins, etc., extensively used in the confectionery trade, etc., for flavouring.

Dry stowage away from odorous goods.

VARNISHES		1.53/1.67 1.67/1.81	Cases Drums

See IMDG Code.

VASELINE	 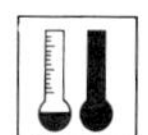	1.62/1.73 1.45/1.50	Drums Cases

A yellowish substance obtained from petroleum, practically tasteless, used as a lubricant, etc. As it melts with heat, must be stowed in a cool place and apart from dry goods.

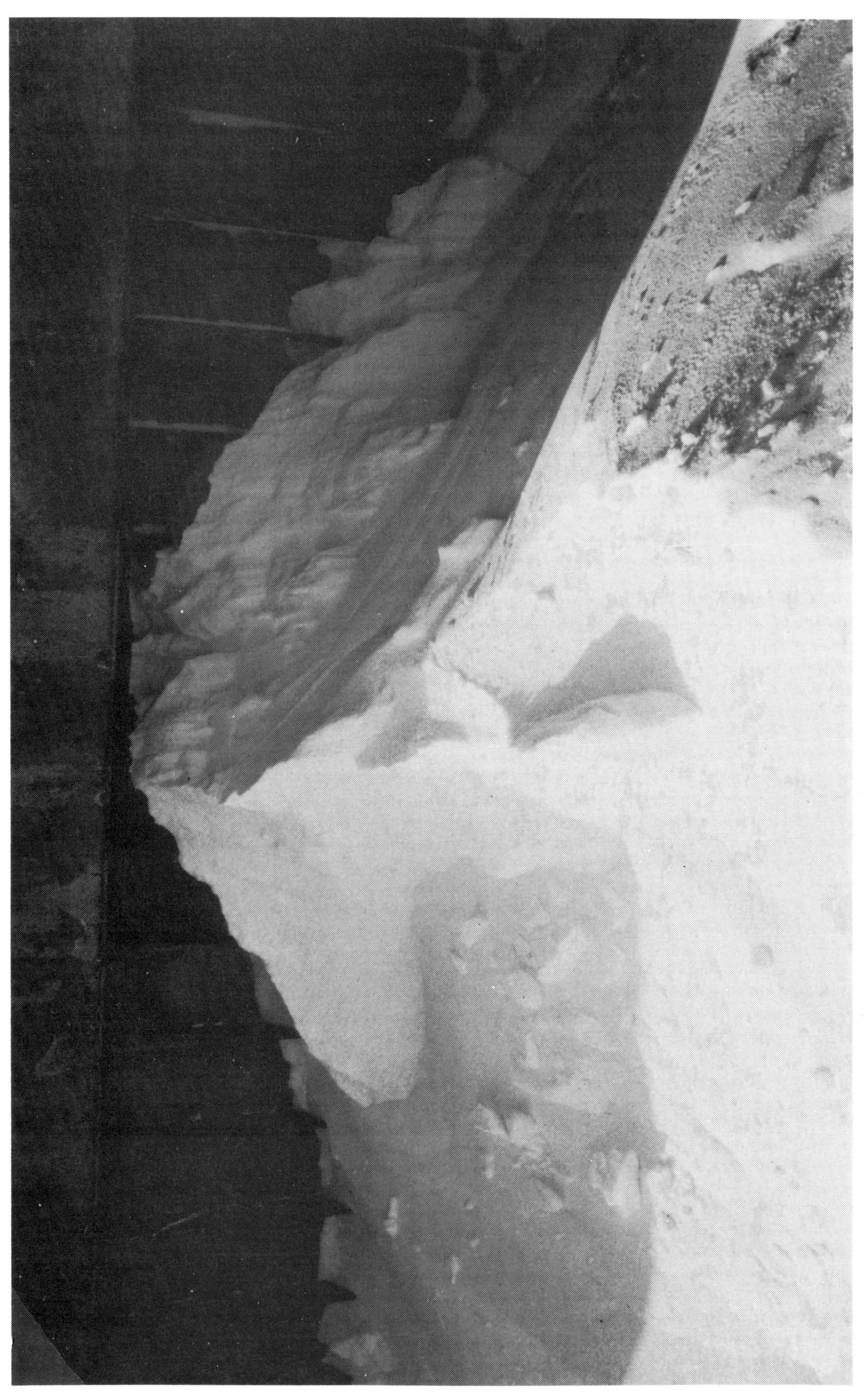

Bulk urea.

Commodity	*Characteristics*	*M3 per Tonne*	*Packaging*
VEAL	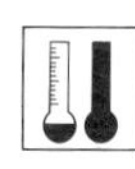	1.80/2.20	Cartons

VEGETABLE FATS   

Are the semi-solidified oils extracted from the nuts and fruits of certain tropical trees. The principal kinds e.g. coconut palm, palm seed, kapok oils, etc., are shipped in large quantities in bulk.

In lesser quantities, the undermentioned are shipped in metal containers or in casks and demand vigilance to avoid receiving leaky recepticles on board and the appropriate clausing of B/L to protect the ship from claims for leakage:—

Avocado oil, Chinese vegetable tallow, Cacao butter, Macassar oil, Mahwa butter, Nutmeg oil, or Oil of Mace, Piney tallow, Shea butter.

These fats or oils are largely in demand for the manufacture of butter and cream substitutes, soap, medical stores, etc., some of which are usually referred to as oils, the only difference between fats and oils being that of the temperature at which they solidify; those which solidify at ordinary temperatures are usually referred to as fats.

The principal vegetable fats or oils are referred to herein under their respective names. Wet stowage, in a cool place and remote from edible or delicate goods liable to be tainted by fat odours. See Vegetable Oils and Vegetable Oils in Bulk, also Vegetable and Animal Oil Constants for Stowage Factors.

VEGETABLE FIBRE	  	4.46/5.30	Bales

Keep dry. Liable to spontaneous combustion. See IMDG Code.

VEGETABLE IVORY 

See Corozo and Coquilla Nuts.

VEGETABLE OILS Bulk		1.67	Drums

The fruit and seeds of all plants yield oil or fat; the oil yield varies greatly for different seeds, that of the castor bean being as high as 50 to 53, while in the case of maize it is as low as 5 to 10 per cent.

The difference between vegetable oils and fats is only that of the temperature at which they solidify, those which solidify at ordinary temperatures being usually referred to as fats.

The principal vegetable oils of commerce are dealt with herein under their respective names and a table giving the specific gravities; the temperature at which they solidify, and the cubic feet space occupied by one tonne of oil in bulk, is given in Appendix 12.

The following are amongst the most important Vegetable oils that move in large volumes — mostly in bulk, viz.:—

Bean Oil, China Wood (Tung) Oil, Coconut Oil, Cotton Seed Oil, Kapok Seed Oil, Linseed Oil, Palm Oil and Palm Nut Oil, the characteristics of which are given in the Alphabetical Section.

Some oils are shipped in barrels, recoopered and second-hand receptacles and even in old kerosene cans, especially coastwise. In such cases considerable leakage is to be apprehended and guarded against.

See Part 2, Oils and Fats.

VEGETABLES			Crates, Bags, Cartons Open Mesh Sacks

May be carried on short hauls with good ventilation. Several factors must be taken into consideration when vegetables are presented for shipment. Temperature of product, length of voyage

Commodity | *Characteristics* | *M3 per Tonne* | *Packaging*

and climatic condition expected on voyage. Vegetables breathe which results in the production of heat and, left alone, it will ultimately rot. On long voyages it is essential that this heating process is retarded by the application of refrigeration. If the product is not pre-cooled for shipment, it is essential that after loading the temperature is brought down as quickly as possible.

Load and carry according to shippers instructions and temperature of carriage.

See Part 2, Refrigerated Cargoes.

Commodity	Characteristics	M3 per Tonne	Packaging
VEGETABLE WAX	 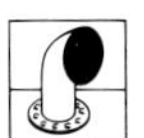 	1.28/1.39	Boxes
		1.53/1.67	Bags
		1.81/1.95	Boxes, Africa

A plant secretion which is collected from certain trees, their berries or fruits, in the form of a greasy powder which is shipped from Japan and certain ports of Africa. Cool stowage in well-ventilated space.

Commodity	Characteristics	M3 per Tonne	Packaging
VEGETABLES, FRESH	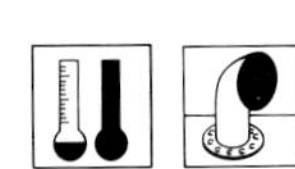		Crates, Bags, Cartons, etc.

Other than Potatoes and Tomatoes, which are referred to separately, vegetables such as Beets, Cabbage, Cauliflower, Carrots, Celery, Lettuce, Onions, Parsnips, Peas, Radishes and Turnips all carry well at 0.5/1.0 degrees C. Cucumber, seed and green Beans should be carried at a temperature of about 4.5 degrees C.

Adequate floor dunnage is required and care must be taken to avoid crushing in stow. Most vegetables can be stowed together but onions should be given separate stowage and kept well away from cargoes liable to take taint.

There is a continuing growth in the shipment of vegetables and fruit under quick frozen conditions. Carrying temperature should be − 18 degrees/−15 degrees C. Provided that all shipments are fully packaged, there should be no possibility of taint and therefore mixed stowages can be obtained. Normal floor dunnage should be provided. Avoid deep stowages. Refrigerated containers are particularly suitable for this class of cargo. See Part 2, Techniques, Refrigeration.

VELLUM

A superior and finer kind of parchment prepared from the skins of kids, lambs, calves, etc. A valuable commodity, usually made up into rolls and packed in airtight cases. Special stowage.

Commodity	Characteristics	M3 per Tonne	Packaging
VELLS	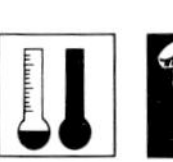 	2.93	Cases

Animal intestines which give off an offensive odour. When kept in a hard frozen condition, this cargo is odourless. Ordinary frozen stowage as for case goods, but should not be stowed in the same compartment as butter.

See Part 2, Techniques, Refrigeration.

Commodity	Characteristics	M3 per Tonne	Packaging
VENEERS		3.34/4.18	Crates

Stow so as to avoid uneven weight distribution. Distortion may occur if damp. Do not stow with moisture inherent commodities.

Commodity	*Characteristics*	*M3 per Tonne*	*Packaging*
VERDIGRIS		1.56/2.17	Bags

The rust of diacetate of copper, used as a green pigment, also for medicinal preparations. Poisonous. Dry stowage away from edible goods.

VERMECELLI 

A form of macaroni, which see.

VERMICULITE		1.37	Bulk

A mineral of the Mica group. Used in insulation and fire proofing. Angle of repose 36 degrees.

VERMILLION		1.23/1.28	Cases

A pigment — the bright red sulphide of mercury which is obtained from cinnabar. Dry stowage. See Cinnabar.

VINEGAR		1.73/1.81 1.84/1.95	Drums Demijohns

A form of acetic acid obtained from cider, inferior wines, etc., by fermentation.
A corrosive liquid. Wet storage.
See Part 2, Techniques, Drums and Demijohns.

VITRIOL, OIL OF (SULPHURIC ACID) 

See Part 2, Dangerous Goods. See IMDG Code.

VITRIOL, GREEN 

See Copperas.

VOMICA NUTS 

See Nux Vomica.

WALLBOARD 

Stow as for Plywood. See also Beaver Board.

WALNUTS (Unshelled)		3.07/3.62	Bags

See Nuts.

Commodity	*Characteristics*	*M3 per Tonne*	*Packaging*
WALNUT MEAT (Shelled)	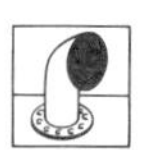	2.23/2.51	Cases
		4.18	Bags

Must be stowed in a cool well-ventilated space, well removed from engines and boilers.

WASTE	 	4.74/5.57	Bales

Cotton thread refuse. Unless guaranteed to be entirely free of oil or grease, this is a dangerous cargo, as oily waste is peculiarly liable to spontaneous combustion and should only be carried as deck cargo. See INDG Code.

WATER

See Fluids, Weight, etc. Appendix 7.

WATTLE		1.25/1.39	Bags

A gum shipped from E. Africa in bags. Dunnage and sawdust between tiers, or the stow will set solid. See Gums.

WAX	 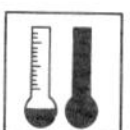	1.48/1.53	Cases, White
		1.62/2.01	Bag
		1.39/1.48	Cases
		1.95/2.09	Drums, Paraffin
		1.39/1.53	Bag, Paraffin

Melts and runs with heat. Stow in a cool place. See Vegetable Wax.

WEEDKILLER	  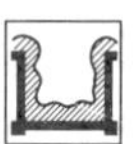		Bags

Hessian, Polylined. Poisonous. Do not stow with foodstuffs. Line *ISO container* floor against sifting. See IMDG Code.

WET HIDES	 	See Hides	

WHALEBONE		1.81/1.95	Cases
		3.07/3.21	Bundles

A very valuable commodity, which requires careful tallying and watching to avoid pilferage. Stow in special locker away from acids, greases and oils.

WHALE MEAL	 	1.95	Bags

Odorous cargo. Dry stowage.

Commodity	*Characteristics*	*M3 per Tonne*	*Packaging*
WHALE MEAT	 		

Carried hard frozen.

Shipped in cartons, bags or slabs. Whale meat should not be stowed with mutton or lamb and should be closely watched for drip if pre-cooled.

See Part 2, Techniques, Refrigeration.

Commodity	*Characteristics*	*M3 per Tonne*	*Packaging*
WHALE OIL	 	2.09	Drums
		1.14	Bulk

See Fish Oil. See Part 2, Oils and Fats.

Commodity	*Characteristics*	*M3 per Tonne*	*Packaging*
WHEAT		1.18/1.34	Bulk
		1.34/1.50	Bags

See Grain.

See Part 2 for Carriage in Bulk.

WHEELS
(Waggon Wheels Axle Mounted)

By nesting the wheels and axles in two directions (i.e. at right angles to one another) great saving of space (up to 25%) may be obtained. This further reduces the requirements of securing, which is necessary if only shipped one tier high. Typical weights and dimensions of axle mounted wheels would be: length 2,350 mm, 1,000 mm diameter, weight 1,422 kg. 2,350 mm × 1,092 mm diameter, weight 1,524 kg.

Great care must be taken to ensure that the permitted axle weight of the lower tier is not exceeded when nesting and increasing the height of stow.

Commodity	*Characteristics*	*M3 per Tonne*	*Packaging*
WHISKY/WHISKEY		1.84/2.01	Barrels
		1.67/1.81	Cases

The utmost vigilance is necessary to avoid pilferage of this and other kinds of spirits. When receiving, a sharp lookout should be kept for light cases, i.e. cases from which bottles have been abstracted on shore. Suspicious-looking cases should be opened or rejected.

An officer or a reliable watchman should remain in the hold all the time the handling of this cargo is in progress. Cases of whisky should preferably be stowed in special cargo locker — and not in the open hold. See Part 2, Techniques, "Specials".

Barrels of whisky should be covered over with other cargo as soon as possible.

For sizes and measures see Appendix 6. See IMDG Code.

Commodity	*Characteristics*	*M3 per Tonne*	*Packaging*
WHITE QUARTZ		0.61	Bulk

Various sized lumps. Angle of repose 42 to 45 degrees.

Commodity	*Characteristics*	*M3 per Tonne*	*Packaging*
WHITE ROCK		See Lime	
WHITE WOOD OIL		See Eucalyptus Oil	

Commodity	*Characteristics*	*M3 per Tonne*	*Packaging*
WHITENING		1.11/1.25	Drums
WHITING		1.06/1.11	Bags

A chalky substance which, when mixed with water, is used as a covering medium. Ordinary dry stowage. See Part 2, Techniques, Obnoxious Cargoes.

WINES		1.62/1.78	Casks
		1.67/1.95	Cases
			Tanks
			Flexibags

The bulk of wines are carried in casks which, according to their capacity, are termed tuns, pipes or butts, puncheons, hogsheads, tierces, barrels, etc., but case wine is in much evidence and cheaper varieties of wine travel well in portable tanks, flexitanks or in bulk tanks.

Wines Bs/L should protect ship from liability for leakage, ullage and spills.

Loss of wine through leakage is often very considerable and, notwithstanding ships are not responsible for leakage, claims for loss are often made on the ground of improper stowage, which, if true, may be due either to imperfect bedding, chocking or dunnaging; to stowing too many heights, or to the rough handling of casks — all of which should be guarded against. For stowage of casks, see Part 2, Techniques, Drums, Barrels, etc.

With small shipments of wine in barrels, it is now largely the practice to stow the barrels on end. An adequate floor should be made between each tier and the barrels should not be stowed more than three in height. This form of stowage has the advantage in that chocking off is reduced and the process of loading and discharging is quickened.

Leakage is often due to poorly-made, old or ill-conditioned casks, so that it is necessary for each cask carefully to be scrutinised to ensure that no stained, leaky, partially empty, spilled, re-coopered or damaged casks are received. Clean receipts should never be issued for repaired, patched, spilled or stained casks and, if a properly qualified receipt for such is not acceptable to the shipper, the casks should be rejected if claims are to be avoided.

If landing leaky casks to Custom store, lighter, etc., or casks which are alleged to be short of contents, the ullage should be taken or cask weighed and particulars agreed upon with the party to whom delivery is made.

Wine deteriorates with heat and is readily reduced in value or even rendered unfit for consumption by strong smells, so that it should be kept well apart from cargo which is liable to heat, from engines and boiler space and from oils and strong-smelling commodities.

Wine in bottles should be stowed and the same precautions taken as for Whiskey, which see.

In certain trades, vessels are now equipped to carry wine in specially constructed tanks in the vessel. In addition, wine is now sometimes shipped in portable tanks which are carried either on deck or in 'tween decks. Care should be taken to see that there are no leakages during loading of portable tanks and no other cargo should be stowed above the tanks.

ISO containers — wines and fortified wines may be successfully carried in tank containers. Specific Gravity 1.1. Suitable for carriage in bulk bags in General Purpose closed box containers (see Part 2, Techniques, Containers), though care must be taken to ascertain that the bag material is compatible with the product, i.e. that no risk of taint exists and de-lamination will not occur.

WIRE, BARBED		1.56/1.67	Reels

Barbed wire is very useful for filling broken stowage amongst steel and iron goods, heavy packages of machinery, etc. When used for that purpose, the different marks, lots of consignments should be stowed well apart — in different compartments, if quantities shipped and other conditions permit — in

Commodity	*Characteristics*	*M3 per Tonne*	*Packaging*

order to avoid mixing and consequent expense of sorting out and delay in delivery at destination. See Separations.

	Plain wire, black or galvanised	0.98/1.06	Coils
	Fencing wire	0.98/1.06	Coils
	Netting wire	1.81/2.51	Rolls
	Copper wire	0.95/1.00	Cases

Should not be overstowed with heavy goods.

WITHERITE  0.50/0.56 Bulk

A heavy ore — a native carbonate of baryta — also called Barolyte. See Ores.

WOLFRAM  See Tungsten Ore

WOOD OIL See China Wood Oil

WOOD PULP See Paper Pulp

Avoid contamination with foreign substances, rope yarns, etc.

WOODCHIPS 

		2.93	Soft Bulk
		2.29/2.31	Hard Bulk

Now loaded in bulk in considerable quantities. Low angle of repose. As this cargo is frequently left in the open for some time before loading, it is liable to have a substantial moisture content. May oxidise and cause oxygen depletion. Space/hold must be tested prior to entry.

See also Part 2, Techniques, Neobulk and Bulk Cargoes. See IMDG Code. (See IMO Code of Safe Practice for Solid Bulk Cargoes).

WOOD CHIP PELLETS  3.05 Bulk

May oxidise and cause oxygen depletion. Space/hold must be tested prior to entry. See IMDG Code. (See IMO Code of Safe Practice for Solid Bulk Cargoes).

WOOL  

The "Stowage factor" varies enormously for different countries and grades of wool.

Slipe wool is wool from sheep killed for mutton. Shipped in bales, kiln or sun dried, the latter kind being liable to spontaneous combustion. Pie wool is chemically removed wool from mutton and is the residue that cannot be removed with a slipping knife. Both slipe and pie wool must be especially marked. Incipient fires in Pie Wool can often be detected early by an oily smell akin to burning fat.

Some Pie wool may only be carried on deck at shippers' risk. See local regulations and recommendations (e.g. New Zealand D.S.R.) for up to date information on carriage requirements, marking of bales, etc.

In addition to meaning the fleece of the sheep, the term wool includes the finer hair of a number of other animals.

Commodity	*Characteristics*	*M3 per Tonne*	*Packaging*

The principal wool exporting countries are Australia, New Zealand, S. America, U.S.A., Far East and S. Africa.

Wool shipments are generally divided into two classes, i.e. wool in the grease (greasy or unscoured) and clean or scoured wool. Formerly the two kinds were not stowed together for fear the unscoured would damage the scoured wool.

To avoid claims for rust damage, bales of wool should not be stowed on iron-bound cases or bales without overlaying them with dunnage wood.

Wool is packed for shipment in bales, covered with burlap or gunny, pressed to varying densities. The size and weight of bales vary considerably for different countries as the following will show:—

Australian single dump varies from 160 to 200 kg.
New Zealand bales vary from 128 to 190 kg.
S. Africa bales vary from 136 to 408 kg.
Plate bales vary from 317 to 453 kg.
India bales vary from 136 to 181 kg.

Typical space requirements (M3 per bale)

Australian wool:—

	Conventional	Random (Bulker)
Undumped	0.85 M3	0.99 M3
Dumped 3 Band	0.54 M3	0.59 M3
Medium Density	0.48 M3	0.54 M3
High Density	0.42 M3	0.48 M3

Medium density bales may stow up to 0.62 M3 per bale depending on hold configuration.

New Zealand wool:—

Ordinary Bales	0.45-0.51 M3	170 kg
High Density	0.31-0.37 M3	170 kg
Tripak Dump	0.75-0.87 M3	510 kg

Typical stowage factors (M3 per tonne) for Low density dumped New Zealand wool:—

	Lower Holds	'Tween Decks
Greasy	3.6	4.5
Scoured	4.6	5.7

Up to nine bales may be strapped together, into which stack is secured a suitable lifting device such as a T-bar. Given the right type of ship, high rates of loading and discharge can be obtained, but such a method has severe limitations in vessels with long runs from hatch to ships' wings.

ISO containers — closed box containers are suitable for greasy or scoured wool. Typical payloads for 20 ft containers:—

Normal dump	56-63 bales	7.2-10.5 tonnes
Medium density	66-81 bales	10.5-14 tonnes
High density	92-100+ bales	14.5-17 tonnes
Triple dry bales	36 (×3) bales	18-21.5 tonnes

Precautions: spontaneous combustion can take place in wool, pie being the most susceptible. See IMDG Code.

Wool should not be stowed with oil, fat, tallow, tow or flax; or in contact with packages containing such products; or in contact with other material more readily combustible than wool itself.

The risk of fire being so great when wool cargoes are carried, the greatest care should be taken to guard against same, especially when hatches are open.

Wet or damp wool should be rejected; wool should not be stowed on top of ore, moist or oily goods, without adequate separation of planking, etc., neither should wool be stowed with or above maize or other cargo liable to heat and throw off moisture, many claims having been paid in the past because of sweat damage to wool arising as a result of that class of stowage. Dunnage and mat well for wool cargoes.

Commodity	*Characteristics*	*M3 per Tonne*	*Packaging*
WOOL GREASE (Lanolin)		1.67/1.78	Barrels or Drums

Contrary to what is sometime believed to be the case, wool grease is not liable to spontaneous combustion, but like all other oily substances it maintains and increases combustion. Wet stowage. See also Part 2, "Obnoxious Cargoes".

XYLENE (See Benzene). See Part 2, Chemical Cargoes.

YAMS		2.37/2.79	Bags and Crates

Frequently shipped in second-hand bags, stowage factor varies. Care should be taken to ensure that shipment is not over-ripe, as yams rapidly decompose. They are a sweet potato type vegetable and when deteriorating give off a staining liquid. Must be stowed on their own ground and not overstowed. Cool dry stowage. Good ventilation required; this can be augmented by vertical breeze holes at hatch corners. 'Tween deck stowage to avoid crushing; stow up to about eight high. Cocoa yams should be restricted to five high. Liable to sweat, stow well away from logs, etc. May be carried under refrigeration.

YARN (COIR)		See Coir	
YARROW SEED		2.51	Sacks

See Seed Grass.

YEAST		1.67 1.70/1.95	Bags Drums

Very susceptible to damage by variations in carriage temperature. Such variations may cause irrepairable damage. May be carried under chilled or hard frozen conditions as requested by the shipper.

See Part 2, Techniques, Refrigeration.

YELLOW OCHRE		See Ochre Puree	
YELLOW SULPHUR		See Copper Pyrites	
YESO			Cartons

Blackboard chalk. Fragile. See Chalk.

YOLK		See Egg Yolk Powder; Eggs Liquid	
ZINC		0.22/0.33	Ingots

See Spelter.

Commodity | *Characteristics* | *M3 per Tonne* | *Packaging*

ZINC ASH    1.11 Bags
Bulk

Very liable to heat if wet or damp. Reject all wet or damp bags, stow in a dry place, remove from goods liable to throw off moisture. In bulk form must be kept dry as contact with water may produce toxic and flammable gases. See IMDG Code. (See IMO Code of Safe Practice for Solid Bulk Cargoes).

ZINC CONCENTRATES  0.56/0.61 Bulk

May liquify. Check physical properties before loading. See Ores and IMDG Code.
See Appendix 13.

ZINC DROSS   0.56/0.61 Cases
Bulk

In bulk form must be kept dry as contact with water may produce toxic and flammable gases. See Ores and IMDG Code. (See IMO Code of Safe Practice for Solid Bulk Cargoes).

ZINC DUST   0.56/0.70 Cases or Drums

(Zinc flue dust)

Inflammable on account of its very fine particles and liable to spontaneous ignition when damp. Largely used in the manufacture of certain coal tar dyestuff intermediates, also in cyanide mills as a precipitant of gold and silver. Packed in hermetically-sealed lined cases or metal drums. Dangerous cargo. Stow where easily accessible in a dry place and clear of all acids. If explosives are carried, the engine and boiler spaces must separate same from zinc dust. See IMDG Code.

ZINC ORE  See Blende, also Ores

See Appendix 13.

ZINC WHITE  0.56/0.70 Drums or Kegs

Treat as wet cargo. Stow on end, chock off well and avoid over stowing with weighty cargo.

ZIRCON SAND  0.70/0.84 Bags. Bulk.

Used in the process of hardening steel, etc. Shipped from Australia and the Far East. Cool, dry stowage. See also Sand.

ZIRCONIUM ORE  0.56
0.50 Bags
Bulk

Used in the manufacture of certain steels. See ores.

PART 4

DAMAGE AND CLAIMS

CHAFE

Damage of this nature arises from the to and fro movement imparted to packages via the motion of the vessel, or by the vibration and movement of the vehicle on land, causing them to rub and chafe against sharp projections, rough surfaces, other packages or themselves.

Chafing occurs, not only during heavy weather when a vessel is pitching and rolling, but under fine weather conditions when the pulsating movement of the ship is taken up by the packages which are not securely and firmly stowed on a solid floor. This applies in a similar way to container loads which suffer the vibration of road and rail movement.

Baled goods are particularly susceptible to this type of damage, particularly cotton piece goods and other textiles, skins, etc.

Goods which are shipped without the protection of packaging may be directly affected by chafe; e.g. reels of cable, which, if the protective insulation coating is affected, may lead to very high claims; items with reduced packaging and incorrectly stuffed in a container.

Cartons, if loosely stowed, may rub against each other during transport and remove a thin layer of the carton surface. This may reduce the value of the consignment, since it will also have removed marks, advertisements, etc.

TEMPERATURE

Some commodities may be affected by temperature fluctuations (which may cause condensation, or accelerate deterioration — see Part 2, Techniques, Refrigerated Cargoes) or extremes of temperature which may shorten the "shelf life" of the commodity, or even change its chemical composition.
shorten the "shelf life" of the commodity, or even change its chemical composition.

Some commodities are so susceptible to temperature change or the degree of temperature at which they may be carried, that controlled temperature may be a requirement.

High temperature may affect various cargoes in the following way:

1. Softening of chocolates, fats, coatings, etc., either temporarily or permanently changing their appearance.
2. Cause partial or full decomposition, e.g. reduce the shelf life of perishables.
3. Provoke a chemical change that will render the commodity valueless, e.g. in the case of drugs.
4. Take the commodity up to its own flash point and cause it to be an explosive risk.
5. Cause expansion (liquids and gases) so that packaging materials may be strained to bursting point.
6. Aggravate any self heating tendencies.
7. Cause distortion and cracking, e.g. in the case of linoleum.

Extremely low temperatures may also affect some commodities. For instance:

1. Freezing (and spoiling) of fresh fruit, cheese, vegetables, etc.
2. Freezing of bottled liquids, canned liquids, etc., with subsequent bursting of the receptacles.
3. Change in the chemical state of some goods (e.g. drugs) making them useless and sometimes dangerous.

4. Hardening and cracking of certain textiles and cushioning materials.
5. Separation of emulsions.

The following commodities (the list is by no means exhaustive) may take damage from extremes of heat:

Abrasive based papers, beeswax, biscuits (chocolate coated, cream filled), butter (canned), candles, cellophane and film, celluloid goods, chemicals (some), chocolate and confectionery, cocoa butter, cream of tartar, cube gambier, drugs, furniture, gelatine, grass seeds, gum (certain types), hides (wet salted), hops, jams, kernels, lard, linoleum, nails, meals (certain types), middlings, pork, ham (canned), peas, pulps, photographic material, photographic film (unexposed), rubber merchandise, seed (certain types), shellac, soap, tallow, tobacco (certain types), vinyl film, waxes.

Some of the commodities which may suffer if exposed to extremes of cold include:

Beers, wine, chemicals (some), chocolate, cork, drugs, fruit (canned), fruit (fresh), jams, latex, motor cars (if radiators, etc., are not drained), paints (emulsions), rubber merchandise, vegetables (canned), vegetables (fresh).

Some commodities are liable to spontaneous heating, which would include:

1. Coal with accompanying escape of marsh gas increases the risk of explosion, also loss of calorific value.
2. Rice, oats, maize, oil seeds—and other grain especially if shipped in an unripe condition, give off a great deal of moisture, causing sweat damage to the grain as well as to other goods in the vicinity—and may cause the seeds to germinate and deteriorate—also the possibility of loss of weight.
3. Fine seeds—heat and germinate if stowed in a badly ventilated space or where moisture gets at them.
4. Jute and certain fibres—give off a great deal of moisture, resulting in sweat damage, rotting and loss of weight.
5. Seed Cake.
6. Hay, wool, pepper, cocoa—if wet or damp, very liable to spontaneous combustion, sweat damage and deterioration.
7. Gambier, jelatong, copra—give off a great deal of heat and moisture thus affecting other commodities in the vicinity—also loss of weight.
8. Nuts, beans—peculiarly liable to heat, sweat and deterioration.

Commodities which are subject to damage due to heating (see above) should be stowed away from local sources of heat, i.e. away from engine and boiler room bulkheads and casings, and apart from wet goods to avoid increasing the evaporation of the liquid contents of the latter. If the cargo is in containers, stowage on board and ashore should be protected from direct sunlight. N.B. This latter requirement for containers may also include inland transport, e.g. lorry parks, etc.

Where vegetable oils, molasses, etc., which require to be liquefied by steam heat to facilitate discharge by pumping are carried, care must be exercised to avoid stowing cargo liable to melt or to damage by heat near the deep tank bulkheads. Temperature increase to pumping viscosity must be gradual to avoid burning.

Ventilation may be required to reduce the effects of heat build up (see Part 2, Techniques, Ventilation).

Whilst stowage pressure may aggravate the problem of heating, such commodities should be carried in 'tween decks. Certain kinds of nut cargoes are so given to heating that they should be stowed in bins and turned over from time to time during the voyage.

Damp or wet packages of all the above types of cargo should be carefully examined, and if in doubt the commodities rejected—otherwise the chance of poor out-turn will be increased.

CARGO MIXTURES

Claims of this class are mostly in respect of bulk grain and seed cargoes, but heavy claims may be paid for other mixtures such as china clay and silver sand, seeds with jaggery, oil with ore, charcoal with sugar, broken and unbroken coke, pulp with fibres, plastic granules with rice, etc. Such mixtures may also occur in spite of packaging, e.g. plastic granules infiltrating bags of rice.

To avoid damage and loss of this kind, careful regard should be given to:

1. Overstowing: goods should be selected for stowing on or over others with a view to eliminating or minimising the risk of such mixtures, avoiding where possible, such stowage as bagged seeds over jaggery, powdery goods over sugar or seeds, oils over ores, plastic granules over bagged rice, etc.
2. Separations: when different lots of bulk grain, etc., or bagged goods are carried, the contents of which are liable to mix as a result of torn packages or of sifting, they should be separated in such a manner as to preclude mixture and facilitate the collection of "sweepings" from the top cargo before disturbing the stow below.
3. Separations within holds should be at charterer's risk and expense and stipulated within the charter party.

When receiving considerable quantities of sugar it is often necessary to separate one parcel from another.

Separation of different grades or types of grain in bulk may be best achieved by trimming the cargo and covering with plastic sheets. These sheets should be anchored by pieces of timber at the shipside ullages taken from top of/or underside of hatch coaming should be taken and noted at loadport. These measurements should be applied by ship's staff or stevedores at discharge port. This will avoid the separation being broken through by grab or the plastic being sucked into vacuvators. The last foot or so should be hand shovelled until the separation is clear.

The mixing of different types of cargo, e.g. unitised cargo and break bulk cargo, may result in mechanical damage to the latter, as fork lift trucks and other mechanical handling equipment operate to load and discharge the units (see Part 2, Techniques, Unitised Cargo).

The mixture of general cargo, to reduce broken stowage, and the mixture of consignment of bills of lading may also lead to claims for lost or overcarried cargo. Refrigerated cargoes frequently require separation. Nets or coloured tapes are usual for carcass and quarter beef; colour wash for cartons and boxes. Correct separation must be carried out as described in Part 2, Techniques, Refrigerated Cargoes.

DUST AND STAIN

When loading dusty commodities, other cargo in the vicinity should be well covered up, but, circumstances permitting, the safest way is to take in the dusty cargo first, then sweep down and make adjacent areas good for subsequent cargo. The same precaution should be observed during discharge.

Dusty goods, or those from which sifted contents may occur (e.g. plumbago), should not be stowed over goods which are susceptible to damage by dust, or on commodities whose packaging would be stained by the dust without first laying an efficient separation between them (see also Part 2, Techniques, Obnoxious Cargoes).

Packages forming the bottom tier of a stow in the lower hold may get into direct contact with oil (where tank top sheathing is not laid) if there is leakage from the double bottoms. The tank top sheathing itself may become oil stained by capillary action. Any such leakage will not only cause staining, but will also provide a source of taint—and even inflammable fumes—with particular types of oil. Fumes and vapours from diesel oil are the most likely to cause taint.

Wooden bulkheads may need to be constructed on occasions to separate cargoes, and where necessary (e.g. in the case of coal) the bulkheads must be rendered dust-tight. Wooden bulkheads which are constructed by laying the boards athwart provided they are well supported by stout upright shores, placed at each side, in pairs, at reasonable distances apart, are much more easily kept dust-tight than with the boards upright—bulkheads should be made of wide planks in order to reduce the number of seams.

In the case of vessels engaged in the bulk grain trade, better results would be obtained by placing the bulkhead on the cargo side of beams and frames, with the pressure of cargo keeping the planking bearing hard on the beam. They should be well supported to prevent bulging under pressure, the seams especially at the sides, caulked and the hold covered on the cargo side with thick paper generously overlapping, held in position by laths placed fairly closely. The seam at the sides should be protected by pasting paper over the same. The use of mats, separation cloths, or old torn tarpaulins for this purpose is not attended with good results.

Limber boards—Considerable damage may occur if these are not maintained in a grain tight condition. It is important to prevent grain, sugar, dust and other debris from percolating into the bilges with resultant detrimental effect on bilge suctions.

RUST DAMAGE

Due mainly to moisture, rain, fresh or salt water, and sweat (and the presence of oxygen) rust is a corrosion producing red discolouration and, in certain circumstances, heavy pitting.

The moisture causing the rust may be introduced by: leakage, other cargo, packaging, green timber (dunnage), rain (when ashore, or when the hatches are open), or even the ventilation itself.

Processed steel may suffer irreparable damage if rust is permitted to gain a hold. Canned goods, spotted with rust, or worse, may lose much of their value to the consumer. Goods liable to damage by rust should not be stowed in the same compartment or container with cargo, packaging or dunnage which is liable to give off moisture. Ventilation may be a requirement, but as mentioned above, it could on occasion aggravate the situation (see Part 2, Techniques, Ventilation).

WET DAMAGE

Condensation, sometimes known as sweat, may be described as tiny beads of moisture settling on a relatively cool surface when the air surrounding that surface can no longer support all the water vapour that it carries.

Moisture thus formed can cause:

rust;
discolouration;
mould;
caking and clogging;
dislodging of labels;
collapse of packages and parts of the stow depending on the commodity, its packaging, the time factor, and other variable conditions.

There are two principle types of condensation affecting cargoes, which may be described as:

cargo sweat;
ship (or container) sweat.
Each is caused by a separate set of circumstances, and each affects the cargo in a different way.
For condensation to appear at all, two conditions must be satisfied:

there must be a source of moisture;
there must be a temperature gradient.

The source of moisture may be:
the cargo itself (or other cargo in the same compartment);
the packaging surrounding the cargo;
the dunnage restraining the cargo;

the pallets or skids;
the timber sheathing, spar ceilings, etc. (or container walls and floor);
the air trapped in at time of loading (or stuffing);
any air introduced in the course of ventilation, opening hatches (or container doors).

The temperature gradient may be caused by any fairly sudden change in the ambient temperature, sea water temperature, or change of cargo temperature—for instance self heating meals.

DISCHARGING FIRE DAMAGED GOODS

Such goods should be discharged under survey, and to facilitate the drawing up of fire damaged reports and average statements, they should be divided into separate lots:

(a) goods damaged by fire and/or smoke;
(b) goods damaged by water or steam used in extinguishing the fire;
(c) goods damaged, broken or otherwise injured in obtaining access to the seat of fire.

PILFERAGE—BROACHING CARGO

In order to protect the ship from blame and responsibility for pilferage occurring on shore the greatest vigilance by the Ship's Officers is necessary. Incoming cargo should be carefully scrutinised—which is best done on the dock or wharf rather than on board—and every unsound or suspicious package rejected until its contents have been ascertained. When discharging, clean receipts for all packages delivered should be demanded except, of course, for those actually found to be short of contents.

To guard against pilferage and broaching on board ship, the Ship's Officers should organise a close watch on the holds and other cargo liable to be broached. When many holds containing broachable cargo are being worked, responsible ship's personnel (and where necessary special shore watchmen) should be employed in watching cargo. Frequent and regular visits to the holds and areas where broachable cargo is being worked by the officer on deck is also a good deterrent.

Where possible broachable cargo should be confined to as few holds as possible. Such cargo should be stowed near the hatch square where it can be more easily seen from on deck while the hatches are being worked. Where hatches with broachable cargo have to be worked at night, it is imperative that good lighting be provided in the holds, on deck, and over side—particularly if working to lighters.

Broachable goods should be overstowed with other suitable cargo, as soon as possible, and the hold watchman should not be withdrawn until that has been done to such an extent as definitely to preclude access to the broachable goods. Hold watchman should be relieved for meals so as not to leave the hold unwatched during meal hours.

Circumstances permitting, all labour should be compelled to leave cargo compartments during meal hours, which is also an additional precaution against fire.

Ventilators, etc., giving access to compartments containing broachable goods should be fitted with iron bars, suitably placed. And reliable locking bars should be fitted to all hatches giving access (however indirect) to broachable goods when covered for the night.

Summarising:

Accurate and reliable tallying should be carried out.
Sweepings and residue should be removed by responsible personnel.
Attractive and high value goods should be stowed in appropriate secure lock-ups, or, at second best, buried in the stow.
Adequate lighting should be provided.
Proper instructions for the safe handling and stowage of the goods should be clearly marked on the packages.
Contents identification and advertising should be avoided.
Proper tamper-proof seals should be used where appropriate.
Correct documentation must be provided and used.

Standards of packaging must be adequate for the commodity and for the necessary handling and stowage.

Strapping and securing must be sufficiently strong.

RATS AND MICE

A rat consumes approximately its own weight in food per week, but the mischief, unfortunately does not end there. Edible goods which have been attacked by rats often have to be condemned as unfit for the purpose for which they were intended. Further, the cost of re-bagging, replacing rat torn bags by new, the loss of bag contents and damage to textile or leather goods, all caused by rats in their search for nest-making materials may amount to heavy sums at times, while the risk of fire, which their propensity for carrying oily rags into remote and ill ventilated corners engenders, is very considerable.

The rat is a carrier of many types of virulent diseases, consideration of which has impelled most governments to spend large sums of money in protecting their ports from invasion by these pests and to enforce strict regulations on vessels frequenting their ports.

Special efforts by Ships Officers are necessary to combat the destructive activities of these pests—they breed so rapidly and prolifically; the quays, warehouses, wharfs, lighters, etc., alongside which the vessels are berthed, provide an apparently inexhaustible reserve from which the ship rat army draws its reinforcements, that any relaxation of effort towards keeping their numbers down will quickly be reflected in increased claims for rat damage.

In the interests of health as well as the preservation of cargo from rat damage the fumigation of holds, peaks and accommodation generally is at times essential. Except when plague is suspected machinery and boiler rooms are not included.

It has to be borne in mind however, that whilst "de-ratisation" by fumigation, or otherwise, may destroy all rats when on board its efficacy is only temporary and that, in the absence of proper safeguards, the ship may quickly become rat infested again. Neither does thorough fumigation eliminate rat harbourage on the ship or their means of livelihood.

SMALLER VERMIN

Certain commodities are liable to introduce vermin in the ship, the presence or ravages of which may prove costly.

Bales of rags, unless thoroughly fumigated, often harbour lice and other insects which, in turn are carriers of certain diseases. This class of cargo should not be received on board unless accompanied by a reliable sanitary certificate, which in most cases requires to be endorsed by the consul of the country of destination.

Certain tropical woods—more especially the kind used for dunnage—are apt to harbour the white ant—a most destructive insect capable of causing serious damage to certain kinds of goods, wood fittings, etc.

Other woods, mostly the hollow bamboo variety and old cord wood, are apt to introduce different species of cockroaches.

Timber dunnage from temperate regions may harbour eggs or larvae, so that some countries (particularly Australasia), have developed strict safeguards and regulations to prevent the import of these insects. The Sirex Woodwasp, in particular, if allowed to become established, is capable of decimating whole sections of forest in those countries where such insects have no natural controlling predators.

Maggots, bred in imperfectly cleaned animal hoofs, horns, bones, skins, etc., if permitted to mix with fine goods give rise to claims and should be safeguarded against.

Certain cereals and their flour, pulses, seeds, copra (kiln and sun dried), rice bran, meal and dust are on occasions found to be insect infested—cereals and flour by weevils, copra by copra bugs.

Dried blood shipped from Argentina, Australasia, etc., is also very prone to infestation by small beetles.

CLEANING AND FUMIGATION

Compartments and containers which have been emptied of cargo, must be made suitable for the reception of the next load of cargo by being thoroughly cleaned (see also Part 2, Techniques, Containers), the dirt and dust collected. Where evidence of infestation exists the compartment or even the hold may have to be fumigated. This would normally be carried out by specialist companies who are fully versed in the safety procedures for such an operation. There are however certain fumigants which are safe to be used by inexperienced operators which may be suitable for fumigating a locker or 'tween deck space. These are usually of the kind that emit smoke, and therefore when the operation is being carried out the ship's smoke detector system should be turned off.

Some cargoes, e.g. bulk grain, may come aboard with remains of the previous fumigation still within the cargo. This may be in the form of small white pellets, and if seen should on no account be approached or touched until the risk has been ascertained. Such pellets could be producing phosphene gas by reaction between the pellet and the moisture in the atmosphere, which ultimately causes the pellet to collapse and become safe.

REGULATIONS

Most countries have very strict quarantine regulations, to prevent the import of unwanted vermin and insects, etc. These regulations usually prohibit the importation of timber with any bark still attached (as is often the case in dunnage), and timber entering the country has to be specially treated.

MECHANICAL DAMAGE

Lowering heavy slings or drafts of cargo too fast on to cargo already in stowage may be responsible for damage, which often goes undetected until discharge. Similarly, forcefully dragging cargo out that is wedged by other cargo or even overstowed, may be another source of damage at the time of discharge.

The use of cargo hooks may be indispensable in the handling of a large variety of break bulk commodities, but with bagged cargo, fine baled goods, hides, furs, rolls of paper and matting, light packages, liquid containers, crates and the like, packages whose contents are exposed or unprotected, the use of cargo hooks may be productive of much mischief and claims, and should be strictly prohibited.

Crow and pinch bars may also be indispensable to the sound stowage or breaking out of many classes of heavy packages, but their use should never be permitted when stowing barrels, other liquid containers, or with any other packages which are not substantial enough to withstand damage from their use. Their use may be reduced or even obviated by the introduction of other techniques, e.g. pre-slinging of cargo.

While special lifting and handling gear may be used (and best suited) for certain types of cargo, the improper use of such equipment may damage the cargo or its packaging. Net slings are most useful with many kinds of small packages but if used with bagged stuff, light cases, etc., a great deal of damage may result. Similarly chain slings are indispensable for certain types of packages and useful for most classes of iron goods but the use of such with light cases, sheet iron, coils of copper piping, sawn logs of valuable timber and other goods liable to buckling, fraying or marking by chain may be productive of damage or claims.

Canvas or man-made fibre slings should be used for slinging bagged flour, coffee and light cargo, while the use of trays for certain classes of goods is much to be preferred to slinging by net or rope.

Too much weight in the draft of a sling endangers the safety of packages situated at the outside edge of bottom and top tiers, into which the sling is liable to be drawn by weight below and compression above.

A draft composed of many packages should taper off on top to prevent springing or crushing the outside upper packages by compression of the sling. Light or fragile packages should not be slung along with heavy packages.

SWEEPINGS

This refers to that part of the original contents of bag or other packages of cargo such as grain, coffee, cocoa, sugar, rice, nuts, etc., that have been lost from their proper packaging. This may be as a result of the bags getting torn in handling; chafed or rotted in stowage; being defectively sewn at the mouth or sides; damaged by vermin, or because of the use of second-hand bags. The condition of second-hand bags often requires to be endorsed in the Mate's Receipt and on the Bill of Lading. A certain amount of sweeping is unavoidable when a large quantity of bagged cargo is carried, but the spillage of such is often unnecessarily large due to rough handling.

Recognising that sweepings will, to some extent, be present on discharge, separation material should, if available, be laid under the cargoes to ensure that the sweepings be clean when collected.

Sweepings are the property of, and, are to be delivered to, the consignees of the cargo of which they form part.

When various consignments of the same commodity are stowed in the same compartment, sweepings from all parcels will be found mixed together, in which case the whole of the sweepings are divided among the consignees pro rata to their respective consignments and shortages of same.

Torn or slack bags or damaged packages with their contents (and even empty bags) are sometimes added to the sweeping pile. This should never be permitted; such packages with their contents preserved by sewing up, etc., should be kept entirely apart from sweepings and delivered as packages, not as sweepings.

Sweepings are not accepted to make up any shortage of bags, etc. The number of bags specifically receipted for must be delivered whole or torn, however badly.

In cases where the cargo is discharged on to a dock or wharf, or is taken into store before delivery, the sweepings which accumulate after the cargo has left the ship's tackle are usually left with ship's sweepings and all apportioned out as above as ship's sweepings—a ship thus getting the blame for loss of bagged contents occurring after delivery.

To avoid that, the ship's sweepings should be bagged, suitably tagged, weighed and delivered as such, and receipt demanded. This, if done, will materially assist in reducing claims, and incidentally, ensure better and more careful handling and more care of the cargo after it has passed beyond the control of the Ship's Officers.

Further to reduce the amount of sweepings, bag sewers should be in constant attendance in the holds whilst loading and discharging are in progress.

PART 5

PROCEDURES

RECEIVING

Goods of packages which are received in unsound condition clearly cannot be delivered in a sound condition. It is essential, therefore, that during loading, a careful watch be continually maintained for any packages which may have been tampered with or which are improperly or inadequately protected, broken, leaky, damaged, repaired, spilled, torn or stained. Such packages should be rejected unless reconditioning is an option. (Care should of course be taken to see that any packages for which Mate's receipts — see below — have been issued but for some reason have to be returned ashore are delivered back to the ship).

MATE'S RECEIPTS

The mate's receipt is a receipt given for goods actually received on board — it is given up to the agent or broker authorised to issue the Bill of Lading.

Such a receipt should therefore be carefully drawn up and any remark as to number or condition of goods, or marks thereon, be incorporated into the Bill of Lading. Receipts should be issued on ship's forms and numbered, receipt books being in triplicate — one copy for the shipper/forwarder, one for the agent and one left in the book.

Disputes should be investigated immediately (while a recount may still be possible). When this is not done, and other means of arriving at a satisfactory solution is not available, the number in dispute should be clearly and unambiguously stated in words on the receipt — for example:

'Received on board 17 packages; three more in dispute' (and not 'Received 20 packages, three in dispute').

Should a shipper or charterer demand that Mate's receipts be issued for cargo alongside but not actually on board, they should be endorsed 'at shipper's risk until actually shipped'. When in doubt as to weight, quality, quantity or condition, Mate's receipts should be claused 'weight, quality, quantity or condition unknown'.

Some charters or shippers declare that nothing but a clean receipt will be accepted; charter parties also often contain a clause to the effect that the Bill of Lading must be signed as presented. It should be borne in mind that neither law nor custom exists which either justifies or compels an officer or Master to issue a receipt for cargo said to have been shipped which careful tallies do not show to have been so shipped, or to certify either quality or condition where they are not able to do so.

BILLS OF LADING

Bills of Lading are internationally respected documents on which Banks and other institutions will rely when advancing large sums of money. If a Master or agent signs a bill of lading knowing or suspecting that its description of the cargo is wrong, or if he deliberately inserts a false date onto a bill, his conduct will probably amount to deceit and this will render worthless any guarantees or letters of indemnity that may have been tendered by the shipper, and expose the ship to liability for any loss that has been suffered.

The Master should be aware of the condition of any cargo loaded to the extent that he can reasonably be expected to have inspected it. If it is impossible for him to properly inspect the cargo, the Bill of Lading should be qualified accordingly. Additionally, the Master will often be under a duty to state the quantity or weight or number of pieces of cargo shipped, and once he has signed bills of lading to that effect it will be very difficult to claim that a different quantity was shipped.

The intrinsic value represented by a bill of lading tends to give rise to many disputes, and accordingly, if there is any doubt about how a bill should be claused, advice of Owners or P&I agents should be sought.

TALLYING

Where tallymen are employed jointly by shipper and shipowner the tally-clerk's receipt may take the place of the Mate's receipt. Spot checks by ship's staff are therefore advisable.

Tally cards should be preserved and forwarded to the shipowner at the end of the voyage. The ship's tally (both when loading and discharging) is indispensable if claims are to be avoided. A careless tally may however be worse than useless. They should be compared and agreed at the termination of each shift. Disputes should be promptly investigated and if a recount is possible this should never be left over until work ceases when it may be too late.

The ship's tally should be taken on board and never on lighters or the loading area — the ship's responsibility frequently commences and ceases when cargo passes or leaves the ship's rail. During discharge, valuable cargo should be tallied at the rail. This should help relieve the ship of liability for any shortages occurring after it has passed overside.

A truly efficient tally is impossible when an officer or tally clerk is tending to the intake or output of more than one hatch; it is unreasonable to expect that an officer can correctly tally and be relied upon to supervise stowage, etc., at the same time.

Tally books or sheets should be delivered to and retained by the Chief Tally Clerk whenever work ceases.

DELIVERING CARGO

Should there be reasonable grounds for anticipating damage to cargo before opening the hatches, protest should be noted. This protest should be noted as soon as possible and not later than 24 hours after arrival; the protest can be extended and can be made without waiting to sight the damaged cargo, continuing the extension of protest as the extent of damage is revealed. A surveyor's attendance on behalf of ship, whilst discharging is in progress, it is always beneficial. Every reasonable facility should be extended to the surveyors attending on behalf of consignees, but this does not mean that consignees or their representatives have a right to full access to the ship or to examination of documents such as log books.

Where a surveyor is not in constant attendance, a survey should be called at once where damaged cargo is found. In the case of damage by moisture or water, or leakage from casks etc., dunnage and matting should not be disturbed until they have been sighted and positioned by the surveyors.

The following points may be useful:

(i) When damaged cargo is sighted in the stow, a sound practice is to make a sketch or take photographic evidence of the position of the cargo in relation to other cargo in the compartment.

(ii) All packages found broken during discharge should be laid aside and segregated. It may be possible to securely lock such cargo away for further investigation or reconditioning.

(iii) Care should be taken to replace back into position any cases bearing marks and numbers that may have become displaced. Where this is not possible every effort should be made to ascertain the correct marks and numbers which then can be clearly painted on to other packages to avoid confusion which may arise through lack of identification.

(iv) It is worth gathering up and including with other discharged cargo all torn, slack or empty bags or packages. Where Bills of Lading state a specific number of bags or packages, it is best at least to show delivery of that number so far as the instigation of a claim may be concerned.

(v) Cargo should never be delivered except on production of the original Bill of Lading properly stamped and endorsed and it should be exchanged for the Master's own signed copy. When the cargo is consigned to order, the Bill of Lading should bear the shipper's endorsement and also that of the merchant to whom it has been transferred. Cargo should never be delivered against invoices, letters of guarantee or indemnity or whatever, and if in doubt, advice of Owners or the P&I correspondents should be sought.

PORT MARKING

Marking of cargo is of course a function performed ashore and prior to loading but it is worth remembering that the marking of cargo with the port of destination and other relevant information is a valuable aid to good stowage and efficient out-turn.

STOWAGE PLANS

Plans showing the disposition of all cargo loaded should always be prepared. The value of full and accurate plans cannot be over emphasised. Inaccurate planning may cause delay and costly, unproductive employment of stevedores. During the passage cargo plans should be carefully checked — any error likely to affect the intended discharge pattern can always be radioed ahead.

The plan should be large and, whilst not necessarily to scale, should be sufficiently approximate so as to indicate the comparative volume of a stow in any compartment. This may be the only means of conveying to agents and others at discharging ports the proportion and quantity of cargo in any hold destined for that port. All pillars, beams, lockers and positions of doors should be shown and it is good practice to indicate the number and type of derricks/cranes of each hatch and the length and breadth of each unless the ship is well known to the agent or stevedore. The position of cargo should be shown accurately in relation to these features. Suitable details should be advised: whether cargo is on pallets, whether stowed by forklift trucks, whether or not pre-slung etc. Where mechanical handling equipment has been used the exact flow of traffic used to achieve the stow may be useful to facilitate discharge. Notes on how any particularly awkward cargoes have been loaded can be of considerable assistance during discharge.

With a highly mixed general cargo it will not of course be possible to show the marks of various packages on the plan, but in the case of substantial consignments the marks, quantities and position should be shown. Where practical, identification of individual Bills of Lading in the stow can be extremely useful. Fuller details, particularly of tonnages, should be given in relation to overstowed cargo, always indicating whether such tonnages are weight or measurement.

A copy of the plan should, where possible, reach the discharging ports ahead of the vessel.

APPENDIX NO. 1

THE TON AND TONNAGE

Originally the "tun", a wine container or cask, the capacity of which by an Act of 1423 in England, was not to be less than 252 gallons.

According to Sir Wm. White, the term "tonnage" probably came into use in connection with the levying of dues on wine carrying ships, the dues being based on the number of tuns of wine which could be accommodated, this according a convenient, if not very accurate, method of comparing the sizes of different vessels.

It is said that the weight of the tun of wine of 252 gallons was approximately 2,240 lb, the gallon measure, then in use, differing somewhat from the imperial gallon; so that the tun, while originally a measure of internal capacity, became also one of deadweight capacity and a weight unit; the term "poundage" which represented ship dues levied on the deadweight capacity thus becoming practically synonymous with "tonnage", the latter surviving in use to these days.

The ton as a unit of measurement is variously used, the following being the principal examples:

WEIGHT TONS

English or long ton avoirdupois = 2240 lb or 20 cwt or 112 lb
= 1.12 short or American tons
= 1016.05 kilos or 1.01605 metric tons

American or short ton = 2,000 lb
= 0.892857 long tons
= 0.90718 metric tons

Metric or tonne = 1,000 kilos or 2204.621 lb
= 0.98421 long tons
= 1.10231 American or short tons

Tonnage Conversion Factors

	Long ton	*Short ton*	*Metric tonne*
1 long ton	× 1	× 1.12	× 1.016
1 short ton	× 0.893	× 1	× 0.907
1 metric tonne	× 0.984	× 1.102	× 1

MEASUREMENT TONS

Shipping Ton—equals 40 cu. ft, i.e. 31.16 Imperial and 32.14 U.S.A. bushels, 1.1327 m^3.

Cargo measuring 1 cu. metre per tonne (1,000 kg) or 40 cu. ft—per long ton (2,240 lb) and upwards per ton is termed "measurement cargo"; that which is under 1 cu. metre or 40 cu. ft is termed "deadweight cargo".

Freight ton conversion factors—

	Freight ton @ 40 cu. ft	*Freight tonne @ 1 cu. metre*
1 freight ton @ 40 cu. ft	× 1	× 1.333
1 freight tonne @ 1 cu. metre	× 0.883	× 1

Displacement—is the measure of the volume of water displaced or dislodged by a floating object, and may be measured either in cubic feet, or as is most customary, in tons; 35 cu. ft of saltwater being equal to or weighing one ton of 2,240 lb.

Tons Net Displacement—is the weight of vessel fully equipped including water in main boilers, but without cargo, fuel, stores, other water, crew, etc.

Tons Gross Displacement—is the corresponding weight of the vessel when fully laden, including cargo, fuel, stores, water, crew, etc., as above.

Tons Deadweight or Burthen Cargo Tonnage—is the weight a vessel can carry when fully laden, i.e. the combined weight of cargo, stores, water (including that in donkey boiler if any), crew, etc., and is the difference between net and gross displacements.

Gross Register Tonnage—is the total of underdeck tonnage; 'tween deck tonnage (if any) and tonnage of closed in spaces above upper deck, less certain exemptions.

Net Register Tonnage—is designed to represent the earning capacity of vessel for cargo and passengers, being the gross tonnage less the non-earning spaces occupied by: machinery, permanent bunkers, water ballast (except deep tanks), Master and crew spaces.

Most dock dues and port charges are levied on this tonnage; some, however, being based on gross tonnage.

Suez and Panama Tonnages—are arrived at along the same lines as the foregoing, but each of them is, as a rule, larger than the British tonnage because of the inclusion of space which, under British rule, is exempted.

APPENDIX NO. 2

THE STOWAGE FACTOR

The stowage factor of any commodity is the figure which expresses the number of cubic metres per tonne (or cubic feet per ton of 2,240 lb), will occupy in stowage—not the actual cubic measurement of a ton—and should include a proper allowance for broken stowage and dunnage which, as in the case of barrels or goods of irregular form and size, enter largely into the composition of the stowage factor and similarly cargoes carried on pallets.

Stowage factors necessarily are based on the assumption that the stowage will be such as to ensure proper regard being given to the second essential of "good stowage", i.e. the economy of cargo space.

From this it follows that the ordinary stowage factor of certain commodities, which are carried under both ordinary and refrigerated conditions will have to be increased substantially in order that it may serve as a useful guide in dealing with refrigerated goods when so much space is reserved for air circulation, battens, etc.

The most carefully determined stowage factor is not absolute—at best it can only serve as a guide—but a useful one—inasmuch that the ratio of broken stowage varies according to whether the compartment is an end or a body compartment—wide or narrow, deep or shallow. It also varies for the same commodity for different countries and ports according to the methods of packing, the degree of density to which the goods are pressed, whether the bags are full and well rounded, or slack, in which case they 'fill solid". It also varies according to the extent to which the goods have been seasoned or ripened as well as the quality of the crop, and whether the loading has proceeded at a normal rate, or the cargo "rushed in". In the latter case, any figure is apt to be misleading.

Notwithstanding the foregoing, a knowledge of the stowage factor, intelligently applied, is very useful to the Ship's Officer in arranging his stowage, and equally so to the freight manager in estimating and checking freight earnings, especially with low measurement goods.

Another method used in estimating what measurement cargo a vessel can stow, is that of applying an allowance for "broken stowage" to the actual cubic measurement of a ton—weight or scale as the case may be.

Liquids of course will fill the tank into which they are put. For this reason the number of tonnes per cubic metre (specific gravity) for the temperatures of the liquid is usually considered in preference to the stowage factor.

Grain and Bale Measurement

These measurements are to be found on ships capacity plans and are given in cubic feet or metres.

Bale measurement is the space in the compartment or hold measured from the inside of the frames or spar ceiling. Normally used when calculating box or bale cargoes etc.

Grain measurement is taken to the ships side (i.e. includes space between frames) and is used in calculating capacity for bulk cargoes such as grain.

APPENDIX NO. 3

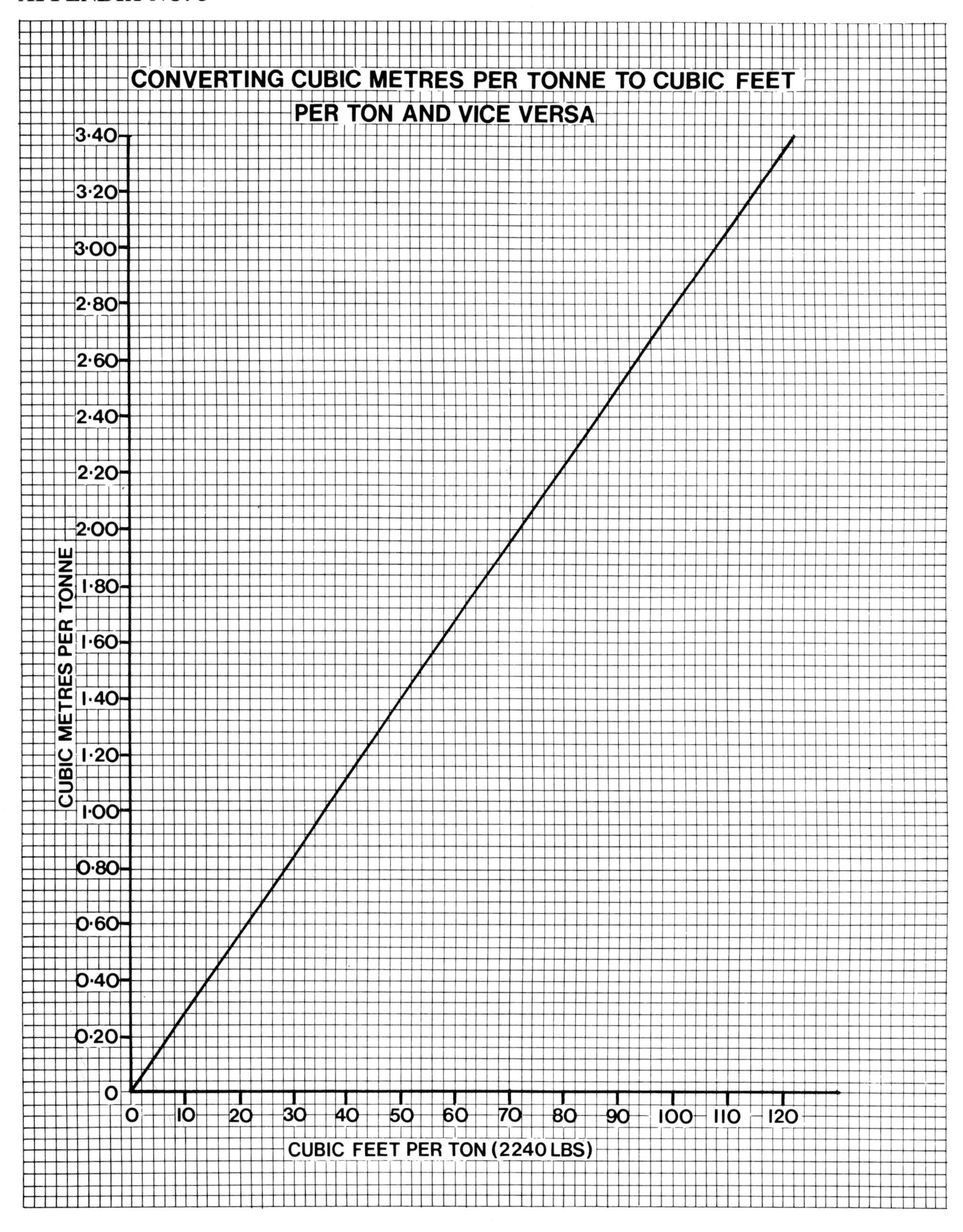

APPENDIX NO. 4

CONVERSION OF LONG TONS TO METRIC TONNES

(1 metric tonne = 2,204.622 lb; 1 long ton = 2,240 lb)

To obtain metric tonnes add the number in the "*Add*" column to the appropriate number within the range shown in the "*Long tons*" column.

Long tons	*Add*	*Long tons*	*Add*	*Long tons*	*Add*	*Long tons*	*Add*	*Long tons*	*Add*
0–31	—	1216–1277	20	2462–2523	40	3708–3770	60	4955–5016	80
32–93	1	1278–1339	21	2524–2586	41	3771–3832	61	5017–5078	81
94–155	2	1340–1402	22	2587–2648	42	3833–3894	62	5079–5141	82
156–218	3	1403–1464	23	2649–2710	43	3895–3957	63	5142–5203	83
219–280	4	1465–1526	24	2711–2773	44	3958–4019	64	5204–5265	84
281–342	5	1527–1589	25	2774–2835	45	4020–4081	65	5266–5328	85
343–405	6	1590–1651	26	2836–2897	46	4082–4144	66	5329–5390	86
406–467	7	1652–1713	27	2898–2960	47	4145–4206	67	5391–5452	87
468–529	8	1714–1776	28	2961–3022	48	4207–4268	68	5453–5514	88
530–592	9	1777–1838	29	3023–3084	49	4269–4330	69	5515–5577	89
593–654	10	1839–1900	30	3085–3146	50	4331–4393	70	5578–5639	90
655–716	11	1901–1962	31	3147–3209	51	4394–4455	71	5640–5701	91
717–778	12	1963–2025	32	3210–3271	52	4456–4517	72	5702–5764	92
779–841	13	2026–2087	33	3272–3333	53	4518–4580	73	5765–5826	93
842–903	14	2088–2149	34	3334–3396	54	4581–4642	74	5827–5888	94
904–965	15	2150–2212	35	3397–3458	55	4643–4704	75	5889–5951	95
966–1028	16	2213–2274	36	3459–3520	56	4705–4767	76	5952–6013	96
1029–1090	17	2275–2336	37	3521–3583	57	4768–4829	77	6014–6075	97
1091–1152	18	2337–2399	38	3584–3645	58	4830–4891	78	6076–6138	98
1153–1215	19	2400–2461	39	3646–3707	59	4892–4954	79	6139–6200	99
								6201–6262	100

APPENDIX NO. 5

TEMPERATURE CONVERSIONS

FAHRENHEIT TO CENTIGRADE; CENTIGRADE TO FAHRENHEIT

Using the centre figure in each column the conversion to Centigrade or Farenheit may be read off as appropriate.

°C		°F	°C		°F	°C		°F	°C		°F
−34.4	−30	−22.0	−15.0	5	41.0	4.4	40	104.0	23.9	75	167.0
−33.9	−29	−20.2	−14.4	6	42.8	5.0	41	105.8	24.4	76	168.8
−33.3	−28	−18.4	−13.9	7	44.6	5.6	42	107.7	25.0	77	170.6
−32.8	−27	−16.6	−13.3	8	46.4	6.1	43	109.4	25.6	78	172.4
−32.2	−26	−14.8	−12.8	9	48.2	6.7	44	111.2	26.1	79	174.2
−31.7	−25	−13.0	−12.2	10	50.0	7.2	45	113.0	26.7	80	176.0
−31.1	−24	−11.2	−11.7	11	51.8	7.8	46	114.8	27.2	81	177.8
−30.6	−23	−9.4	−11.1	12	53.6	8.3	47	116.6	27.8	82	179.6
−30.0	−22	−7.6	−10.6	13	55.4	8.9	48	118.4	28.3	83	181.4
−29.4	−21	−5.8	−10.0	14	57.2	9.4	49	120.2	28.9	84	183.2
−28.9	−20	−4.0	−9.4	15	59.0	10.0	50	122.0	29.4	85	185.0
−28.3	−19	−2.2	−8.9	16	60.3	10.6	51	123.8	30.0	86	186.8
−27.8	−18	−0.4	−8.3	17	62.6	11.1	52	125.6	30.6	87	188.6
−27.2	−17	1.4	−7.8	18	64.4	11.7	53	127.4	31.1	88	190.4
−26.7	−16	3.2	−7.2	19	66.2	12.2	54	129.2	31.7	89	192.2
−26.1	−15	5.0	−6.7	20	68.0	12.8	55	131.0	32.2	90	194.0
−25.6	−14	6.8	−6.1	21	69.8	13.3	56	132.8	32.8	91	195.8
−25.0	−13	8.0	−5.6	22	71.6	13.9	57	134.6	33.3	92	197.6
−24.4	−12	10.4	−5.0	23	73.4	14.4	58	136.4	33.9	93	199.4
−23.9	−11	12.2	−4.4	24	75.2	15.0	59	138.2	34.4	94	201.2
−23.3	−10	14.0	−3.9	25	77.0	15.6	60	140.0	35.0	95	203.0
−22.8	−9	15.8	−3.3	26	78.8	16.1	61	141.8	35.6	96	204.6
−22.2	−8	17.6	−2.8	27	80.6	16.7	62	143.6	36.1	97	206.6
−21.7	−7	19.4	−2.2	28	82.4	17.2	63	145.4	36.7	98	208.4
−21.1	−6	21.2	−1.7	29	84.2	17.8	64	147.2	37.2	99	210.2
−20.6	−5	23.0	−1.1	30	86.0	18.3	65	149.0	37.8	100	212.0
−20.0	−4	24.3	−0.6	31	87.8	18.9	66	150.9	38.3	101	213.8
−19.4	−3	26.6	0	32	89.6	19.4	67	152.6	38.9	102	215.6
−18.9	−2	28.4	0.6	33	91.4	20.0	68	154.4	39.4	103	217.4
−18.3	−1	30.2	1.1	34	93.2	20.6	69	156.2	40.0	104	219.2
−17.8	0	32.0	1.7	35	95.0	21.1	70	158.0	40.6	105	221.0
−17.2	1	33.8	2.2	36	96.8	21.7	71	159.8	41.1	106	222.8
−16.7	2	35.6	2.8	37	98.6	22.2	72	161.6	41.7	107	224.6
−16.1	3	37.4	3.3	38	100.4	22.8	73	163.4	42.2	108	226.4
−15.6	4	39.2	3.9	39	102.2	23.3	74	165.2	42.8	109	228.2

APPENDIX NO. 6

FEATURES AND CAPACITIES OF CASKS, ETC.

FEATURES OF CASKS

A. Head	The head staves are placed so that they are vertical when the bung is up.
B. Chime	The circle formed by ends of staves. Chimes should always meet fair in stowage.
C. Bilge	The bulge or greatest circumference. This part should never bear the weight of the cask, and so should always be "free".
D. Quarter	Is the strongest part, which should carry the weight of the tier above, etc.
E. Bung	Should always be on top to avoid leakage and ensure that the head staves are up and down.
F. Hoops	The hoop rivets are always in line with the bung.
G. Cantline	The space between the upper part of two casks when placed together side by side.
H. Beds	Short pieces of soft wood, 12/14″ × 2″ × 2″ for placing under the quarters to ensure that the bilge is "free".
I. Quoins	Wedge or other shaped pieces of wood placed each side to keep the barrel from moving on the beds, etc.
J. Longer	A row of barrels in the athwart direction.
X. Tier	A horizontal layer. When stowing bilge and cantline depth of tier equals a little more than diameter of head.

The term "Barrel" nowadays seems to be displacing that of "Cask", this probably arising from the fact that the "Barrel" type of "Cask" is most generally in use.

TABLES OF CAPACITY OF VARIOUS CASKS

According to British Liquid Measure

Firkin	equal to	$\frac{1}{4}$	barrel(s),	9 Imperial gallons
Anker	" "			10 " "
Kilderkin or Rundlet	" "	$\frac{1}{2}$	"	18 " "
Barrel	" "			36 " "
Tierce	" "			42 " "
Hogshead	" "	$1\frac{1}{2}$	"	54 " "
Puncheon	" "	2	"	72 " "
Butt or Pipe	" "	3	"	108 " "
Tun	" "			252 gallons (see Ton)

The capacities of the various classes of casks differ considerably as for different countries or trades, there being no strict uniformity of practice in regard to same.

The U.S.A. standard barrel = 5.62 cu. ft = 42 U.S. gallons
= 35 Imperial gallons
(outside vol. approx. 12 cu. ft)
The Imperial gallon = 277.274 cu. in = 0.164 cu. ft
The U.S.A. gallon = 231.00 cu. in = 0.1337 cu. ft

TO ASCERTAIN CAPACITY OF A CASK

Add area in square inches of interior circle at bung to area in square inches of interior circle at head. Divide by 2 and multiply by interior length in inches, which equals cubic inches capacity, which, divided by 277.274, equals capacity in Imperial gallons, or divided by 231.00 equals capacity in U.S.A. gallons.

TO ASCERTAIN BUOYANCY OF A CASK

Actual buoyancy, in poinds, equals capacity in gallons $\times$ 10, less weight of cask. For a practical operation reduce result by $\frac{1}{10}$.

APPENDIX NO. 7

FLUIDS, WEIGHTS AND CUBIC CAPACITIES

CONVERSION OF DATA, ETC.

Fresh Water (distilled)	Specific gravity is	1.000
	1 cu. ft weighs	1,000 oz = $62\frac{1}{2}$ lb
	1 ton occupies	35.84 cu. ft
	1 ton	= 224 Imperial gallons
		= 269 U.S.A. gallons
		= 1 cu. metre, or kilolitre, nearly.
Salt Water (standard)	Specific gravity is	1.025
	1 cu. ft weighs	1,025 oz = 64 lb
	1 ton occupies	35 cu. ft
	1 ton	= 218.536 Imperial gallons
		= 262.418 U.S.A. gallons

The Imperial Gallon	= 277.28 cu. in
	0.1604 cu. ft
	4.54 litres
	1.20 U.S.A. gallons
	10.00 lb of fresh water
	10.25 lb of sea water
The U.S.A. Gallon	= 231.00 cu. in
	0.1337 cu. ft
	3.785 litres
	0.833 Imperial gallons
	8.328 lb of fresh water
	8.536 lb of sea water

TO ASCERTAIN

Lb fluid per Imperial gallon;	Multiply specific gravity by 10.
Lb fluid per U.S.A. gallon;	Multiply specific gravity by 8.328.
Lb fluid per cu. ft;	Divide specific gravity by 16.

$$\text{Gallons per ton} = \frac{\text{Cu. ft per ton}}{0.16}.$$

$$\text{Cu. ft occupied by 1 ton of fluid} = \frac{\text{2,240 or 2,000 lb}}{\text{Specific gravity} \times 10} \times 0.16045 = \text{cu. ft per ton.}$$

$$\text{Oil capacity of water ballast tank} = \frac{\text{Tons capacity (s.w.)} \times 35}{\text{cu. ft occupied by ton of oil at } 60°\text{F}} - \text{expansion allowance*}$$

To ascertain minimum permissible ullage of ISO tank container:

Gross wt. of cargo and container = 20 tons (for 20 ft tank)
Net wt. of cargo = 20 − tare weight of tank (displayed on tank)

$$\text{Net gallons of cargo} = \frac{\text{Net weight} \times 224}{\text{Specific gravity at } 60°\text{F}}$$

$$\frac{\text{Capacity of tank (gallons)} - \text{net gallons of cargo} \times 100}{\text{Capacity of tank (gallons)}} = \%\ \text{ullage.}$$

The capacities of the various classes of casks differ considerably as for different countries or trades, there being no strict uniformity of practice in regard to same.

The U.S.A. standard barrel = 5.62 cu. ft = 42 U. S. gallons
= 35 Imperial gallons
(outside vol. approx. 12 cu. ft)
The Imperial gallon = 277.274 cu. in = 0.164 cu. ft
The U.S.A. gallon = 231.00 cu. in = 0.1337 cu. ft

* Allow 1% for every 10° rise of temperature above 15.5°C, it is estimated will occur between shipment and discharge.

APPENDIX NO. 8

STANDARD GRAIN, WEIGHT PER BUSHEL AND BUSHELS PER TON

Imperial Bushel = 1.2837 cu. ft = 1.0315 U.S.A. or Struck Winchester Bushel.
8 Bushels = 1 Quarter = 10.27 cu. ft.
Sack = 3 Bushels.
U.S.A. or Struck Winchester Bushel = 1.2445 cu. ft = 0.9694 Imperial Bushel.

Barley ..	48 lb	= 46.66 Bushels per ton
Beans ..	$63\frac{1}{2}$ lb	= 35.28 " " "
Buckwheat ..	50 lb	= 44.8 " " "
Clover ..	63 lb	= 35.5 " " "
Linseed ..	50 lb	= 44.8 " " "
Maize ..	56 lb	= 40.0 " " "
Oats ..	32 lb	= 70.0 " " "
Rye ..	56 lb	= 40.0 " " "
Wheat ..	60 lb	= 37.3 " " "

APPROXIMATE WEIGHT PER BUSHEL OF SEEDS, ETC.

Bran	20 lb	Lentils	60 to 62 lb	Rape	48 to 53 lb
Canary Seed	53 to 61 lb	Millet	56 to 64 lb	Salt	46 lb
Clover	62 to 64 lb	Onions	60 lb	Timothy	45 lb
Cornmeal	50 lb	Onion Seed	36 to 38 lb	Potatoes	54 to 58 lb
Flour (Wheat)	56 lb	Peas	63 to 65 lb		
Hemp	42 to 44 lb	Poppy	47 to 49 lb		

TO CONVERT GRAIN CAPACITY INTO BUSHEL CAPACITY, ETC.

$$\frac{\text{Capacity cu. ft}}{10} \times 8 \text{ bushels, add 1 bushel to every 40} = \text{capacity in bushels.}$$

$$\frac{\text{Bushel capacity}}{8} = \text{capacity in quarters;} \qquad \frac{\text{Bushel capacity}}{8{,}000} = \text{capacity in (U.S.A.) loads.}$$

BRITISH CORN MEASURE AND METRIC EQUIVALENTS

1 Bushel =	8 Gallons	= 3.637 Dekalitres.	1 Dekalitre =	2.20 Gallons
1 Quarter =	8 Bushels	= 2.900 Hectolitres.	1 Hectolitre =	2.75 Bushels
Load =	5 Quarters	= 14.545 Hectolitres		
Last =	10 Quarters	= 29.090 Hectolitres		

N.B.—U.S.A. load = 8,000 bushels (*see* preceding page).

To find number of 2-bushel bags necessary to secure grain cargo in any compartment:

$$\frac{\text{Length of hold}}{3} \times \frac{\text{Breadth of hold}}{2} = \text{bags for } \textit{one} \text{ tier.}$$

FOREIGN MEASURES WHICH MAY BE USED IN CONNECTION WITH GRAIN, ETC., AND THEIR BRITISH EQUIVALENTS

Argentine	Arroba = 25.35 lb
	Fanega = $1\frac{1}{2}$ Imperial bushels
Bolivia	Quintal = 100 libras = 101.44 lb
	Arroba = 25.36 lb
	Arroba of Wine, etc. = 6.7 Imperial gallons
Brazil	Arroba = 32.38 lb
	Quintal = 129.54 lb
Canada	British Imperial weights and measures are in use, with the exception that a cwt = 100 lb and the ton = 2,000 lb as in the United States
Chile	Arroba = 25.36 lb
	Quintal = 101.44 lb; 20 quintals = 1 tonnelada
China	Tael = 1,333 oz = 37.78 grammes
	Catty = 1.333 lb = 604.53 grammes
	Picul = 133.333 lb = 60.453 kilogrammes
Crete	Oke = 2.8 lb
Denmark	Ceutner = 110.23 lb
	Tolnde (Toime) = 3.827 bushels
Egypt	Ardeb = 5.44739 bushels = 43.579 gallons
	Oke = 2.7513 lb
	Cantar = 100 rotts or 36 okes = 0.884 cwt = 99.0492 lb
	Cantar of Alexandria = 112 okes = 2.7514 cwt
	Hernl = 200 okes
	100 ardebs of wheat = $62\frac{1}{2}$ quarters; 100 ardebs of beans = $65\frac{1}{2}$ quarters
Greece	Oke = 2.832 lb; 791 okes = 1 ton
	Cantar = 124.6 lb = 44 okes
	2,128 great venetian lb = 2,240 lb (a currant measure)
Honduras	Fanega = $1\frac{1}{2}$ Imperial bushels
Hong Kong	Tael = $1\frac{1}{3}$ oz
	Catty = $1\frac{1}{3}$ lb
	Picul = $133\frac{1}{3}$ lb
Italy	Tonnellata = 2.200 lb
	Ettolitro = 2.75 Imperial bushels
	288 hectolitres wheat = 100 quarters
Japan	Catty = 1.322 lb
	Picul = 132.27 lb
	Koku = 4.9629 bushels
Mexico	Arroba = 25.357 lb = 25 libras
Indonesia	Catty = 1.36 lb
	Picul = 136 lb
Paraguay	Arroba = 25.35 lb
	Fanega = $1\frac{1}{2}$ Imperial bushels
Russia	Pood = 36.113 lb; 100 poods = 1.6121 tons
	Chetvert = 5.77 Imperial bushels
Archangel	100 chetvert wheat = 70 quarters; 100 chetvert oats = 68 quarters
Odessa	100 chetvert wheat = 72 quarters; 100 chetvert linseed = 83 quarters
Singapore	Picul = $133\frac{1}{3}$ lb = 100 kati
Turkey	Oke = 2.8283 lb
	Almud = 1.151 Imperial gallons
	Kileh = 0.912 Imperial bushel = 0.36 Imperial quarter
	100 kilch = 12.128 Imperial quarters = 35.266 hectolitres

United States Same as British weights and measures, with the following exceptions:

Bushel = 0.9692 Imperial bushel
Hundredweight = 100 lb
Ton = 2,000 lb
Load = 8,000 bushels
Bag (New York) holds 200 lb wheat, stows about 4 ft 6 in–4 ft 9 in

APPENDIX NO. 9

TO CALCULATE BOARD MEASURE (B.M.)

$$\frac{\text{Thickness} \times \text{width (in inches)}}{12} \times \text{length in feet} = \text{board measure feet, or square feet of planking } 1'' \text{ thick.}$$

TO ESTIMATE UNDER-DECK LUMBER CAPACITY

$$\frac{\text{Bale capacity} \times 100}{12} = \text{approx. under-deck capacity in board feet, which divided by 1980}$$

= capacity in Petrograd Standards.

TIMBER WEIGHTS—AVERAGE

Wood	*Lb per C.F.*	*Cubic Meas. of 1 Ton*	*Wood*	*Lb per C.F.*	*Cubic Meas. of 1 Ton*
Ash	48	46.6	Lime	35	64.0
Beech	48	46.6	Mahogany, Honduras	35	64.0
Birch	45	49.8	" Spanish	53	42.2
Box, English	61	36.7	Maple	42	53.3
„ French	83	27.0	Oak, African	62	36.0
Cedar, American	35	64.0	" American, White	49	45.7
" Lebanon	30	74.6	" " Red	53	42.2
" W. Indies	47	47.6	" English	52	43.0
Chestnut	38	59.0	Pine, Scotch	41	54.6
Cork	15	149.0	" Red	34	66.0
Cottonwood, Black; Green	46	48.7	" Longleaf	42	53.3
" Dry	24	93.3	" Shortleaf	38	59.0
Cypress	27	83.0	" Loblolly	38	59.0
Deal, Christiania	43	52.0	" Norway	34	66.0
" English	29	77.0	" Pitch	38	59.0
" Scotch	31	72.0	" Western White	28	80.0
Ebony	74	30.0	" Northern White	27	83.0
Elm, Canadian	45	49.8	" Yellow	32	70.0
" English	35	64.0	" White	30	74.6
Fir, Douglas	38	59.0	Redwood California, Green	60	37.3
„ Larch	43	70.0	" Dry	27	83.0
„ Spruce	35	64.0	Sycamore	37	60.5
Greenheart	71	31.5	Teak, African	60	37.3
Hornbeam	47	47.6	" Burma	54	41.5
Ironwood	71	31.5	" Indian	46	48.7
Junglewood	57	39.3	Yew	50	44.8
Lignum Vitae	83	27.0			

Clear lumber being cut from the outside of logs is heavier than the coarser grades.

COMPOSITION OF VARIOUS "STANDARDS" AND OTHER UNITS OF TIMBER MEASUREMENTS

Unit	*Pieces*	*Inches*	*Feet*	*Board Feet*	*Cubic Feet*
Petrograd Standard	120	$1\frac{1}{2} \times 11$	$\times$ 12	= 1,980	= 165
Christiania Standard	120	$1\frac{1}{4} \times 9$	$\times$ 11	= $1{,}237\frac{1}{2}$	= 10.312
London or Irish Standard	120	3×9	$\times$ 12	= 3,240	= 270
Quebec Standard	100	$2\frac{1}{2} \times 11$	$\times$ 12	= 2,750	= 229.2
Drammen Standard	120	$2\frac{1}{2} \times 6\frac{1}{2}$	$\times$ 12	= $1{,}462\frac{1}{2}$	= 121.9
Drontheim	of Sawn Deals			= 2,376	= 198
Drontheim Standard	of Square Timber			= 2,160	= 180
Drontheim Standard	of Round Timber			= 1,728	= 144
Wyburg Standard	of Sawn Deals			= 2,160	= 180
Wyburg Standard	of Square Timber			= 1,963	= 163.5
Wyburg Standard	of Round Timber			= 1,560	= 130
A Petrograd Standard Deal = 1 piece 6′ 0″ × 3″ × 11″					

THE AMERICAN TIMBER MEASURES

A Board or superficial foot = 1 sq. foot of lumber 1 inch thick
A Square = 100 Board feet
A Mille = 1,000 Board feet = $83\frac{1}{2}$ cu. ft; slightly more than half a Petrograd Standard

The Mille is the unit of measurement in the American Timber Trade.

THE METRIC TIMBER MEASURES

A Stere = 35.314 cu. ft
= 0.2759 Cords
= 0.1635 Fathoms
= 423.77 Board feet
3.624 Steres = 1 Cord
0.028317 Steres = 1 Cubic foot
A Metric Ton = 0.9842 Long Tons
= 1.1023 Short Tons

APPENDIX NO. 10 *(Refer I.M.D.G. Code for Sea Carriage)*

CONSOLIDATED EDITION 1992

SPECIFIC GRAVITIES OF SOME COMMODITIES SUITABLE FOR CARRIAGE IN BULK BAGS IN ISO CONTAINERS

Commodity	*Specific Gravity*
Aluminium Chlorohydrate	1.336
Antifreeze	1.1
Apple Juice	1.35
Beer	1.005
Cashew Shell Nut Liquid (irritant if unrefined)	0.92
Castor Oil	0.945–0.965
Crude Cod Liver Oil	0.918–0.927
Crude Oil	0.8–0.9
Decanol	0.82
Detergent Concentrate	0.98
Diethylene Glycol	1.12
Dobane—DDB	0.856
Dodecyl Benzine	0.87
Ferric Chloride Solution—corrosive	2.90
Furferaldehyde	1.156
Glycerine	1.25
Glycol	0.93
Glyoxal	1.26
Grape Concentrate	1.33
Heavy Liquid Paraffin	0.88–0.915
Isophorene	0.923
Iso Propanol (ISO Propyl Alcohol)	
Inflammable Liquid	0.786
Lactic Acid	1.2
Latex—natural	0.94
synthetic	0.98
Light Liquid Paraffin	0.88–0.915
Linseed Fatty Acids	0.93
Maple Syrup	1.313 approx.
Medicinal White Oil	0.88–0.915
Methyl Ethoxol	0.964–0.969
Nylon Salt Solution	1.09
Oleic Acid	0.897
Orange Juice	1.35
Palm Acid Oil	0.952
Plasticiser	0.90
Polyethylene Glycol	1.127–1.130
Protein based fire fighting foam	1.25
Refined Linseed Oil	0.931–0.936
Refined Soya Oil	0.924–0.929
Sodium Silicate	1.5–1.7
Sorbitol	1.47
Soya Bean Oil	0.924–0.929
Sugar Solution	1.47
Technical White Oil	0.90
Transformer Oil	0.895
Turpentine Oil	0.860–0.875
Vegetable Oils	0.98
Wine	1.1

APPENDIX NO. 11

VEGETABLE AND ANIMAL OILS, CONSTANTS

At Standard Temperature, viz. 60°F = 15°C

Oil	*Specific Gravity at 60°F = 15°C*	*Solidifying Point C*	*M³/Tonne*
Arachis (Ground Nut)	0.915 to 0.920	−6.7/−3.3	1.09
Bean, Soya	0.926 „ about	−17.8/−11.9	1.08
Castor	0.960 „ 0.970	−16.7	1.04
China Wood or Tung	0.936 „ 0.940	3.1	1.07
Colza (Rape)	0.913 „ 0.916	−6.7/−4.4	1.07
Coconut	0.925 „ 0.931	15/21.1	1.08
Cottonseed	0.923 „ about	(11.7 Partial) 1.1/10	1.08
Kapok Seed	0.915 „ about	11.7 about	1.09
Lard	0.912 „ about	−3.9/5.6	1.10
Lardine	0.967 „ 0.980		1.03
Linseed (Hemp)	0.932 „ about	−15/−8.3	1.07
Menhaden (Fish)	0.927 „ about	−3.9	1.08
Neatsfoot	0.912 „ 0.914	0/10	1.10
Nigerseed	0.924 „ 0.928		1.08
Olive	0.915 „ 0.919	−5/−1.1	1.09
Palm	0.920 „ 0.926	23.9/40.6	1.08
Palm Nut	0.952 „ about	25.6/30	1.05
Seal	0.924 „ 0.929	−3.1	1.08
Sesame (Gingili)	0.923 „ about		1.08
Sperm	0.880 „ 0.884	0	1.14
Sunflower	0.918 „ 0.922		1.09
Tallow Oil	0.911 „ 0.915		1.10
Whale	0.920 „ 0.931	−1.1	1.08

APPROXIMATE SPECIFIC GRAVITY AND DENSITY OF OILS AT COMMON HANDLING TEMPERATURES

Oil	*Temp. °F*	*Specific Gravity Ref.—Water 60°F*
Castor	65	0.962
Coconut	104	0.909
Cottonseed	65	0.921
Groundnut	65	0.916
Lard	105	0.899
Linseed	65	0.930
Menhaden	75	0.924
Palm	120	0.892
Palm Kernel	120	0.900
Rapeseed	65	0.914
Soya Bean	65	0.922
Tallow	122	0.891
Teaseed	65	0.915
Whale	89.6	0.912
Wood	85	0.932

Change in S.G. of Liquid Oil per 1 degree C = 0.00063

All oils are inflammable. The difference between oils, fats and greases is one of temperatures only.

Specific Gravities of Oils—There is no Specific Gravity indicating the invariable weight of oils.

Specific Gravity will not indicate what an oil is, but what it is not; i.e. it will not prove an oil to be pure but it will indicate when it is not so.

Temperature—Being an important factor in relation to Specific Gravity of liquids it is necessary that the temperature of oil be taken at the same time as gravity tests are made.

Specific Gravity decreases as the temperature increases and vice versa, the ratio of change being:

0.00035 for each degree above or below 60 degrees F
0.00063 for each degree above or below 15 degrees C

Expansion—An allowance for expansion of about 1 per cent for every 14 degrees C expected increase of temperature between shipping temperature and that likely to be experienced in the course of the voyage or after heating for discharge, should be made when carrying oils in bulk.

APPENDIX NO. 12

DRY ICE (SOLID CO_2)

Calculating the quantity required:

$$\text{Quantity of Dry Ice required (lb)} = \frac{\text{T.D.} \times \text{Time in hours} \times \text{Box Heat Gain}}{\text{Solid } CO_2 \text{ (Dry Ice) B.T.U. factor}}$$

T.D. (Temperature Difference) is the difference in temperature between the average ambient temperature during transit and the required carriage temperature of the cargo.

Time in hours is the total time (allowing for possible delays) that the commodity will be required to be held at the carriage temperature, e.g. the total transit time plus say 15 per cent to allow for delays, etc.

Box Heat Gain is the gain in B.T.U.s based on the rated insulation value of the container—the average to maximum heat gain in a modern insulated container is 45 to 50 B.T.U.s/hour/degrees F/T.D.

CO_2 B.T.U. factor: 1 lb of solid CO_2 (dry ice) will absorb 260 B.T.U.s at 32 degrees F.

Example

If the cargo carriage temperature is -10 degrees F
Ambient temperature at night: 35 degrees F. During day: 65 degrees F. Average: 50 degrees F.
T.D. is therefore 50 degrees F $-$ 10 degrees F $=$ 60 degrees F.
Time (expected) of transit: say 42 hours. Plus 15 per cent for delays $=$ 48 hours total.
Therefore:

$$\frac{\text{T.D. (60)} \times \text{Time in hours (48)} \times \text{Box Heat Gain (say 45)}}{CO_2 \text{ B.T.U. factor (260)}} = \text{lb Dry Ice required.}$$

$$\frac{60 \times 48 \times 45}{260} = 498.4 \text{ lb. Say 500 lb, or } 20 \times 25 \text{ lb blocks of Dry Ice.}$$

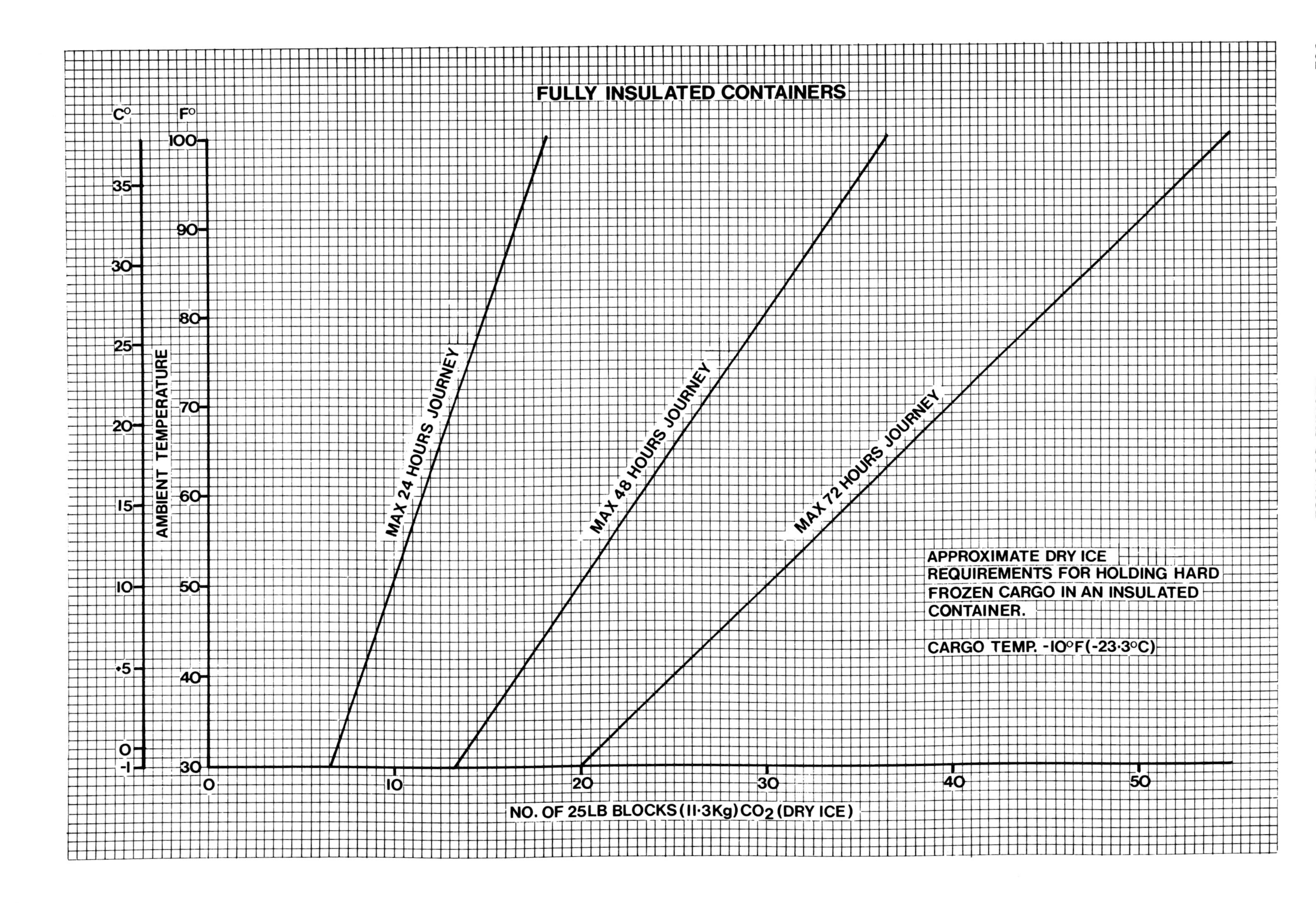
FULLY INSULATED CONTAINERS
C°
F°
AMBIENT TEMPERATURE
100
90
80
70
60
50
40
30
35
30
25
20
15
10
+5
0
-1
MAX 24 HOURS JOURNEY
MAX 48 HOURS JOURNEY
MAX 72 HOURS JOURNEY
APPROXIMATE DRY ICE REQUIREMENTS FOR HOLDING HARD FROZEN CARGO IN AN INSULATED CONTAINER.
CARGO TEMP. -10°F (-23·3°C)
0
10
20
30
40
50
NO. OF 25LB BLOCKS (11·3Kg) CO2 (DRY ICE)

APPENDIX NO. 13

BULK SOLID CARGOES WHICH MAY HAVE HAZARDOUS PROPERTIES OR WHICH MAY LIQUEFY

Cargoes Which May Liquefy

Materials which if shipped "wet", may shift transversely during the voyage due to the effects of moisture migration.

It should be noted that the list of materials is not exhaustive and that the physical and/or chemical properties attributed to them are for guidance only. Consequently, whenever the shipment of a bulk cargo is contemplated, it is essential to obtain currently valid information about its physical properties prior to loading.

Varying terminology exists to describe mineral concentrates. All known terms are listed below, but the list is not exhaustive. The stowage factor for these materials is generally low: from 0.33 to 0.57 M^3/tonne.

Blende (zinc sulphide)
Chalco Pyrite
Copper Ore Concentrate
Copper Nickel
Copper Precipitates
Galena (lead)
Ilmenite ("dry" and "moist")
Iron Ore Concentrate
Iron Ore (magnetite)
Iron Ore (pellet feed)
Iron Ore (sinter feed)
Iron Pyrite
Lead and Zinc Calcines (mixed)
Lead and Zinc Middlings
Lead Ore Residue
Lead Silver Ore
Lead Sulphide
Lead Sulphide (galena)
Magnetite
Magnetite-taconite
Manganic Concentrates (manganese)
Nefelin Syenite (mineral)
Nickel Ore Concentrate
Pentahydrate Crude
Pyrite
Pyrites (cupreous)
Pyrites (fine)
Pyrites (floatation)
Pyrites (sulphur)
Pyritic Ashes (iron)
Pyritic Cinders
Silver Lead Ore Concentrate
Slag (iron ore)
Zinc and Lead Calcines
Zinc and Lead Middlings
Zinc Ore (burnt ore)
Zinc Ore (calamine)
Zinc Ore (crude)
Zinc Sinter
Zinc Sludge
Zinc Sulphide
Zinc Sulphide (blende)

The list below contains only materials (other than cargoes which may liquefy) that have been reported as capable of attaining a flow state and is not exhaustive:

Material	*Approx. M^3/tonne*
Coal (fine particled)	
Coal Slurry (watery silt, material normally under 1 mm in size)	0.98 to 1.15
Coke Breeze	1.8

Cargoes Which May Have Hazardous Properties

Aluminium Dross
Aluminium Ferrosilicon, powder
Aluminium Silicon, powder, uncoated
Ammonium Nitrate Fertilisers
Antimony Ore (Stibnite) and Residue
Barium Nitrate (Briquettes and Pellets)
Castor Beans
Cereals and Cereal Products
  Bakery Materials
  Barley Malt Pellets
  Beet Pulp Pellets
  Bran Pellets
  Brewers' Grain Pellets
  Citrus Pulp Pellets
  Gluten Pellets
  Maize
  Milfeed Pellets
  Pollard Pellets
  Rice Bran Pellets
  Rice Broken
  Strussa Pellets
  Toasted Meals
Charcoal (Briquettes)
Chromium Ore
Coal
Concentrates (Ore)
Copra, dry
Direct Reduced Iron (High Iron)
Ferrophosphorus
Ferrosilicon
  Alfalfa Pellets
Fishmeal, fishscrap
Fluorspar (Calcium Fluoride)
Iron Oxide, spent
Iron Sponge, spent
Iron Swarf, Steel Swarf (Nitrate)
Lead Nitrate
Low Specific Activity Substance (LSA) 1 (Radioactive)
Magnesium Nitrate (containing 8 per cent or more moisture)
Pencil Pitch
Petroleum Coke
Pitch Prill, Prilled Coal Tar
Potassium Nitrate (Saltpetre)
Sawdust
Seed Cake, containing Vegetable Oil
Meal, oily, Oil Cake, Seed Expellers, oily
Sodium Nitrate (Chilean Natural)
Sodium Nitrate and Potassium Nitrate mixture (Chilean Natural Potassic Nitrate)
Sulphur
Tankage
  Garbage Tankage
  Rough Ammonia Tankage (containing 7 per cent or more moisture)
  Tankage Fertiliser (containing 8 per cent or more moisture)
Vandium Ore
Woodchips
Wood Pulp Pellets
Zinc Ashes

See Commodities section Part 3 or further information and details on the above list may be obtained from the IMO Code of Safe Practice for Solid Bulk Cargoes, Latest Edition (see also IMDG Code).

EPILOGUE

"A ship is not a slave. You must make her easy in a seaway. You must never forget that you owe her the fullest share of your thought, of your skill, of your self-love. If you remember that obligation, naturally and without effort, as if it were an instinctive feeling of your inner life, she will sail, stay, run for you as long as she is able, or like a sea-bird going to rest upon the angry waves, she will lay out the heaviest gale that ever made you doubt living long enough to see another sunrise.

The hurry of the times, the loading and discharging organization of the docks, the use of hoisting machinery which works quickly and will not wait, the cry for prompt dispatch, the very size of his ship, stand nowadays between the modern seaman and the thorough knowledge of his craft . . .

Stevedoring, which had been a skilled labour, is fast becoming a labour without the skill. The modern steamship with her many holds is not loaded within the sailorlike meaning of the word. She is filled up. Her cargo is not stowed in any sense: it is simply dumped into her through six hatchways, more or less . . .".

Extracts taken from *The Mirror of the Sea* by Joseph Conrad.

INDEX

C

Q

R

S

T